The COMPLETE SALT AND PEPPER SHAKER BOOK

Mike Schneider

77 Lower Valley Road, Atglen, PA 19310

Dedication

To Robert Dupee, a hard-working public servant who has spent the last twenty years making the village in which I reside a much better place to live, work, play, raise kids, and write.

Pheasants have always been a popular subject with salt and pepper shaker manufacturers, and with collectors, too. The pair in the back, whose tails act as handles, are 4-½ inches high. They are not marked. Neither is the pair in the front, which is 2-¾ inches high. The pair on the sides bear an inkstamp, "H608," along with a Norcrest paper label. *Thornburg Collection.*

Library of Congress Catalog Number: 92-63108.

Printed in the United States of America
ISBN: 0-88740-494-4

Published by Schiffer Publishing, Ltd.
77 Lower Valley Road
Atglen, PA 19310
Please write for a free catalog.
This book may be purchased from the publisher.
Please include $2.95 postage.
Try your bookstore first.

We are interested in hearing from authors
with book ideas on related subjects.

Title Page

The pair of Spanish dancers shown on the title page were made by Ceramic Arts Studio, of Madison, Wisconsin, and are so marked. Each is 7-¾ inches high. This particular set won 1st Place in the Most Beautiful Shakers Contest at the annual convention of the Novelty Salt and Pepper Shakers Club in Memphis, Tennessee, in 1990. The set was also made in forest green and maroon, and perhaps other colors. *Thornburg Collection.*

Acknowledgments

This book was made possible by two things that no amount of money can buy. The first was the help and cooperation of my wife, Cindy. In previous books I have generally credited Cindy with copying information during photo sessions, but I want you to know that she contributes in many other ways. On this project, for example, she helped me label the pictures, proofread and edited copy, and ran numerous time-consuming errands—from dropping off film and picking up prints to copying information at the library. For a time, she even worked two jobs so that I would be able to stay home and write. Simply put, without Cindy's help there would not have been a book.

The second thing that made this book possible was the willingness of collectors to open their homes to us and share their collections. Many people gave up precious hours—or even days—so that their shakers could be photographed. As with *Animal Figures* and *The Complete Cookie Jar Book*, good friends Betty and Floyd Carson supported the project from the beginning. All totaled, I am sure that we spent at least five or six days at their house photographing their extensive collection of salt and pepper shakers. We also picked up much valuable information there, and had some of the best chili that has ever been served in northeastern Ohio.

Good fortune also smiled upon us when we met Irene and Jim Thornburg of Michigan. Irene routinely refers to her 11,000 sets of salt and pepper shakers as a "collection." But if you were to see it, housed in the dust-free environment of the finished and carpeted pole barn the couple built especially for it, I believe that you would agree with me that it is nothing short of a museum. We spent two days there taking pictures, and wish it could have been two weeks. Anyone would be hard pressed to find more hospitable hosts, or more enthusiastic collectors.

A couple of days with Ohio collectors Richard and Susan Oravitz, and Georginia and Harry Lilly yielded not only many good pictures, but also lots of laughs.

Others who contributed in various ways are Joyce Beulah, Heidi Calhoun, Trina Cloud, Cloud's Antique Mall, Dorothy Coates, Jim and Betsy Coughenour, Mercedes DiRenzo, Lesia Dobriansky, Millie Gisler, Dawn Glow, Jack Glow, Henrietta Hillabrand, Joyce Hollen, Sharon Isaacson, Jazz'e Junque Shop, Betty Laney, Allen and Michelle Naylor, Bob Sandoff, Dan Schneider, Virginia Sell, Robert Smith, Carol Sparacio, Lee and Jean Stipes, Bev Urry, and Stan and Karen Zera.

Additionally, I would like to thank publisher Peter Schiffer for suggesting the book, and for the trust he put in me to do it. Peter has never put any arbitrary restrictions on me as far as size, content or anything else, and I believe that it is that type of approach that has been responsible for the reputation of Schiffer Publishing offering the best books in the antiques and collectibles field today. I also want to thank Douglas Congdon-Martin and Nancy Schiffer for their excellent editing and Ellen J. (Sue) Taylor for her superior work in laying out the book.

Both these sets of fish were made by Rosemeade. The swordfish are 4-¼ inches high. The paper label on the shaker at right center is the only mark on any of them. *Thornburg Collection.*

Introduction

This is where most writers welcome the reader to the world of the collectible their book is about. Since I do not have a collection of salt and pepper shakers, and your purchase of the book at a significantly higher price than other books on the subject indicates you are already an ardent collector, I will instead thank you for welcoming me to your world.

I also want to take a moment to talk a bit about why the book is set up the way it is, and the most efficient way of using it.

Three years ago, when I began photographing collections for this volume, I had no idea of how the finished product was going to be organized. Probably like all the others, I thought. Then my wife and I went into the antique business and began setting up at shows every weekend. With salt and pepper shakers being no small part of our business, it didn't take long to devise a book format that would benefit collectors and dealers alike. What I discovered was that in other books on shakers, while beautifully photographed and logically arranged, too many subjects show up in too many different places. This necessitates the time consuming task of checking practically every page to see whether the shakers you are looking up are in the book.

Take bears, for example. In a salt and pepper shaker book picked at random, I found them in six different sections. Shakers depicting people are a worse problem. In a different book, also picked at random, people appear in seventeen sections, many of which, judging by their titles, have nothing to do with people.

Now let's go back to bears to show how this book is different. In the book you are reading now, if you want to look up a pair of bear shakers, you will find them in the section on bears in Chapter 6: Animal Life. It doesn't make any difference whether they are ceramic bears, plastic bears, miniature bears, advertising bears, Goebel bears, nesting bears, Smokey Bears or Yogi Bears. If the shakers are bears, they will be in the section on bears in Chapter 6. For the most part. As stated earlier, when I began photographing the book I had no idea of what format it would take. Early on some shakers were shot in groupings that didn't conform to this type of layout. There is also some natural overlap with go-togethers, shakers that present two different subjects that go together such as a bear and a beehive, or a bear holding a couple salmon. For these cases a complete cross-referenced index has been provided.

So, if you have a pair of bears you want to look up, first check the section on bears in Chapter 6. Then check the other page numbers listed under bears in the index. Work the same way with other subjects. Two simple steps and you know whether or not the pair of shakers you have are in the book. And you do not have to scan every page to find out.

The index also lists shakers according to media other than ceramic (such as plastic, Bakelite, etc.), according to company (such as Ceramic Arts Studio or Rosemeade), tells what page will give you background information on various companies, and lets you know on which pages you will find pictures of specific company marks and paper labels.

In many cases measurements of shakers are provided, too. They have been included for two reasons. First, when shakers are presented as individual pairs (as many are in this book), or in groups of two or three pairs, without a measurement it is impossible to get a lock on what size shakers you would look for if you wanted to acquire a certain set. Second, the same shakers were often made in different sizes, which just as often dictates different values. While the difference between some miniature, table size and range size shakers can often be readily apparent, even to beginning collectors, other size differences are more subtle. The Shawnee fruit stacks come to mind right away. If you are standing in an antique shop looking at only one pair of Shawnee fruit stacks and you are not that familiar with them, it would be hard to tell whether they were the small ones or large ones. The same goes for shakers such as Hull's Little Red Riding Hood, and the Van Tellingen love bugs. If you have only seen the somewhat common small love bugs, how would you know if a larger pair presenting itself to you is the medium or large size if there was no measurement to go by? You wouldn't.

The measurements might also prove handy in the future should some enterprising potters decide to reproduce some of the more valuable shakers by using the shakers themselves as mold cores. The resulting shrinkage would be readily apparent just by knowing the size of the original. Recently cookie jar collectors have had to wrestle a proliferation of reproduction jars flooding the market. As the values of salt and pepper shakers continue to rise, it may be only a matter of time before shaker collectors will be facing the same problem.

The location of the price guide at the back of the book is a genuine nuisance. While I would much prefer the price of each set of shakers to appear in the caption below its picture, the realities of publishing in today's economy dictate a condensed price guide in a separate location. Otherwise, revising the prices in future editions would be cost-prohibitive. When books I constantly refer to in my personal library have price guides like this, I generally find myself photocopying the price guide pages, then going through the book and writing the prices on or adjacent to the pictures. It takes an hour or two, but saves a lot of time and frustration in the long run. To those who say this defaces and devalues a book, I say don't worry about it. A book on collectibles is only as valuable as the information you can obtain from it, and the quicker and easier you can obtain that information, the more valuable it is regardless of whether or not the pages have writing on them.

Last, let me use this introduction to introduce you to Schiffer Publishing. Although not as well known as some other companies, Schiffer publishes today's best and most lavishly illustrated books in the field of antiques and collectibles. The company offers more books on collectible pottery, glass, metal, jewelry, furniture, lighting, toys, clocks, watches, decoys, fountain pens, and American Indian arts and crafts than any other publisher. Plus an assortment of books on general collectibles covering everything from spoons to telephones. I strongly urge you to fill out the enclosed postcard, place a stamp on it and drop it in the nearest mailbox to receive Schiffer's 48+ page color catalog. Whether you are a collector, a dealer, or both I guarantee you will be pleasantly surprised when it arrives.

That's it for the introduction. Thanks again for welcoming me to your world.

Contents

Acknowledgments 3

Introduction 4

Section I: Salt and Pepper Shaker Collecting

Chapter 1 A Short History 7
Chapter 2 Classifications of Salt and Pepper Shakers 10
Chapter 3 Building and Maintaining a Collection 22
Chapter 4 Join the Club! 23

Section II: Figural Salt and Pepper Shakers

Chapter 5 People 26
Occupations and Avocations 26
Native Americans 37
Foreign Costumes 42
Brides and Grooms, Married Couples, and Seniors 50
Babies and Children 53
Bums and Hillbillies 60
R- and X-Rated 62
Miscellaneous People and Pieces Parts 64
Characters and Personalities—Fact and Fiction 66
Santa and Mrs. Claus 72
Almost Human 75
Chapter 6 Animal Life 81
Mammals 81
Birds 127
Amphibians and Reptiles 146
Insects and Other Little Creatures 150
Creatures of the Deep 154
Chapter 7 Plant Life 163
Plants 163
Fruits 167
Fruits and Vegetables 175
Vegetables 177
Chapter 8 Transportation and Fuel 187
Vehicles 187
Fuel 192
Chapter 9 Structures 196
Chapter 10 Appliances and Machines 201
Chapter 11 Containers 207
Chapter 12 Miscellaneous Figural Subjects 217
Athletics and Games 217
Clothing, Apparel and Accessories 219
Eating Utensils 221
Food 221
Lighting 224
Music 226
Potluck 229
Snow People 235
States 236
Tools and Utensils 239

Section III: Non-Figural Salt and Pepper Shakers

Chapter 13 Non-Figurals 241
Condiment Sets 241
China Painting 243
Known American and Foreign Potteries 243
Miscellaneous Non-Figurals 245
Chapter 14 Glass 252
Art Glass 252
Miscellaneous Glass 255

Section IV: Origins

Chapter 15 Manufacturers, Importers, Distributors, Sellers and Artists 257

Appendix I: Sources 276
Appendix II: Beswick Model Numbers by Year of Introduction 277
Appendix III: Parkcraft States 277
Appendix IV: United States Invention Patent Numbers 278
Appendix V: United States Design Patent Numbers 278
Bibliography 279
Index 281
Price Guide 293

Perfect for holiday entertaining, the Christmas trees are 3-½ inches high. Their only mark is an inkstamp, "054/96N," written on one line. *Courtesy of Cindy Schneider.*

Section I: Salt and Pepper Shaker Collecting

The purpose of this section is to explore the history of figural and novelty salt and pepper shakers, present the different types that are available, and point out the best methods of acquiring them.

Since the main thrust of this book emphasizes figural and novelty salt and pepper shakers, we should first attempt to define them. A figural set is one made to resemble something else, such as the Ceramarte advertising shakers shown below, which look more like bears, specifically Hamm's Beer bears, than salt and pepper shakers. Nearly every shaker in the book is a figural.

While the definition of figural shakers is pretty straightforward, putting a handle on novelty shakers is a little tougher as different people define them different ways. Some collectors I interviewed believe novelty shakers are any shakers that are designed to be more attractive than the nondescript clear glass shakers found on all super market and discount store shelves, and most kitchen tables. This definition would include all figural shakers. Other collectors, however, saw novelty sets as something a little different. Novelty shakers, they said, are shakers that are made simply for show, that are not functional for holding, pouring or sprinkling salt or pepper. Under this definition not all figural sets would be included. It would encompass only those that are impractical for one reason or another. For instance, sometimes the holes in the tops may be so large that sprinkling salt through them would create a miniature snowstorm. Other times they may be too small. Sometimes the shakers themselves may be too tiny to be practical. Or too big. Or improperly balanced. Or unsanitary. Or just plain unhandy. Under this definition the ink bottle and spilled ink shown below would have to be considered strictly novelty. The puddle of ink is so flat that it is hard to pick up, and so thin that it would probably have to be refilled before each meal if actually used.

That said, you may choose whichever definition you feel is correct.

Chapter 1

A Short History

In the broadest sense, in order for a future collectible to make a successful debut, two very basic prerequisites must be met. One is that the technology necessary for the manufacture and use of the potential collectible must be developed. The other is that circumstances must be correct for the public to accept and acquire the future collectible on a widespread basis. For figural and novelty salt and pepper shakers, these two requirements began to mesh during the 1920s.

From a manufacturing standpoint, mankind had the technology to make figural salt and pepper shakers in several different media since ancient times. Pottery has been molded for several millennia, glass has been around at least 5000 years, and metalworking goes back even farther. And who would care to guess how long wood has been whittled? So it wasn't the inability to make figural salt and pepper shakers that prevented their manufacture. Rather, it was the inability to use them.

Most of us living today have little if any appreciation for Morton International's familiar trademark depicting a little girl walking in the rain, an umbrella in one hand, a box with salt spilling from it in the other, and the slogan, "When it rains it pours." But our great-grandparents knew exactly what it meant. Prior to the early 1900s, table salt did not pour satisfactorily. It could not be sprinkled from a shaker with any consistency. The slightest bit of humidity would cause salt to cake.

According to Arthur G. Peterson in *Glass Salt Shakers,* now out of print, this problem was first addressed in 1863 when a United States patent was issued for a shaker containing a built-in agitator to break up caked salt. Eight years later, the first of many patents to prevent the caking of salt was granted. The earliest patented processes with this aim depended upon removing moisture-absorbing elements from salt. After a short time, however, researchers changed direction. They began looking for an additive that would neutralize those elements, a "filler" that would coat each grain of salt and absorb moisture

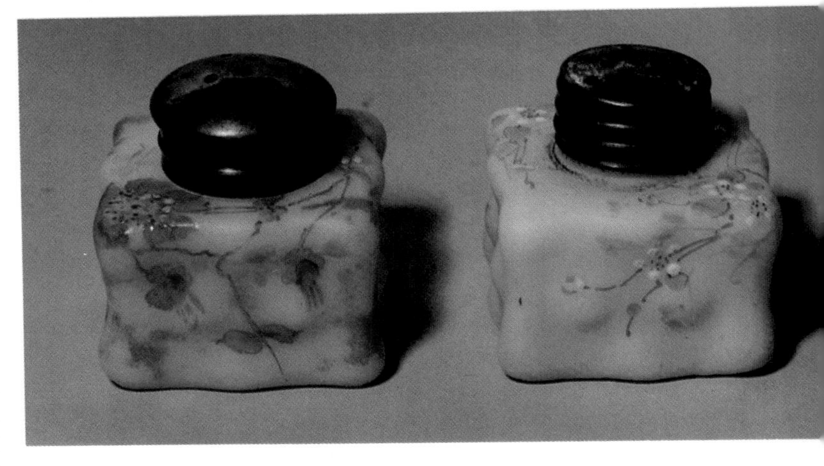

When salt was finally able to be poured with some reliability, art glass and cut glass shakers became the rage among the upper classes. These Wavecrest shakers (billow pattern) would have been a product of the C.F. Monroe Company, a glass decorating studio that operated between 1882 and 1916. Generically known as opalware, the unmarked shakers are 1-¾ inches high. *Courtesy of Cloud's Antique Mall, Castalia, Ohio.*

As their name implies, figural salt and pepper shakers are designed to represent figures of something, in this case the Hamm's Beer bear. These shakers are 5 inches high. They are a Ceramarte product; mark is shown. *Thornburg Collection.*

Mark of Ceramarte Hamm's Beer bears.

Novelty salt and pepper shakers, according to some collectors, can be defined as figural shakers that are impractical for table use, that were made with only decorators or collectors in mind. The ink bottle and puddle of ink fit this description not only because of the problems encountered with handling the pepper (puddle), but also because it was poured very thin and would be prone to breaking. The puddle is 3-3/8 inches across; neither piece is marked. *Carson Collection.*

during periods of high humidity, then release it back into the atmosphere at times of low humidity.

The initial patent along this line was granted to Rhode Islander Charles A. Catlin in 1883. Catlin's filler was bicarbonate of soda. It was not very effective. Several other fillers were used with varying degrees of success over the next 20 or so years. While truly non-caking salt would not become a reality until well into the 20th century, the process of adding moisture-absorbing fillers must have been fairly well along by 1911. That was the year that Joy Morton, then owner of the Morton Salt Company, apparently felt confident enough about the pourability of her salt to debut the now universally recognized umbrella girl. Today sodium silico-aluminate is the most commonly used filler.

The other half of the equation—circumstances that would allow public acceptance and acquisition of figural and novelty salt and pepper shakers—took a bit longer to work out, but during the late 1920s everything gelled. By then America's changeover from an agricultural economy to an industrial economy was nearing completion. In the cities, people had more stores in which to shop for trifles such as salt and pepper shakers, more time to shop for them and more money to spend on them than ever before. Previously, the long hours and generally low income of labor-intensive rural life would not allow the time, money nor energy to engage in such frivolous practices, even if there had been places to do so, which there weren't. Additionally, a few years earlier escalating costs of labor and materials had knocked the brilliant cut glass industry out of existence, and had severely crippled the output of colored art glass. Open salt dishes made of cut crystal, and salt shakers made of cranberry, burmese, agata and other forms of colored glass were largely unavailable, which opened the door for the acceptance of figural shakers.

Although the record is not totally clear, indications are that the first figural shakers to be manufactured on a fairly large basis were probably made during the mid-to late-1920s. Goebel, a fine German pottery that later became famous for its Hummel figurines, marketed its first three sets of figural shakers in 1923. By the end of the decade 14 more pairs of Goebel figural shakers were on store shelves. Then the world economy got hit by the most severe depression it had ever seen, and that helped in a big way to popularize figural shakers. Potteries, both domestic and foreign, were forced to turn out less expensive products than in the past because people no longer had money to spend on high dollar items. But for those who were lucky enough to remain employed, 10 or 15 cents for a pair of salt and pepper shakers wasn't all that expensive. It was even cheap enough that they could start collecting them if they did not overdo it.

Japan appears to have exported a good number of figural shakers to America during the late 1920s and throughout the 1930s, and again in the late 1940s and 1950s following a hiatus for World War II. That hiatus worked to today's salt and pepper shaker collector's advantage. It gave American potteries a chance to get a firm grip on the market. Lacking competition from foreign potteries which had always enjoyed much lower labor costs, during World War II small domestic companies such as Rosemeade and Ceramic Arts Studio, and larger ones such as Shawnee, created and marketed some of the finest figural shakers that have ever been made. Another aspect of World War II that played into the shaker collector's hand was the severing of our supply lines for raw materials, most notably rubber from The Philippines. The ensuing rush to produce synthetic replacements for rubber resulted in major advances in the development of plastics. Not surprisingly, some of today's most highly prized plastic sets were made by enterprising businessmen within the first few years after the fighting ended.

Today things are back to about the same way they were prior to World War II. The United States is largely on the outside looking in concerning shaker production. Few are made here. The Pacific-Asian region is again the king of the salt and pepper shaker mountain. But two things are different. During this round Japan has been forced to share the market with neighbors such as China, The Philippines, Korea, and Malaysia. This has led to increased competition resulting in higher quality and lower prices. The other difference is that many of the shakers entering the country from abroad today have been designed in the United States and made according to American specifications. Again, this has resulted in improved quality. If you are a purist who collects only old shakers and doesn't want anything to do with newer "junk," next time you see a pair marked Vandor, Fitz and Floyd, Otagari or another current importer's name, inspect them closely. You will likely be surprised by the workmanship you find.

Now that we take pourable salt for granted, few people appreciate this Morton International, Inc. trademark. But when it first appeared back in 1911, there seems little doubt that nearly everyone would have understood its significance.

Chapter 2

Classifications of Salt and Pepper Shakers

Figural salt and pepper shakers are generally classified three ways—by size, media, and the way one shaker of a pair interacts with its mate, or how both shakers interact with another component, or components, in sets that include more than two pieces.

SIZE

Figural shakers are made in three basic sizes—range, table and miniature.

Range sets are the largest, with heights usually running from about 5 to 8 inches. While certain table sets often reach or exceed these heights to achieve a comical or dramatic effect, range sets will generally have larger diameters which make them easy to distinguish.

Range sets are exactly what their name implies, salt and pepper shakers that are used on the kitchen range, where the quantities of salt and pepper required in cooking are generally greater than the quantities we sprinkle on our food to suit individual tastes. Many, but not all, range sets have matching table sets. Some complement cookie jars, canister sets or other kitchen artware. Most range sets have been made to be functional.

The table set is by far the most common size of salt and pepper shaker found on both the secondary and primary markets. They run from about 2 inches to 5 inches in height, but sometimes they may be shorter if they have large diameters. Table sets are used to flavor individual servings. The majority of table sets have been designed to be functional, at least to the extent they could be used safely until people tired of the novelty. Others exhibit design flaws in balance, handling, sanitation and other areas that indicate their manufacturers had only collectors and decorators in mind when they made them.

This is a miniature set made by Sandy Srp (pronounced like syrup without the u), of Seven Hills, Ohio, who specializes in miniatures. The little eskimo is 1 inch high, the igloo is approximately ½ inch high. Although Srp makes all her miniatures to be functional, there really would not be much sense in using shakers so small. The address of Sandy Srp appears in Appendix I Sources. *Carson Collection.*

Unmarked Shawnee sunflower shakers in both range and table size. The range shakers are 5-⅛ inches high, the ones for the table stand 3-⅛ inches. Note that the range shakers are not only taller but also much bigger around so they may hold the larger amounts of salt and pepper required for cooking. *Courtesy of Robert Smith.*

Miniatures are usually no more than 1-½ inches high. Few people actually employ them to hold salt and pepper. Most use them for decoration, or simply to collect. The majority of miniatures, in fact, were not made to be functional. A few, such as those produced by Ohio ceramic artist Sandy Srp, are made so they can be used. Nevertheless, they seldom are. Miniatures are currently hot items on the secondary market. Their prices often exceed the prices of table size shakers, and sometimes range size shakers, too.

MEDIA

Far and away the most popular medium used to make figural and novelty salt and pepper shakers has been clay. When clay is fired at a high temperature it changes to pottery, or ceramic. How hot the firing temperature is, along with the specific mixture of the clay, determines what type of ceramic emerges from the kiln. Pottery can be divided into four basic categories—earthenware, stoneware, bone china and hard paste porcelain. (There is a fifth, soft paste porcelain, but as far as I have seen it has never been used to any great extent for figural and novelty salt and pepper shakers.) Each type of ceramic is fired at successively higher temperatures. As pottery moves up the temperature scale, it also moves up the hardness scale—the hotter it is fired, the harder it is. Hardness also runs hand in hand with price. Earthenware is the least expensive fired ceramic, porcelain is the most expensive.

Earthenware is the most common form of pottery used for making shakers, and for just about anything else. Its abundance is attributable to it being the easiest and least expensive to make. It is also the softest ceramic material, and the most easily damaged. Earthenware is porous. Unglazed earthenware will absorb water. Earthenware is fired at 1470 to 1830 degrees F. It is made in several different forms including terra cotta (like common flower pots), redware, yellow ware, and whiteware. The first three identify themselves loudly and clearly by color. You can identify white earthenware shakers by the fact that they are usually poured thicker than those made of bone china or porcelain, and are therefore heavier in weight.

Stoneware is the least used of the four types of ceramic for salt and pepper shakers. Stoneware is often thicker and heavier than earthenware. It is fired at 2280 to 2370 degrees F. That temperature is hot enough to make stoneware non-porous. Unglazed stoneware will not absorb moisture. Stoneware is most often related to crockery and, if I may coin a badly needed word, juggery. One form of stoneware often overlooked is jasperware, perfected and made famous by Josiah Wedgwood in 1774.

A good number of shakers are made of bone china, which falls between stoneware and hard paste porcelain on the firing temperature scale. Because it is harder than earthenware, it can be poured thinner than earthenware. Bone china got its name legitimately. The slip, or liquified clay used for pouring bone china articles, contains up to 50 percent animal bone ash depending upon the particular recipe that is followed. Bone china is fired at about 2240 degrees F. Like jasperware, it is a fairly recent innovation in ceramics, developed by Josiah Spode in England in the 1790s. By comparison, Germany began dabbling in stoneware during the 16th century, the Chinese have been turning out porcelain since the T'ang Dynasty (618 to 906), and earthenware has been made since prehistoric times.

Hard paste porcelain, commonly referred to simply as porcelain, is the strongest, most durable and most expensive type of ceramic. Those three characteristics lumped together enable it to be, and dictate that it be, poured thinner than earthenware and stoneware, which is one of the ways to identify it. Like stoneware and bone china, porcelain is nonporous. Unglazed porcelain may be immersed in water without damage. Porcelain is fired at 2550 degrees F., which is hot enough to vitrify it and make it translucent. Hold a piece of porcelain up to a strong light and you can see the brightness penetrate it, which serves as a second means of identification.

This rather lewd Biblical pair is made of terra cotta, which is also earthenware. Height is 4-3/8 inches. They have an Enesco paper label. *Carson Collection.*

Redware, of which these black cats are made, is a form of earthenware. All of the black cat shakers and other black cat accessories I have seen have been made of redware. The cats on the outside are 4 inches high. They are not marked. The inside pair is 2-7/8 inches high, have a "Japan" inkstamp. Shafford is the company that is best known for selling black cat pieces, but Enesco has marketed some, too, and possibly other companies. *Carson Collection.*

Salt and pepper shakers made by the W. Goebel Company, of Germany, are made of porcelain, the hottest fired, hardest and most durable form of ceramic. The Goebel poodle on the left is 3-3/8 inches high. The shakers are marked with an inkstamp, "Goebel / W Germany," on two lines, and also have an impressed "P 151" on the smaller shaker, "151A" on the larger one. In addition to the pink shown, these shakers have also been made in black and white. According to Hubert and Clara McHugh, in their excellent and highly recommended book, *Goebel Salt & Pepper Shakers—Identification and Value Guide* (Vintage Machinery Company, 1992), this pair of shakers was modelled by Goebel artist Arthur Moller in 1953. *Carson Collection.*

These two sets of dogs are easily identifiable as bone china because they have a paper label that says so—"Bone / China / Japan" on three lines. The dog at far left is 1-¾ inches high. *Carson Collection.*

Many salt and pepper shaker collectors favor sets made of plastic, such as this MAPCO oyster and pearls. The open shell is 3-¾ inches high, 4 inches wide. The removable pearls serve as the shakers. Mark of the oyster is shown. *Courtesy of Allen and Michelle Naylor.*

Mark of the MAPCO oyster shell.

For the purposes of this book, most of the sets that are made of clay are referred to as ceramic regardless of what type of pottery they are because it is often difficult to accurately determine the composition of small pieces. Where the difference is readily apparent, such as with redware, yellow ware, jasperware and other easily identifiable types of ceramic, the exact names are given. The same is true where shaker material is identified with a paper label, as is the case with many bone china sets.

After clay, there are seven other basic media from which figural salt and pepper shakers are made. The most popular with collectors is plastic. Plastic is a catchall word that includes Bakelite, Catalin, Lucite, Burrite and many other trade names. Examples of most of them appear in this book.

Many metal salt and pepper shakers are made of a copper-lead alloy commonly referred to as pot metal. Pot metal is not very strong, but for objects as small as salt and pepper shakers it is a satisfactory material. If you have any metal shakers, be sure to remove all salt from them, lest its corrosive properties will cause the shakers the same fate as the log cabin shown below. As an extra measure, rinse the inside of the metal shakers with warm water to remove all salt residue, then air dry, and do not reinstall the corks. Cork can absorb humidity and release it into the shaker, ruining your collectible from the inside out.

Some salt and pepper shakers are made of wood. Most common are the cedar figural shakers that are sold as souvenirs in various tourist areas. Also quite common are the carved shakers, usually painted red and green with some bare wood exposed, that are made in Mexico. They seem to be in every collection, and at every antique show. Although some are quite clever and attractive, wood shakers represent a rung that is pretty close to the bottom of the figural salt and pepper shaker collector's desirability ladder.

Closely related to wood is wood composition, or simply composition, the basic recipe for which is sawdust, glue and a spattering of other materials. Composition is an acceptable medium for shakers, but like wood, most sets made of it are not considered as collectible as those made of ceramic or plastic.

Glass, of course, is used for salt and pepper shakers, but not as extensively for figural and novelty shakers as for restaurant ware or as accessories for elegant crystal place settings. However, it has been used for many bottle and light bulb sets. Examples of both are shown. Older glass shakers, and art glass shakers, are shown in Chapter 14 Glass.

Chalkware is a plaster mixture somewhat like ceramic except that it is not fired. It is very soft. Chalkware is always painted, never glazed. Chalkware shakers that are in good condition can be very attractive. Those in poor condition can be very ugly.

Cardboard is another material that is sometimes used, especially for small advertising shakers. Cardboard is usually used for the body of a shaker, the bottom and top being made of metal or plastic.

Several other media such as bone, ivory and nuts have been used to make shakers, but only to a small extent.

This is the unfortunate result of salt being left inside a metal shaker. Actually, it is a good idea to remove left over salt and pepper from all of your shakers regardless of medium.

Metal shakers with a paper wrap for advertising. The miniature cans are 1-½ inches high. *Courtesy of Allen and Michelle Naylor.*

Here's a nonfigural wood pair whose clean design and contrasting colors add some class where it usually isn't seen. Made of oak and walnut, they are 3 inches high. The mark on their caps is shown below. Written in pencil on the bottom was, "Sweeden Sept. 1976." *Carson Collection.*

Impressed mark of the wooden shakers above, which helps confirm the information the original owner had scribbled on the bottom.

While glass has generally not been used very prolifically in the production of figural salt and pepper shakers, beer advertising shakers are an exception to the rule. Heights here are 4 inches and 3 inches. The more desirable of the two sets is the one on the right which has metal caps instead of plastic, an indicator that it is older. Mark is shown. *Carson Collection.*

This is the mark that appears on the bottom of most beer bottle shakers. Somewhat hard to read, the raised printing says, "Muth & Sons / Pat. / Pend. / Buffalo NY," on four lines.

Some shakers are made out of heavier metals such as steel or iron. This set of shakers, showing the Ford Rotunda from the 1933 Chicago World's Fair, was cast in a foundry, then planed and threaded in a machine shop. They have a height of 1-5/16 inches. Diameter is 1-⅝ inches. Printing on the rims of the bases reads, "Ford Exposition A Century of Progress 1934." *Lilly Collection.*

Wood shakers don't command much respect from most collectors but they do have their place. These figurals, for instance, might fit in nicely in a cabin or a cottage. Both sets are made of cedar. The pair on the inside is 3-⅛ inches high, the pair on the outside is 2-½ inches. Neither set is marked except to say they were souvenirs of Gettysburg (outside) and Sequoia and Kings Canyon National Park (inside). *Carson Collection.*

Another place glass was used extensively was in light bulb shakers. Apparently some were made overseas, some in America. The mark of the tall brown bulbs in the rear, which are 5-¾ inches high, is "Made in Hong Kong," while the mark of the clear bulbs in front, 3-½ inches high, is "Rite Lite Corporation Clarendon PA Watt Fun." The shorter brown bulbs at the sides are 4-½ inches in height, marked "Made in Taiwan." *Carson Collection.*

Disposable cardboard shakers, a symbol of our "throwaway" society. Morton shakers are still made today, marketed through retail grocery stores for picnics and lunch boxes. They are 1-½ inches high; tops are plastic. The Northwest Airlines shakers apparently came with in-flight meals. Their cardboard tops would seem to indicate they are much older than the Morton shakers. *Carson Collection.*

Not all figural condiments have a tray, this set of pixies being one example. It has a figural base instead, into which the shakers fit. The stump, or mustard, is 4-¾ inches long, has a "Made in Japan" inkstamp. The pixies, or shakers, are 3-¼ inches high, have "Japan" inkstamps. *Oravitz Collection.*

TYPES OF SETS

After size and media, shakers are most often divided into several different categories according to design. Each category has its share of aficionados who specialize in it.

Condiment sets include a third receptacle, a mustard, and all three pieces are usually housed on a tray. The mustard is a carryover from the days when dry mustard was a popular spice. Most condiment sets include a small spoon that fits in the mustard, usually made of the same medium as the rest of the set, and usually missing. Condiment sets may be figural or nonfigural. Three condiment sets are shown below. Many others are shown throughout the book.

This condiment set is called Shakespeare's house. It was made by Beswick, an English pottery. It was designed in 1934 by a Mr. White, a freelance modeler whose first name has been lost to history, but who worked for Beswick between 1934 and 1941. The company sold the set for 37 years, withdrawing it in 1971. The tray is 6-¾ inches long, the mustard 3 inches high. The mark is shown below. *Private Collection.*

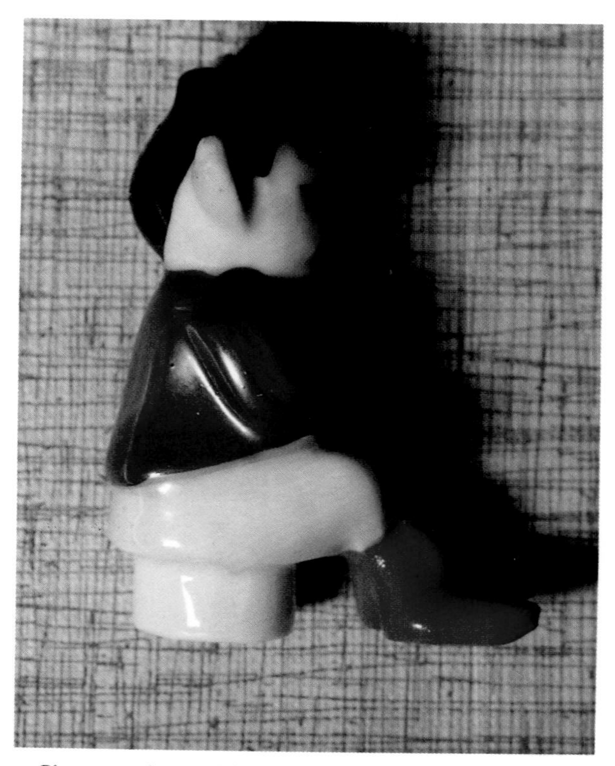

Close up of one of the above pixies showing an extension that keeps it from sliding off the base. If the extension on this type of shaker were to break, the possibility would exist for an unscrupulous dealer to grind it flat and sell it as a sitter.

Inkstamp mark on the bottom of the tray of the Shakespeare's house condiment set. Impressed lettering barely visible is part of "Beswick England" in a semicircle. Also impressed is the set's model number, "253." Beswick has made several condiment sets, each one marked with a model number. By comparing the number with the chart in Appendix II you can determine what year each set was designed.

This is a hand-painted non-figural condiment set. So what is it doing in this book? The answer is that from what I have observed, there are few collectors of figural shakers that do not let non-figural condiments creep into their collections. The tray of this set is 7-⅝ inches long. It has a Noritake mark. *Carson Collection.*

Most authorities use the word nester to define any set of shakers where one sits within or atop the other, or where a third piece holds the shakers as though grasping them. But this broad category might be better broken down into three subcategories, sitters, nesters and holders.

For the purposes of this book, a sitter will be defined as any set of shakers in which one shaker sits atop of the other as though it were sitting on a platform. The monkey sitting on the stump, shown below, is a perfect example.

Here is a basic sitter, the monkey sitting atop the stump as though on a platform. Together the shakers are 4-½ inches high. The pieces are marked with, "Made in Japan," inkstamps. *Carson Collection.*

Sometimes sitters straddle, such as a cowboy riding a horse, or this performer riding a crocodile. The croc is 4-¾ inches long, the woman is 4-¼ inches high. The crocodile has an "Empress" paper label. *Oravitz Collection.*

A nester, on the other hand, would be similar to a sitter except that the pepper shaker (nearly always the smaller of the two) rests deeper inside the salt shaker. Examples of this would be the Ceramic Arts Studio bears, kangaroos and gorillas shown below.

The characteristics of a holder are much more clear cut than the sometimes murky distinction between sitters and nesters. With a holder, a third piece, sometimes functional as a mustard but usually only decorative, holds the shakers by actually grasping them. The butler below is an example of a holder.

The pair of little gray bears hanging onto the tree limbs is an example of a hanger. Hangers seem to be self-explanatory.

These are also sitters, but by strictest definition, I prefer to call them nesters. Note that the simple platform is absent, and that the smaller component "nests" in a cavity in the larger component. All three of these sets were made by Ceramic Arts Studio of Madison, Wisconsin, and are so marked. To give an idea of size, the mother bear is 4-½ inches high. The bears were also made in white as polar bears. *Carson Collection.*

A hanging set, the tree of which is 4-¾ inches high. The mark indicates Japanese origin. *Carson Collection.*

Go-togethers and go-withs are terms used to describe shaker sets in which the salt and peppers represent two different objects with a common theme. The cup and saucer shown below is a go-with. So is the miniature telephone and telephone book. Many go-withs seem like they could have been inspired by psychologists' word association tests—ball and bat, love and marriage (heart and scroll), pen and ink, etc. Think of two such objects that go together and it is likely that you will eventually run onto a pair of salt and pepper shakers that represent them.

This set is an example of a holder, the butler actually holding the bottles which are the shakers. The butler is 6 inches high, the bottles are 2-½ inches high. Each bottle has a "Japan" inkstamp. Inkstamp on the bottom of the butler is, "Hand Decorated Naguya Japan," in a circle with "Trico China" written across the center. *Carson Collection.*

A go-together, two different subjects with a common theme. The cup is 1-⅜ inches high, the saucer is 3 inches in diameter. The set is marked only with a "Japan" paper label. The back of the cup displays hand-painted flowers and leaves. *Lilly Collection.*

This go-with qualifies as a miniature, the phone being but 1-¾ inches high, the phone book much smaller. Neither piece is marked. *Carson Collection.*

Huggers, like the lovebugs below, hug each other, or at least appear to hug each other. The most popular huggers, again like the lovebugs, were designed by Ruth Van Tellingen (later Ruth Van Tellingen Bendel), and made by the Regal China Company.

The pair of skulls with gold trim are nodders. Nodders nod. The long round chambers which hold the contents of the shakers extend down into the base and act as counterweights which make for a slow nodding motion at the slightest jiggle. When several people are sitting at a table eating, nodders never stop moving. The most common nodders are those shown below in which the figural shakers sit in a nonfigural base. They come in three different sizes, and also as condiments with the mustard located between the two shakers. Nearly all have "Patent TT," or a similar designation, impressed in the bottom. Rarer and usually more expensive are nodders which have a figural base. Some are old, some are new. The elephant shown below with the clowns riding it is an old figural-base nodder. The rattlesnake in the southwestern style basket is a recent figural-base nodder.

Here is an example of an older nodder with a figural base, probably from the 1930s or 1940s. Both the clown and the dog nod. The elephant is 3-¼ x 5-¼, has a "Made in Japan" inkstamp. *Thornburg Collection.*

The most desirable huggers are those modelled by Ruth Van Telligen (later Ruth Van Tellingen Bendel) for the Regal China Company, of which this set of love bugs is an example. Made in three different sizes, this pair at 3-⅝ inches in height is the smallest. Impressed in a circle around the very short tail of each of these love bugs is © Bendel Love Bug." You can just see it on the one on the right. "Pat. No. 2560755" (1951) is impressed in the bottom of each shaker. *Oravitz Collection.*

And here is a newer nodder, just a few years old. The base of the set is 4 inches high. It has "© 1988 / Vandor" impressed on two lines. In comparing recent nodders to older nodders, you will find the vintage models were made with more precision and, consequently, display more motion. *Thornburg Collection.*

The base of this set of nodders is 3-¼ inches long. It is marked in three ways. "Patent TT" appears in raised letters, it is ink-stamped "B602," and also has a paper label that reads, "A Quality / Product / Japan," on three lines. *Courtesy of Cloud's Antique Mall, Castalia, Ohio.*

Advertising sets, such as the plastic Esso gas pumps, were and are used to help sell a company's products. Some were given with purchases, others were premiums that could be sent for through the mail. Some advertising sets were given away free of charge. Others required one or more proofs of purchase, sometimes accompanied by a small fee. Certain advertising sets, for instance the pair below depicting Nipper the RCA Victor dog, can be misleading because they illustrate a symbol, trademark or product instead of naming the company they represent.

Here are a couple more advertising sets that are straight forward about their message. The Harvestore silos are 4-5/8 inches high, while the Red Brand fence shakers stand 3-3/8 inches. Neither set is marked. *Thornburg Collection.*

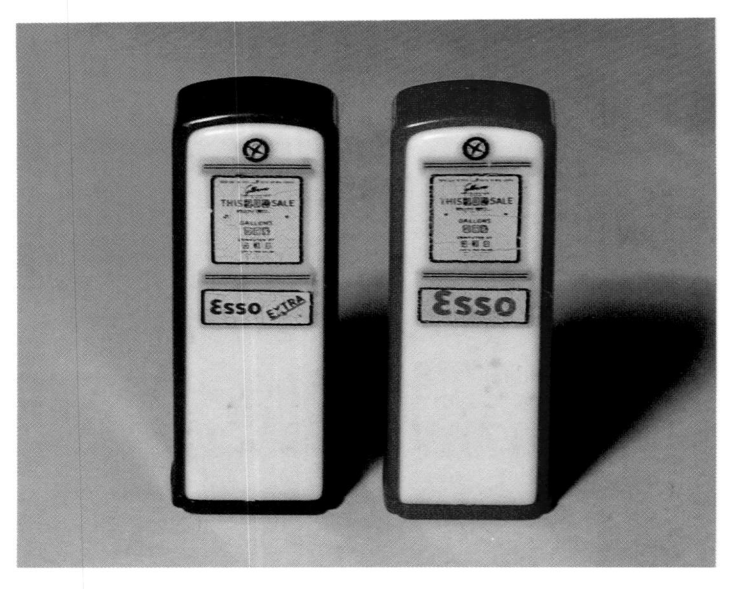

Here is advertising in it's purest form, the company's product standing tall for everyone to see. The unmarked pumps are 2-3/4 inches high. As shown below, this set advertised not only the national brand, but also the local seller. *Courtesy of Allen and Michelle Naylor.*

The kissing couples below are bench sets. Like hangers, bench sets seem to be self-explanatory, except to say they may be displayed as shelf sitters if benchs are not available.

Mechanical sets move in some way. When the plastic lawnmower below is rolled across the table, the white salt and pepper shakers move up and down like pistons. Pop-up sets, such as the television in which the shakers pop up when the knob is turned, are also mechanical sets.

Some salt and pepper shakers are made in series, Washington and its apple being one example. They are part of the Parkcraft states series. Parkcraft made two series of states, one about 1957 that had 48 states, another about 1968 that had 50. Many other themes have been used for series.

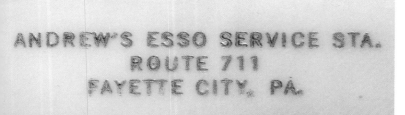

Back of one of the above Esso gas pumps showing the business where their original owner came into possession of them.

These two sets qualify as both bench sets and kissers. Kissers don't necessarily have to sit on benches. Some are standing. The cowboy is 3-3/4 inches high. Neither he nor the cowgirl is marked. Each of the Eskimos is marked "Japan." *Carson Collection.*

This advertising set is a bit more subtle, the subject being Nipper, the symbol or trademark of RCA Victor. Each of these unmarked shakers is 3-3/8 inches high. While this is a vintage set, reproductions of Nipper abound. *Carson Collection.*

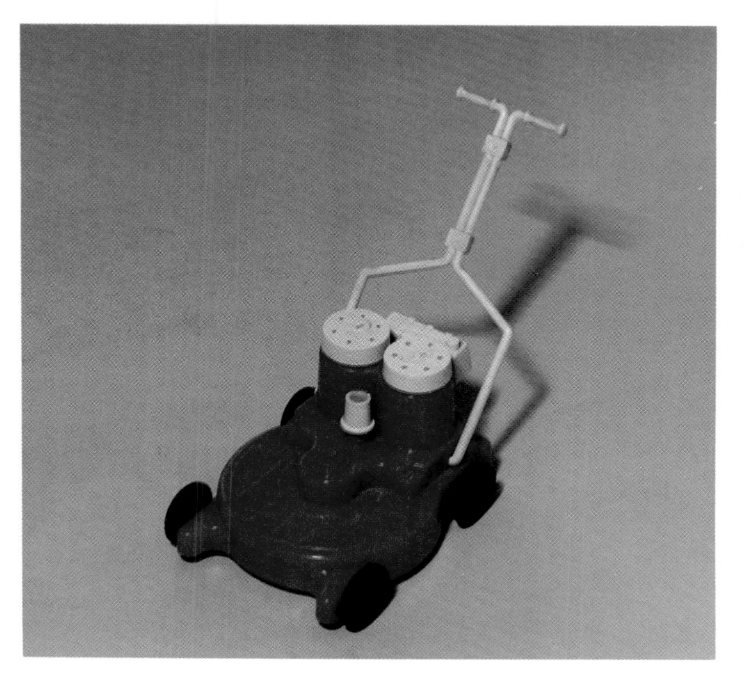

Plastic lawnmower, a mechanical set in which the two pistons (salt and pepper shakers) alternately move up and down when the mower is pushed. The deck of the mower is 4-¼ x 3-¾ inches. Mark information was not recorded. *Carson Collection.*

Here are two members of an obvious series, animals playing drums. The frog is 2-¾ inches high. It is not marked. The monkey is marked "Japan." *Carson Collection.*

Plastic pop-up television, 3 inches high. Turn the knob and the salt and pepper shakers are at your disposal. The set is marked "Made in U.S.A. / Pat. Pend." but with the words backwards. *Carson Collection.*

Probably the best known series is the Parkcraft states. As mentioned above, two sets were made. One had 48 states, the other 50. The apple is 2 inches high, Washington is 3 inches wide. The Parkcraft mark is shown in Chapter 12. *Thornburg Collection.*

Spice sets hold more than just salt and pepper. The Japanese redware dog shown below has four spices in addition to salt and pepper. This set could also be considered a hanger because the small dogs hang on the large dog. The small dogs say, "I am salt," "I am pepper," "I am allspice," etc.

Size is also used as a classification by specialty collectors. Most common here are especially tall sets and especially long sets, generally 7 inches or more in height, 7 to about 13 inches in length. One example would be the long ears of corn shown in Chapter 7 Plant Life.

One-piece salt and peppers are exactly that, one receptacle that holds both salt and pepper in separate chambers. If you see a shaker for sale that looks like it is missing its mate, and you are not familiar with that particular shaker, it pays to inspect it closely because sometimes it turns out to be a one-piece.

A Japanese redware poodle with lots of puppies, better known as a spice set. The large dog is 9-¼ inches long, the puppies are 3-½ inches high. All pieces are ink-stamped, "Japan." *Private Collection.*

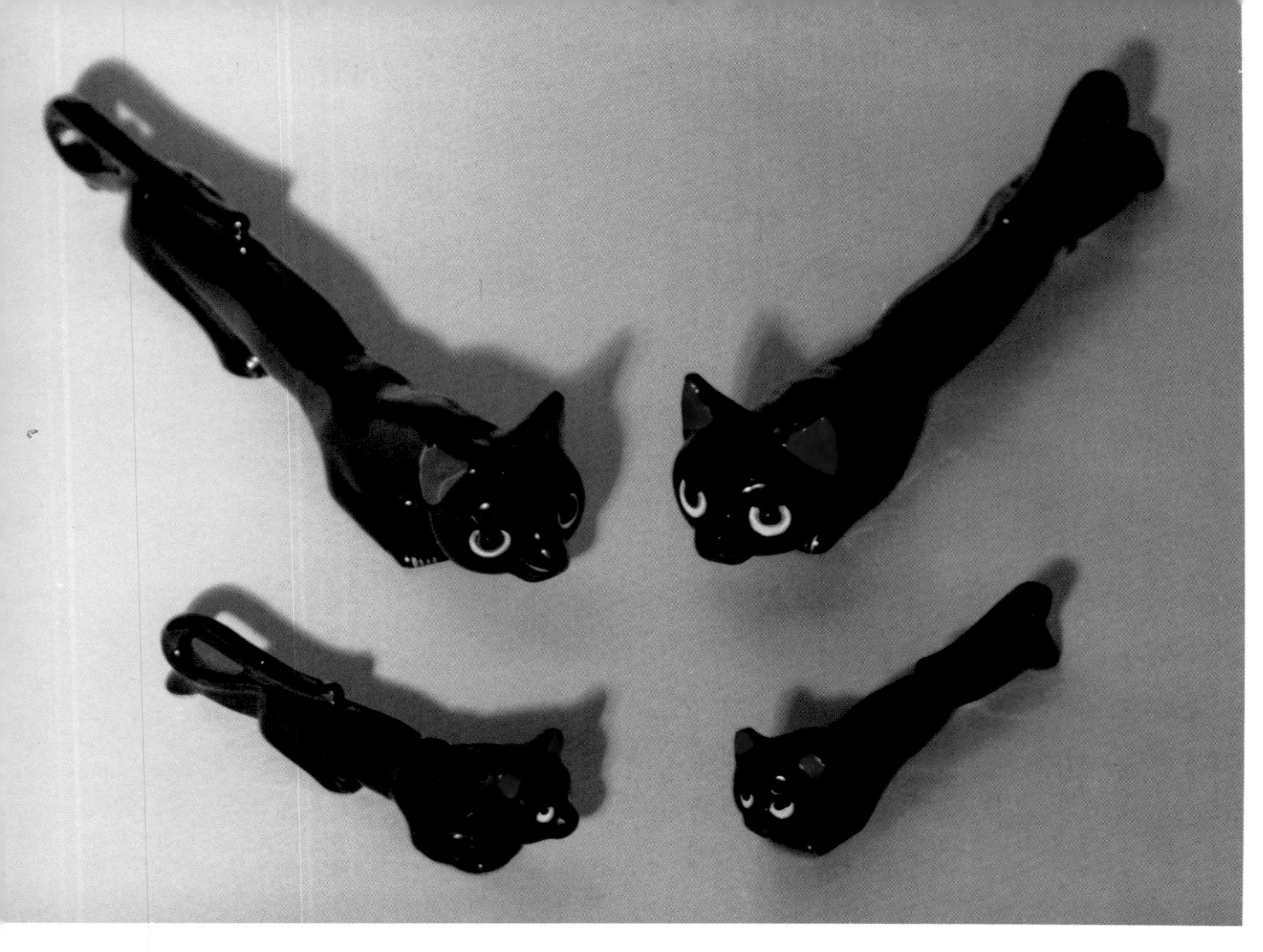

The unmarked black cats in the foreground are only 5 inches long, but those in the background measure 10-¼ inches, which is long enough to gain admittance to any collection of long shakers. They would also qualify for a collection of one-piece shakers because they are not really a pair. Each one contains receptacles for both salt and pepper, salt in the front, pepper in the back. Although black cat collectibles are most often associated with Shafford, an importer of Japanese products, these two lengthy cats were marketed by Enesco, as the paper label below clearly shows. *Carson Collection.*

Enesco paper label of above long black cat one-piece shakers.

Five examples of one-piece shakers. While I failed to record heights and marks of these shakers, it would be a pretty safe bet to say they average around three inches high, and those that are marked show they were made in Japan. *Carson Collection.*

With pushbutton sets, you push a button on the top to allow salt or pepper to sprinkle out of the bottom. The buttons are plastic, and are spring loaded so the opening that releases the flavoring automatically closes when you release pressure. Most pushbutton salt and peppers are one-piece.

Magnetic sets, as the name implies, are held together by magnets. The red skunks and blue pigs are magnetic sets.

Turnabouts are shakers that present a different image than you would expect when they are turned around. The most common theme is before and after the wedding. Several examples are shown in Chapter 5 People.

A category that is not generally acknowledged but perhaps worth mentioning is tray sets. These are pairs of salt and pepper shakers that sit on trays. They are worth mentioning because unless you have seen a picture in a book, an advertisement or the set itself, it is often impossible to tell they are incomplete. The frog set below, for instance, would look just as complete without its tray as it does with the tray to one who did not know a tray was supposed to be included. There is not much that can be done to prevent buying incomplete tray sets except to read all the salt and pepper shaker books and periodical literature you can to become familiar with them. Tray sets aren't necessarily ruined by lack of a tray, but it is always nicer to own something that is complete rather than incomplete. Also, the knowledge that a tray is supposed to be included can sometimes give you bargaining leverage when purchasing salt and peppers on the secondary market.

You don't see a lot of magnetic shakers, but here are two sets. The skunks are 3-¾ inches high, unmarked. Look closely and you will see the magnets at the base of the tails. The height of the pigs when they are together is 3-⅝ inches, also unmarked. *Carson Collection.*

Pushbutton set of black mammy and chef with original box. The shaker is 2-¾ inches high, marked, "Made in Hong Kong." Many salt and pepper shakers qualify for several different specialties. This one for example, might appeal not only to a collector of pushbutton shakers, but also to those whose salt and pepper collections were built around black memorabilia, one-piece or plastic shakers. *Courtesy of Jim and Betsy Coughenour.*

A good looking pair of shakers in themselves, these frogs are 3 inches high and have a "Japan" paper label. Not having the holder would not ruin them, but it is always preferable to have all pieces that were originally included in the set. *Carson Collection.*

Chapter 3
Building and Maintaining a Collection

On a recent summer Sunday my wife and I set up at a large outdoor flea market. While walking the aisles that morning I spied a man with a large suitcase who was pulling salt and pepper shakers from it, unwrapping them and setting them on his table. Upon chatting with the fellow, I found out the shakers had belonged to his father-in-law who had recently passed away. The seller's wife had kept what she wanted from her father's modest collection; now they were selling off what was left at $1 and $2 per pair. I stood there feeling somewhat like a Las Vegas blackjack player, throwing down a buck or two bucks every time a decent pair of shakers hit the table. By the time the man had unloaded the suitcase, I had bought 12 pair for a total of $18. I took the shakers back to our space, priced them reasonably, set them out. By early afternoon only one pair was left, and Cindy and I had taken in $129 for the 11 pair that sold.

What the above story illustrates is that salt and pepper shakers, in the view of the non-collecting public, are one of the few collectibles that command exactly the amount of respect a collector would want them to. Non-collectors generally consider shakers too valuable to pitch in the trash can, but not valuable enough to charge the market price for them, or to even investigate to find out if there is a market price. Consequently, a superb collection can often be built at a very reasonable price simply by attending flea markets and garage sales.

Another thing that salt and pepper shakers have going for them is that they are often collected by people who do not consider themselves collectors, people who do not relate to the value collectors put on things. Another story: a friend of mine met a person whose in-laws had a collection of approximately 700 sets of shakers they wanted to liquidate. They weren't collectors, the person said, but travellers who had bought the shakers as souvenirs of vacations beginning in the late 1940s. My friend, who is a dealer and does not collect salt and pepper shakers, gave the couple $700 for their collection. He said they seemed glad to get that much money for it. Within a week he had sold the choice pairs to collectors and dealers for a total of $800, and had recouped his investment by selling the remainder to a collector for the $700 he had originally paid. Although my friend was a dealer, collectors routinely use the same tactic to build their collections. In this case, a collector could have had $800 worth of choice salt and pepper shakers basically free, just for doing the little bit of work required to sell off those he did not want or need.

But no matter how many shakers you find at garage sales, flea markets or through dispersals, you will eventually want to acquire some that seem to elude you at such places. At that point you have several choices. One is to shop at antique shows. The selection will be greater, but the prices will most often be higher, too.

Another option is to purchase by mail, either by scanning the ads in trade papers and newspapers, or by hooking up with established dealers across the country who either search for what you are looking for after you send them your want list, or who regularly send customers lists or catalogs of what they have available. Buying shakers by either method will likely be more expensive than buying them at antique shows, but you probably will be more apt to find what you want. Names and addresses of mail order dealers appear in Appendix I Sources.

A third option, which is covered in much more detail in the following chapter, is to attend the annual convention of the Novelty Salt and Pepper Shakers Club. While this is probably the most expensive option, especially when you consider travel and convention expenses, it is also without a doubt the place where you will find the most complete inventory. To my knowledge, no one has ever counted the exact number of shakers available at these conventions, but estimates of recent conventions generally range from 50,000 to 70,000 sets. If you can't find what you want there, you probably aren't going to find it anywhere. More information about the Novelty Salt and Pepper Shakers Club appears in Chapter 4, and also in Appendix I Sources.

To briefly recap, salt and pepper shakers are available from a variety of sources. You don't have to be rich to build a fine collection. On the other hand, the deeper you become involved in the hobby the stronger your desire will become to own certain sets, and you will probably end up paying much more for them than you ever would have dreamed when you took up the hobby. And that's true of any collecting hobby you can name.

Chapter 4
Join the Club!

While collectors of many other items are pretty much on their own when it comes to finding other people with the same interest, locating sources to acquire their collectibles and gathering information about them, collectors of figural salt and pepper shakers have a whole army of people to whom they can turn to for help, knowledge or just plain socializing. The Novelty Salt and Pepper Shakers Club was formed in 1987 as a result of a split in the Salt Shakers Collector Club, which at that time was composed of two types of collectors, those who favored figural and novelty shakers, and those who favored art glass shakers.

Today the Novelty Salt and Pepper Shakers Club has more than 1200 members nationwide, and even a few scattered around the world in places such as Canada, England, Australia and New Zealand. It also has several state chapter affiliates that hold quarterly, semi-annual or annual meetings so members may remain active between national conventions.

National conventions are held once a year. During each convention the club takes over a hotel and turns it into a brain trust of salt and pepper shaker knowledge, and a giant salt and pepper shaker store. Respected authorities such as Melva Davern, author of *The Collector's Encyclopedia of Salt and Pepper Shakers—Figural and Novelty, Series I and II,* and Helene Guarnaccia, author of *Salt & Pepper Shakers I, II and III* attend the conventions and impart their vast knowledge to others. So do nationally known collectors and dealers such as Irene Thornburg, Sylvia Tompkins and Larry Carey. As you can see, a convention roster reads like a virtual *Who's Who of Salt and Pepper Shakers*. Estimates of the number of shakers for sale at each convention run as high as 70,000 sets as attendees turn their rooms into miniature shaker shops.

Guest speakers each year provide valuable and often obscure knowledge about the hobby. For example, at the 1988 convention, held in Battle Creek, Michigan, the guest speakers were Bob and Marianne Ahrold. You might not recognize their names, but they were the owners of the Parkcraft Company which produced the popular Parkcraft states series in the 1950s and 1960s. What collector in attendance could ever forget hearing the background on these shakers directly from the people who made and marketed them? Additionally, several contests are held, everything from dioramas featuring shakers to members dressing in costumes of shakers. A club auction is held each year.

As you can see from the pictures on these pages, each convention offers a pair of limited edition shakers of special significance. You don't have to attend the convention to purchase them, but you must be a club member.

But when you join the Novelty Salt and Pepper Shakers Club, you get much more than a yearly convention that you might or might not be able to attend. Quarterly newsletters are well researched and packed with hard to find information in articles written by the top authorities in the field. They can prove very valuable when tracking down information about

The shaker on the left was made for the Salt Shakers Collector's Club 1st Annual Convention, held in Hornell, New York in 1986. It is 3-½ inches high. Only 47 were made, each was numbered. Paperwork indicating the untitled shaker was a limited edition of the Hillbottom Pottery, Alfred Station, New York, is shown below. This stoneware shaker is unusual in that it did not emerge from a mold as most shakers do, but was hand thrown on a potters wheel. The set on the right, "Reach out to Other Collectors," commemorated the 2nd Annual Convention, which took place in Pittsburgh, Pennsylvania, in 1987. The hand is 4-½ inches long. It was made in both white and brown. The world is 2-⅜ inches high. Although unmarked, the set was made by Trish Ceramics, of Pittsburgh, where 250 copies were produced. *Thornburg Collection.*

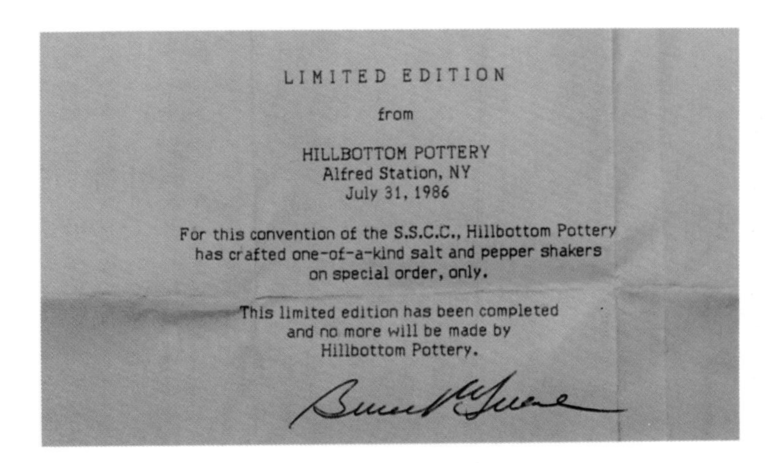
LIMITED EDITION

from

HILLBOTTOM POTTERY
Alfred Station, NY
July 31, 1986

For this convention of the S.S.C.C., Hillbottom Pottery has crafted one-of-a-kind salt and pepper shakers on special order, only.

This limited edition has been completed and no more will be made by Hillbottom Pottery.

Paper accompanying first convention shaker.

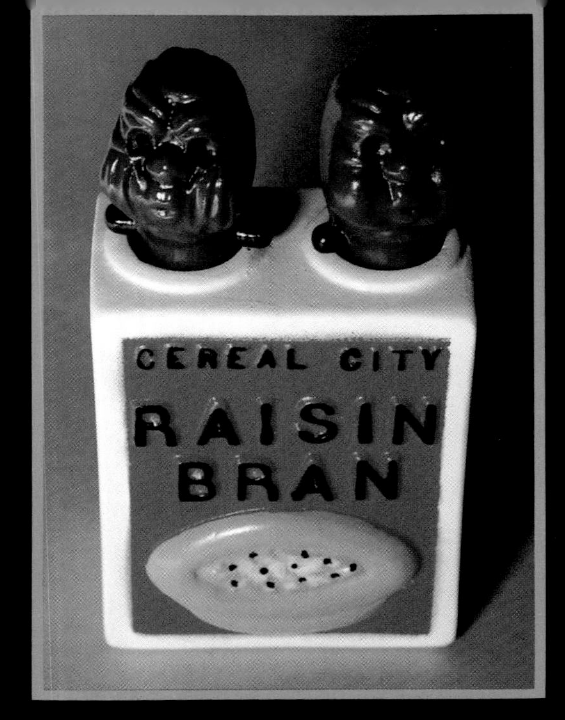

The club's third annual convention was held in Battle Creek, Michigan, home of Kelloggs, so what could be more appropriate than a box of cereal. As you can see, the shakers are nodders. The box is 3-⅝ inches high. Its back is shown below. The set, "Raisin Nodder," was designed by Hubert and Clara McHugh (authors of *Goebel Salt & Pepper Shakers—Identification and Value Guide*), and made by Aura, of Easton, Pennsylvania. One hundred seventy-five were turned out, with the first 100 being numbered. *Thornburg Collection.*

Back of the "Raisin Nodders."

My home state was the site of the fourth annual convention, held in Richfield, Ohio in 1989. A three-piece set titled "Come Travel to Beautiful Ohio," was made by Aura to mark the occasion. They were not numbered. Total production was 225. Under the car and trailer is the word, "Ohio." *Thornburg Collection.*

This set, modelled and completed by Rick Wisecarver of Roseville, Ohio commemorated the 1990 convention, held in Memphis, Tennessee. It is titled "A Float on the Mississippi." The raft is 6-¾ inches long, the black boy is 3-¾ inches high. Four hundred of these were made; marks are shown below. *Carson Collection.*

Same set as above but with the colors of the boys' clothing reversed. *Thornburg Collection.*

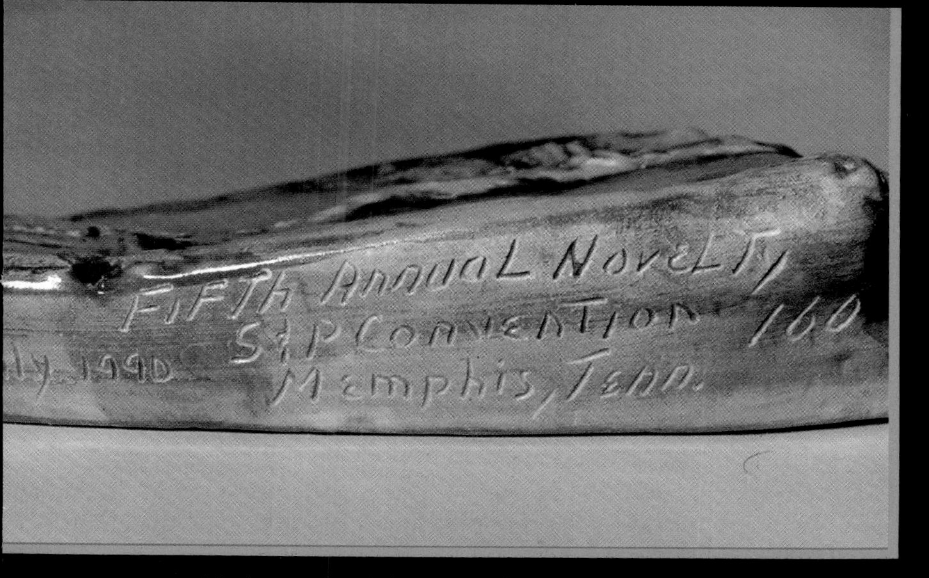

Mark on side of above raft. On the bottom is "Wihoa's Originals by Rick / Wisecarver / R. Simms," on three lines.

Titled "A Nod to Abe," this set commemorated the club's 1991 convention, which was held in Chicago. The designer was Betty Harrington, the creator of many Ceramic Arts Studio shakers and figurines during the 1940s and 1950s. Regal China Company, of Antioch, Illinois, made 600 of the shaker sets. They were not numbered. Lincoln is 5-½ inches high. The children, incidentally, are nodders. Mark is shown below. *Carson Collection.*

Mark of Abraham Lincoln set.

your shakers. But perhaps the greatest resource you get for your nominal dues is the annual club roster. It alphabetically lists all members who wish to be on it (which is the large majority), gives telephone numbers, addresses and special interests. It also lists them by Zip Code so you may quickly locate fellow collectors in your immediate area. In short, the Novelty Salt and Pepper Shakers Club is networking at its finest but I will even go one step further. If the book you are reading belongs to a friend, library or book dealer, and you have decided you want to buy it and also join the Novelty Salt and Pepper Shakers Club but can't afford to do both, join the club first. In the long run, it will become your most valuable asset as a collector. Consider the book for later. Coming from someone whose livelihood depends upon book sales, I think that's the strongest recommendation for the club that you will ever receive.

Section II: Figural Salt and Pepper Shakers

Here is the meat of the book, the large section on the figural shakers we collect and admire. On the surface it is simply a photographic overview of a collecting hobby, a panorama of the little objects that often make our hearts beat faster when we see them at antique shows, and sometimes make others wonder about diminished mental capacity when they see them cluttering our homes.

But look deeper and you will see much more. Collectors of salt and pepper shakers display on their shelves the history of civilization. From Adam and Eve to the space race, from the simplest log house to the Whitehouse, it is all presented in colorful handpainted detail. A record of who we are, where we came from, and how we got here. If an alien from outer space was cruising toward Earth without any knowledge of what it is like, Section II would give him a very good idea of what he could expect to find upon landing.

Chapter 5
People

Salt and pepper shakers depicting people are some of the most popular among collectors, perhaps because they mirror our lives—our jobs, our families, our hobbies. They express joy, sorrow, love, hate, humor, solemnity, strength, frailty and most of the other emotions and conditions that make us unique to the world in which we live.

Most "people" shakers are very detailed in both mold work and glazing or painting. A few are rather plain. As with other types of figural salt and pepper shakers, prices run the gamut from a couple dollars to several hundred dollars or more. Without regard to specific companies or artists, people shakers depicting blacks are *generally* the most expensive. Characters would be second, followed by certain occupation shakers.

Large collections have been built not only on the general subject of people, but also on sub-specialties within the genre such as bench sets, brides and grooms, or athletes. This is a large field; how it is cultivated is up to each individual collector.

Occupations and Avocations

Far and away the most produced sets of occupation shakers over the years have been those made in the images of chefs and cooks. After all, what better subject could there be for kitchen artware than the person who spends most of their time in the kitchen. Clowns probably take second place. But that doesn't mean you will not find many other ways of making a living here, too, everything from the "world's oldest profession," to growing the crops that nourish us so we have the strength to go out and find more shakers.

These ceramic shakers are 2-¾ inches high. Although you can't see it well in the photograph, "Sunshine" is written across the front of their hats. This is believed to be an advertising set of Sunshine Biscuits, Inc. The company was listed in the 1979 *Trade Names Dictionary* as a subsidiary of American Brands, Inc., with headquarters at 245 Park Avenue, New York. *Carson Collection.*

Soft (squeezable) plastic chefs, an advertisement for Tappan stoves, according to Guarnaccia in Book I. They are 4-⅞ inches high, not marked by the manufacturer. As you can see, the picture shows two pepper shakers. Salt shakers are exactly the same. These were also made in red. Sets turn up in single color, and in two colors, so perhaps they were marketed both ways. Falling in between range size and table size, they were apparently made for either as the tops of the hats may be rotated to three positions—small pour holes, large single pour hole, and closed. The legs unscrew for filling. *Carson Collection.*

Milk glass Magic Chef shakers, much older than those above. Height is 3-½ inches. Sides and backs are ribbed to afford the user a sure grip. The caps appear to be made of Bakelite. *Courtesy of Cindy Schneider.*

The Campbell Kids in hard plastic, 4-¼ inches high. On the back they say, "Permission of Campbell's Soup Company." Although not marked as such, most authorities agree this set was made by Fiedler and Fiedler Mold & Die Works (F & F), of Dayton, Ohio in the early 1950s. A nonfigural porcelain set with a picture of 1930s era Campbell Kids—they looked quite a bit different then—was produced in 1983. The Kids themselves date to 1904, when they were created for the Campbell Soup Company by Grace Gebbie Widerseim Drayton (1877-1936), a widely known portrait artist who also created Kuddle Bunny. For a more detailed history of these fascinating trademark symbols, see the excellent article, "The Campbell Kids," by Muriel O'Connor, in the Novelty Salt and Pepper Shakers Club November 1991 newletter. *Thornburg Collection.*

A Japanese condiment set, the shakers being 3-⅛ inches high. Shafford mark is shown below. *Thornburg Collection.*

Mark of Japanese chef condiment set.

Pillsbury Doughboy, 3-½ inches high. Impressed mark reads "© Pillsbury Co." This set was probably made before his name was changed to Poppin' Fresh, in 1972. Davern, in Book I, shows a similar set but it appears to be of a much whiter ceramic. She also states that when the name was changed, a female companion, Poppie Fresh, was added, and shows a picture of a plastic set of the pair. *Courtesy of Trina Cloud.*

More chefs advertising ranges, this time Magic Chef. Made of plastic, the shakers are 5 inches high. They are marked, "Old King Cole / Canton, OH," on two lines. *Thornburg Collection.*

Another pair of kid chefs, this time advertising Hot N Kold Shops. The shaker on the right is 4-⅝ inches high, neither is marked. *Thornburg Collection.*

The style of the refrigerator would seem to date this pair as being manufactured during the 1930s or 1940s. Both pieces are 2-⅜ inches high. The decal across the front of the chef's apron indicates this specific set was a souvenir of Niagara Falls. *Carson Collection.*

Heights left to right are 2-½, 4-¼ and 3 inches. All shakers are ceramic, all sport "Japan" inkstamps, red on the outside sets, green on the inside set. *Carson Collection.*

Meet Mamma and Pappa Pizza, a 4-⅝ inch high unmarked couple who apparently favor Italian food. When I see sets like this I cannot help but wonder if the missing P on the collar of the pepper was intentionally or inadvertently left off. *Carson Collection.*

These three sets measure 6, 4-¼ and 4-½ inches high left to right. The set on the left is marked with a "3031" inkstamp. The set in the middle is unmarked, while the set on the right carries a "Japan" inkstamp. *Carson Collection.*

This set is apparently supposed to be comical, but I fail to understand exactly what is happening. Did the cat kill the bird? Are these two characters so hard up they are preparing to stew their pets? Or is there some other meaning? The shakers are 4-⅝ inches high. This set is unmarked but Guarnaccia, in Book II, pictures the same set with a Shafford mark. *Carson Collection.*

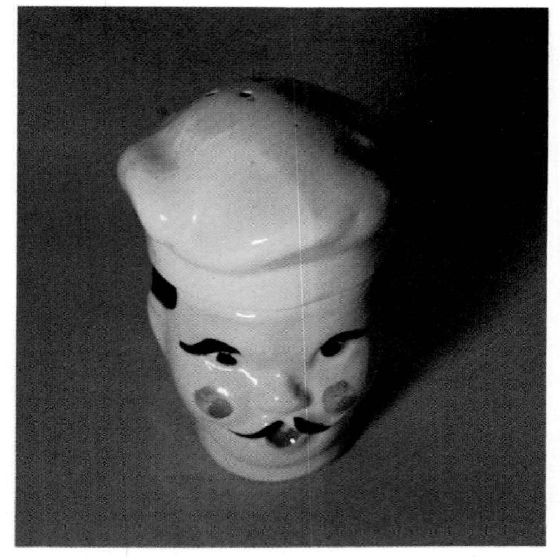

Here is a chef who must have gotten so frustrated he blew his top. The hat, which is the pepper, lifts off to reveal the salt, which is the head. Height of the unmarked set is 3-¼ inches. *Carson Collection.*

At first glance you might miss the finger under the nose of the peppery plastic chef on the left—obviously trying to prevent a sneeze—but it really makes the set. Height is 4-⅞ inches, mark is "U.K. Reg. No. 984,941 / Made in Hong Kong" on two lines. *Carson Collection.*

Kids? Girls? Eggs? In any event, they are wearing chef hats, even if they are red. Height is 3-⅞ inches. The shakers are unmarked. *Carson Collection.*

Two chefs ready to barbecue. The grill and the chefs' hats are plastic, the chefs themselves are glass. The chefs are 2 inches high. Mark on the grill is "© 1958 Rubicon, Inc." *Lilly Collection.*

These unmarked earthenware heads are made of redware, a shaker medium that usually indicates Japanese origin. They are 3 inches high. Although unseen in the picture, somewhere on them, according to my notes, is the phrase, "Souvenir of Hightown Gap, Pa." *Carson Collection.*

Here is another old set that is truly different than most of the others. It is also range size, 6-¾ inches high. *Carson Collection.*

The pair on the left is a real utility set, the salt having a corkscrew on the bottom, the pepper a bottle opener that is barely visible in the picture. Excluding appendages, the shakers are 2-¼ inches high. The shakers of the double set on the right are 2-⅜ inches high. None of the wooden pieces in this picture are marked. *Courtesy of Joyce Hollen.*

Salt, pepper and sugar bowl. The shakers are 5 inches high, the stove 4-½ inches. None of the pieces are marked. Stoves are rare, shakers are fairly common. As evidenced below, they were made in several colors. They also may have been made by several companies as they often show technical differences. For example, some of these shakers have flat bottoms while others have ridges across them. *Oravitz Collection.*

The chef and cook head shakers are each 2-¼ inches high, and have "Japan" inkstamps. The boy and girl just kind of wandered in from the section on children later in this chapter, and couldn't be removed because of the way the picture is arranged. As long as they are here, note that the shakers are marked "Poinsettia Studios California," and that the boy's height is 2-¾ inches. (I know, I know, the chef looks much taller than the boy. The illusion is due to the angle at which the photo was taken. I doubted the measurements, too, so I placed dividers on the chef from the bottom of the shaker to the highest point on the front of his hat. When transferred to the boy, they came only up to his eyes.) *Carson Collection.*

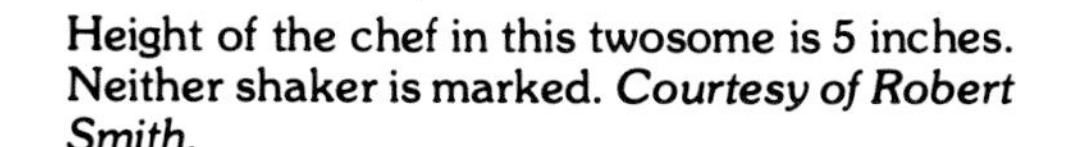

Height of the chef in this twosome is 5 inches. Neither shaker is marked. *Courtesy of Robert Smith.*

Different size here. The shaker on the left is 4-3/8 inches high, the one on the right 4-1/8 inches high. *Carson Collection.*

As with most of the others, not much is known about these three sets. The cook on the far left is 4-1/2 inches high. None of the shakers are marked. *Courtesy of Jim and Betsy Coughenour.*

Wooden black chef and mammy, each 3-1/4 inches high. "Japan" inkstamp. *Carson Collection.*

While the sets above are believed to have been made several decades ago, probably in America, this pair is of recent vintage. Its paper label reads, "Made in Taiwan." Both shakers are 4-3/4 inches high. *Courtesy of Jazz'e Junque Shop, Chicago, Illinois.*

The outside shakers are unmarked, 4-1/2 inches high. The inside set is marked "Japan." *Courtesy of Jim and Betsy Coughenour.*

The main figure of this hanger set is 5 inches high. None of the pieces are marked. *Carson Collection.*

Height of the chef here is 5-1/8 inches. Neither shaker is marked. *Oravitz Collection.*

All of these shakers are unmarked, other than the New Orleans souvenir line on the pair on the left. The chef on the right is 5 inches high. *Courtesy of Jim and Betsy Coughenour.*

It's women only in this southern kitchen of yesteryear. The outside set is 3 inches high, the inside set 2-¼ inches high. None are marked. Davern, in Book II, shows three examples of the outside set, each decorated differently--two white with red trim, the other basically pink. More importantly, one of them in her book is marked Elbee Art, Cleveland, Ohio. *Carson Collection.*

A more recent set, 3 inches high, unmarked. *Lilly Collection.*

A maid and butler, he being 4-¼ inches high. Neither is marked. *Carson Collection.*

Hard plastic Aunt Jemima and Uncle Mose, made for Quaker Oats by Fiedler and Fiedler Mold & Die Works, of Dayton, Ohio. Quaker used the shakers as a premium for its Aunt Jemima Pancake Mix. This stove set is 5 inches high, a smaller table set was also made. In fact, the premium campaign included an entire series of kitchen containers and dispensers—cookie jar, syrup, cream and sugar, plus a spice set for allspice, cinnamon, cloves, ginger, nutmeg, and paprika. The F & F mark is shown below. *Courtesy of Allen and Michelle Naylor.*

F & F mark of Aunt Jemima and Uncle Mose. Somewhat difficult to make out, it reads, "F & F / Mold & Die Works / Dayton, Ohio / Made in U.S.A." on four lines.

Out of the kitchen and on to other occupations. Although this Twin Winton cookie jar dwarfs the company's matching salt and peppers, they are actually quite tall, 6-¼ inches. The cookie jar stands 12 inches. The shakers are unmarked while "Twin Winton Calif. USA © '60," is impressed in the cookie jar. *Courtesy of Joyce Hollen.*

The simple rule here is that green skirts are originals, red skirts are reproductions. The green-skirted shakers were made by F & F for Langniappe of New Orleans to advertise its Luzianne Mammy Coffee. Originally, there were dark vertical stripes on the skirts which have since worn off on this pair. Height is 5 inches. The outside shakers, also plastic, are reproductions. They are worth but a fraction of the originals which are quite rare. *Thornburg Collection.*

Mark of original Luzianne Mammy shakers. Another way to remember which is the original and which is the reproduction is that only the originals have the words, "Luzianne Mammy," on their bottoms.

At the opposite end of the spectrum from the giant cookie jar above is this miniature set of monks, just ⅞ inches high, photographed on dimes to help illustrate their small size. The shakers are made of plastic. They are not marked. *Carson Collection.*

From Canada to England for a pair of bobbies. Height of the pair is 4-½ inches. Neither component is marked. *Oravitz Collection.*

The rebel soldier is 4-⅜ inches high, the cannon is 4 inches long. Paper label says "Made in Japan." *Carson Collection.*

All of these shakers were made by Goebel, all have Goebel marks. The musical pair in the rear is called "Monks with Instruments." They are 4-½ inches and 4-¾ inches high left to right. The other monks dressed in brown are called "Friar Tuck." The pair dressed in red is called "Cardinal Tuck." Monk shakers were first issued by Goebel in 1954 beginning with a condiment set. Some monks were issued in baskets. For a very detailed explanation of all known Goebel monk shakers, see *Goebel Salt & Pepper Shakers—Identification and Value Guide,* by Hubert and Clara McHugh, a book I cannot recommend strongly enough for all shaker collectors. Information on how to obtain it is listed in Appendix I Sources. *Thornburg Collection.*

Meet the hooker and the booker. She appears to be offering an occupational bribe. He is apparently trying to resolve a conflict between hormones and law enforcement. Which one prevailed is unknown. The shaker on the left is 3-1/16 inches high, the one on the right is 3-⅛ inches. Both are made of metal, neither is marked. Guarnaccia, in Book I, shows what appears to be this same policeman with an obvious male thug wearing a red shirt and gray pants. I don't know which, if either, set is correct. Both themes are appropriate; maybe it was marketed both ways. *Carson Collection.*

On to English toy soldiers. The inside pair was made by Goebel. They are 3-⅜ inches high, have a full bee mark. According to the McHugh's, they debuted in 1931. The set on the outside carries an inkstamp, "Handpainted / Japan," on two lines. I remember comparing these when I photographed them, and determining the Goebel shakers were of much higher quality. Looking at them now, however, I think I find the Japanese pair more appealing. *Thornburg Collection.*

Canada's finest, the Royal Canadian Mounted Police, 2-¾ inches high, unmarked. *Carson Collection.*

Civil War soldiers, 5-½ inches high. Each has a "Japan" paper label. I find the subtlety of design fascinating here, the creator of the set doing a masterful job of making his work inoffensive to both sides. Northerners should find the staid composure of the Union soldier appealing because it indicates he is in charge. Southerners can take pride in the fact that the Confederate soldier appears to be the more aggressive of the two. *Carson Collection.*

Old Salty and Cap'n Pepper, 6-¾ inches high. Paper label reads, "Our / Own / Import / Japan" on four lines. *Courtesy of Allen and Michelle Naylor.*

Another Old Salty and Cap'n Pepper. But this time they are shorter, 4-⅞ inches high, and made of plastic. Neither shaker is marked. *Carson Collection.*

A sailor and his dream, a lovely mermaid. This Regal China set is also the dream of many collectors, as you do not see it every day. The sailor is 3-¾ inches high, the mermaid 3-⅞ inches high. Impressed on the back of both shakers is "© Van Tellingen", while "Pat. Pend." is impressed on the bottom of each. *Oravitz Collection.*

Guarnaccia, in Book II, pictures this captain and tugboat as part of a series that includes a car and driver, baker and pastry table, and a fireman and fire engine. There may be others. All are well done and worth searching for. The man here is 3-¼ inches high, the boat is 4 inches long. It has an inkstamp, "H-87," on the bottom. *Courtesy of Allen and Michelle Naylor.*

Busts of Old Salty and the Cap'n made of wood composition. They are 3-⅝ inches high, carry a "Japan" paper label. Printing on the shaker on the left indicates they were purchased as a souvenir of Annapolis, Maryland. *Carson Collection.*

Two singles, not a pair. The correct second member of the set has an open jacket, light color shirt and pants, and is holding a watch. (See Guarnaccia Book II, page 133.) The other member is also slightly shorter than this one, which is 4-⅜ inches high. The example here is marked, "Japan." *Carson Collection.*

This is the Lennox Furnance Company man. He is 4-½ inches high. His impressed mark, showing a date of 1950, appears below. Inkstamp on the bottom reads, "Manufactured by / Juan Products / Columbus, OH" on three lines. *Oravitz Collection.*

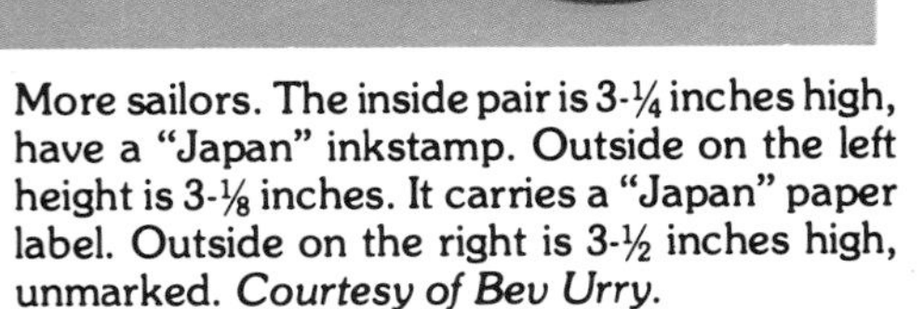

More sailors. The inside pair is 3-¼ inches high, have a "Japan" inkstamp. Outside on the left height is 3-⅛ inches. It carries a "Japan" paper label. Outside on the right is 3-½ inches high, unmarked. *Courtesy of Bev Urry.*

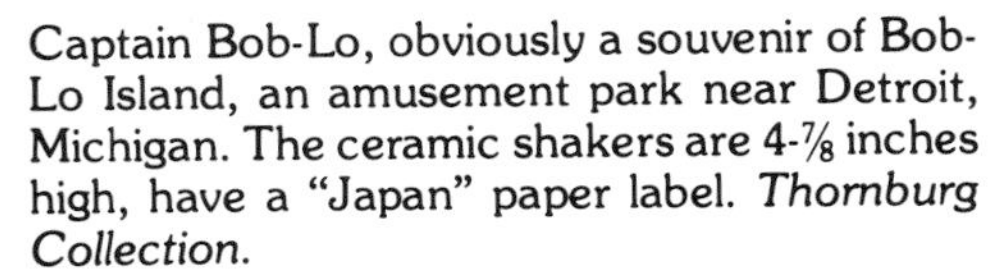

Captain Bob-Lo, obviously a souvenir of Bob-Lo Island, an amusement park near Detroit, Michigan. The ceramic shakers are 4-⅞ inches high, have a "Japan" paper label. *Thornburg Collection.*

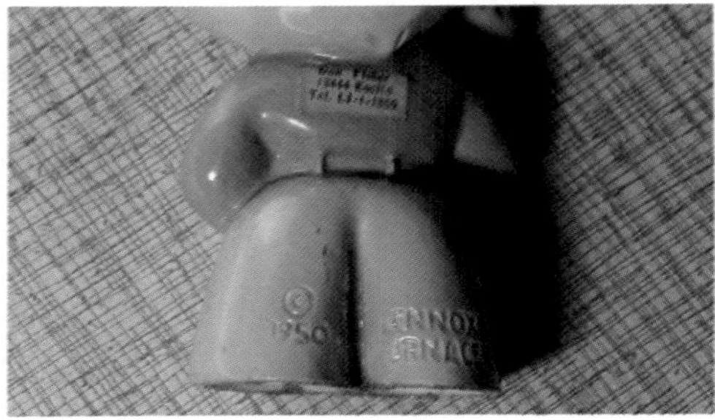

Impressed mark of Lennox Furnance Company man. The decal reads, "Don Fisher / 13444 Euclid / Tel. LI-1-1350."

The fellow on the left is 4 inches high, has a "Japan" inkstamp. The fellow on the right is 4-¾ inches high. He is not marked but each of his bags is inkstamped, "Japan." *Thornburg Collection.*

A farmer and his wife made of chalkware. Height is 2-¾ inches. Like most chalkware sets, the pair is unmarked. *Carson Collection.*

This set is called Ma and Doc. They are 4-¾ inches high. Impressed mark is "No. 97 / Imperial Porcelain / Zanesville" on three lines, indicating they are a product of The Imperial Porcelain Company of Zanesville, Ohio. They are from one of the lines Paul Webb designed for the company. Webb was the artist who developed Lil' Abner for Al Capp. *Carson Collection.*

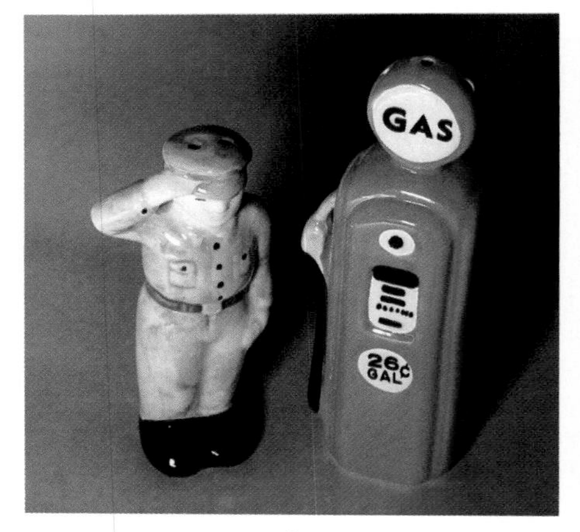

Gasoline pump and service station attendant. The pump is 5 inches high; each piece has a Vandor paper label. While Vandor is a quality company that markets only top of the line imports of its own design, it could mark these nostalgic type shakers with something more permanent than a paper label. Considering the globe and the 26 cents per gallon price that is shown, an unscrupulous dealer could remove the paper labels and attempt to pass the set off to a beginning collector as something old—at a premium price, of course. And a reputable dealer could honestly mistake it for an older set if he got hold of a pair on which the paper labels had fallen off. *Thornburg Collection.*

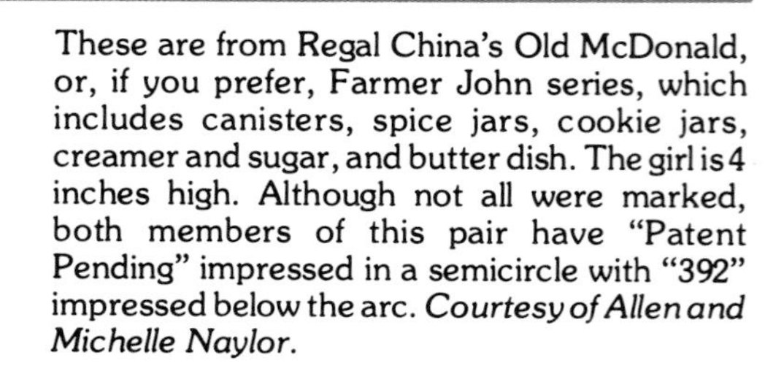

These are from Regal China's Old McDonald, or, if you prefer, Farmer John series, which includes canisters, spice jars, cookie jars, creamer and sugar, and butter dish. The girl is 4 inches high. Although not all were marked, both members of this pair have "Patent Pending" impressed in a semicircle with "392" impressed below the arc. *Courtesy of Allen and Michelle Naylor.*

Chimney sweeps by Goebel. According to the McHugh's, they appeared in 1973. An earlier version, slightly different and a tad bit taller, debuted in 1949. Each shaker is 3-¼ inches high, each has a Goebel paper label, and each has an impressed, "7323308 / 1972," on two lines. The wicker basket is original. The interesting card that accompanied it is shown below. *Thornburg Collection.*

Farmer and wife, 4 and 3-¾ inches high. Paper label reads "Goldammer / Ceramics / San Francisco" on three lines. *Carson Collection.*

Amish farm woman milking a cow. The cow is 3 inches high. Neither of these metal shakers are marked. This set is seen quite often, usually in good to excellent condition, so they are assumed to not be very old. *Courtesy of Allen and Michelle Naylor.*

The card that came with the chimney sweeps above. It is significant because Goebel, the German porcelain manufacturer that is famous for its Hummel figurines, is said never to have made a Hummel salt and pepper shaker set. Yet, from the printing on this card, it appears the company may have been labeling these shakers Hummel.

The cowboy is 3-⅝ inches high. This ceramic set has a "Japan" inkstamp. *Carson Collection.*

Above clowns together.

The clowns on the outside are 3-¼ inches high, have a "Made in Japan" inkstamp. The inside set is not marked. *Carson Collection.*

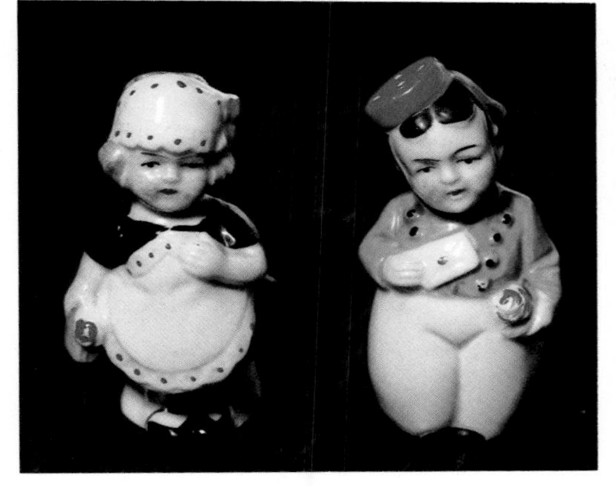

Maid and bellhop marked "Germany." These were not measured but are about 2-½ to 2-¾ inches high. *Private Collection.*

The sitters on the left are 4-½ inches high as shown. They are not marked. The heads in the middle have a "Japan" paper label. The set on the right has a "Japan" inkstamp. Each member is 3-½ inches high. *Thornburg Collection.*

Maids and clowns. When I first saw "Dearie is Weary," my initial thought was that it wasn't a true pair, and I would probably have to look around for a "Jeeves Is Peeved" to make a set. But a closer look quickly revealed that the girls' stances are reversed, that they are carrying different objects, and that the bases are clearly marked "S" and "P." The shakers are 5 inches high, have Enesco paper labels. The clowns are a leapfrog set, sitters, shown together below. They are unmarked except for a sticker that indicates they were originally purchased as a souvenir of Huntington, Indiana. *Carson Collection.*

Not two clowns but a clown and a dog. The clown is 3-¾ inches high, the dog 2-½ inches. Both have marks of Ceramic Arts Studio of Madison, Wisconsin. *Thornburg Collection.*

These clowns have the same inkstamp mark, "Germany," as the maid and bellhop couple above. They are approximately 3 inches high. *Private Collection.*

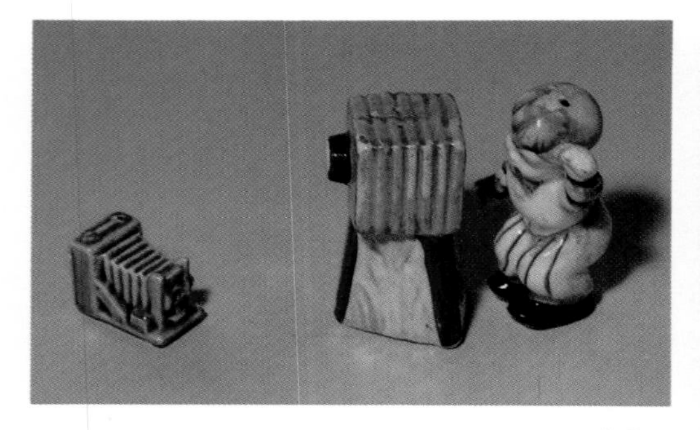

A young photographer and his camera, each 3 inches high. Neither piece is marked. No information was recorded on the single camera. *Carson Collection.*

One could easily think this set was complete if found without the base. The drums are 2-½ inches high. Mark information was not recorded. *Thornburg Collection.*

The bull is 5 inches long, the bullfighter 3-¾ inches high. The bull has a "Japan" paper label, the bullfighter a "PAC" inkstamp. Several sets of bulls and bullfighters have been made, this is one of the more impressive ones due to the natural appearance of the participants. *Carson Collection.*

Two lethargic looking boxers; must be toward the end of a long fight. Each is 3-⅜ inches high, each has a "Japan" paper label. *Carson Collection.*

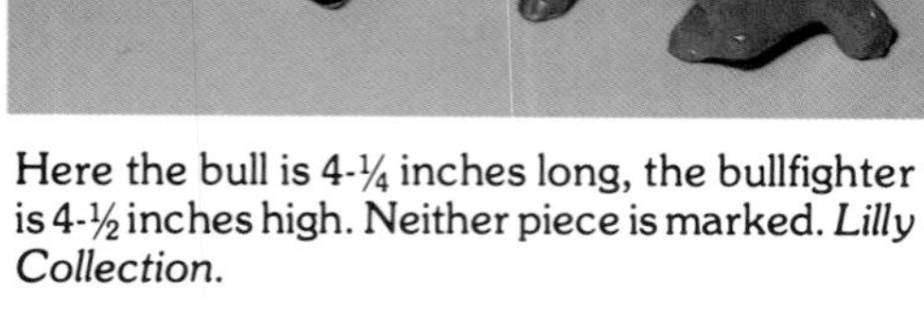

Here the bull is 4-¼ inches long, the bullfighter is 4-½ inches high. Neither piece is marked. *Lilly Collection.*

Here we have an occupation (bartender), an avocation (drinker) and an unfortunate fact of life (horse's patoot). The horse and the drinker lift off as the salt and peppers. The drinker is 4-⅛ inches high. The base has a "Made in Japan" paper label, and the entire set is made of redware. *Oravitz Collection.*

The salt of this set is the surfboard and the wave, the surfer lifts off. The salt is 4 inches long, the surfer 2-⅛ inches high. There are no marks. *Oravitz Collection.*

Condiment nodders of bullfighters. The bases are 4-⅞ inches long. Only mark is "Pat TT" in raised letters. *Thornburg Collection.*

A bit grotesque perhaps, but humorous, too. The shaker on the left is 4-3/16 inches high, the one on the right 3-⅞ inches long. Both shakers sport "Japan" paper labels. *Oravitz Collection.*

Three avocations—fishing, hunting and golfing. Each person in this series is approximately 3-¾ inches high. All have red and gold Enesco paper labels. *Carson Collection.*

Looks like a similar fate befell this chap. The head is 1-¾ inches high, the football is 3 inches long. Neither piece is marked. *Lilly Collection.*

Shutterbug from above series, companion piece is unknown. Height is 4-½ inches, red and gold Enesco paper label. *Carson Collection.*

Native Americans

Two of the darkest clouds casting shadows on the history of our freedom-loving country are the despicable manner in which it has treated native Americans and Afro-Americans. Strangely, salt and pepper shaker manufacturers have handled each race differently. While blacks have often been portrayed in demeaning caricature on shakers, Indians have generally been portrayed with class and dignity. This circumstance becomes even more intriguing when you consider that most of these shakers were made from the 1930s through the 1960s, a period during which Hollywood thrived on B-grade movies in which the cowboys always won and the Indians always lost.

The unmarked pair in front is 2-½ inches high, made of redware. The canoe set, made of a white-bodied ceramic, is also unmarked. The canoe is 7-¼ inches long. *Carson Collection.*

The base holding the drum salt and pepper is 4 inches high. Inkstamp says "Handpainted / Japan" on two lines. The canoe set appears to be the same as above, just painted differently. This one has a "Japan" paper label. *Carson Collection.*

Hiawatha is 5 inches high, Minehaha 7 inches high. In front are the Wee Indians, the girl standing 3-¼ inches high. All have Ceramic Arts Studio marks. *Thornburg Collection.*

The teepee is 3-½ inches high, the Indian 3-¼ inches. The set is marked "Made in Japan" on one line. *Carson Collection.*

If this young brave was a real Indian, he would probably take a lot of razzing about his feminine appearance. Height of the two ceramic pieces together is 5 inches. Neither is marked. *Carson Collection.*

The drums are 3-⅛ inches high, have "Victoria / Made in Japan" paper labels. The back of this set, is shown below. *Carson Collection.*

This is a Sandy Srp miniature set; what you are looking at is probably smaller than you think. The Indian is only 1-⅛ inches high. The buffalo is 1-9/16 inches long. *Carson Collection.*

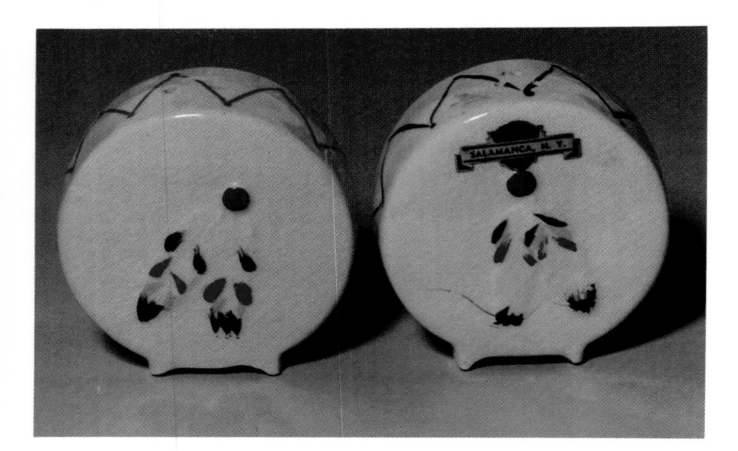

Back of above Indian drums. Souvenir decal says "Salamanca, N.Y.," a city about 50 miles south of Buffalo. *Carson Collection.*

For size reference, the boy in the middle set is 3-½ inches high. The set on the left has a "Japan" paper label, the set on the right a "Japan" inkstamp, while the set in the middle is unmarked. *Carson Collection.*

The girl of this pair is 3-¼ inches high, has a "Japan" inkstamp. *Carson Collection.*

Each of the shakers in this set, a souvenir from the Dells of Wisconsin, is 3-⅛ inches long. Paper label reads "A Quality Product / Japan" on two lines. *Carson Collection.*

The set on the left, 3-⅛ inches high, has a "Japan" inkstamp. The taller of the middle pair is 3-½ inches high, unmarked. Both left and middle are ceramic, the set on the right is chalkware. It is unmarked. *Carson Collection.*

Ceramic Indians or Eskimos, take your pick. Each is 4 inches high, neither is marked. *Thornburg Collection.*

The boy of the kissing couple is 3-⅝ inches high. Inkstamp says "Japan." The shakers on the right are 4-¼ inches high, unmarked, probaby not a pair. *Carson Collection.*

The unmarked Indian children are made of terra cotta bisque. They are 4 inches high. *Carson Collection.*

These kissers, and several others in the book, are part of a Napco series. Each of the pieces is about 3-¾ inches high. They have a Napco inkstamp, plus a paper label that reads "National Potteries Company / Cleveland, O. / Made in Japan" on three lines. *Carson Collection.*

The set on the left is made of chalkware. The shakers are 4 inches high, unmarked. The ceramic set on the right was a souvenir of Conneaut Lake Park, Pennsylvania—note the decal around the top of the girl's dress. *Carson Collection.*

The set on the left has a paper label that says "Souvenir of / Pembine, Wisconsin" but has no information as to origin. The man in the middle set is 4-½ inches high. This set is marked "—Mark L Exclusive / Made in Japan" on two lines. The set on the right is rather unusual, sporting turned wood bottoms with ceramic heads. It is not marked. *Carson Collection.*

Indian nodders. Both bases are 3-½ inches high. The base on the left has "Pat" impressed, plus a Victoria paper label. Paper label of the base on the right reads, "A Quality Product." *Thornburg Collection.*

The busts on the left stand 2-½ inches high, have a "Japan" inkstamp. Center set is 2-⅞ inches high, inkstamp simply says "Handpainted." The set on the right is 3-⅛ inches high, has a "Japan" inkstamp. *Carson Collection.*

More nodders. The base of the set on the right stands 4-¼ inches high. It has a "U435" inkstamp along with a Royal paper label which is shown below. The base on the left has "Patent TT" in raised letters. *Thornburg Collection.*

The whiteware of these two sets is identical, but note the much higher quality of decoration on the pair on the right. Height is 3-⅞ inches. The set on the right has a "Japan" inkstamp, the set on the left is unmarked. As with several other sets in this section, the painter who decorated the set on the left didn't allow the chief very much masculinity. My theory is that painting salt and pepper shakers in Japan, as in America, was probably a rather low paying, menial job that was performed by people with little education who had no knowledge of native Americans. Under those circumstances, a reasonable person might assume that only women would adorn their heads with feathers, and thus paint them that way. *Courtesy of Joyce Hollen.*

Royal paper label of the nodder on the right in the above picture.

The set on the right is same as the two above but, again, painted differently. The shakers have "Japan" inkstamps. The set on the left is 2-¾ inches high. It also has "Japan" inkstamps. Looking at the bases of the set on the left, it appears it might have sat in a tray or on a stand at one time. *Carson Collection.*

The chief of the pair on the inside is 3-¼ inches high. Mark is "Quality Product Japan." The outside set has a "Japan" inkstamp, but is it a pair? Read on. *Carson Collection.*

Same chief as above but this time with a totem pole, which is probably correct considering that the yellow matches, and that Guarnaccia shows it this way in Book I. Acquiring bogus pairs is a pitfall that will haunt you as long as you collect salt and pepper shakers, and there is little that can be done to avoid it. The totem pole is 3-½ inches high; the set has a "Made in Japan" inkstamp. *Lilly Collection.*

The totem is 3-¾ inches high, carries a "Made in Japan" paper label. *Carson Collection.*

The copper plating on the metal brave has just about worn off, a pretty good indication that he served as the salt since we tend to use more salt than pepper when eating our food. The brave is 2-¾ inches high. Neither piece is marked. *Carson Collection.*

Unmarked plastic set with paper lithograph. The shakers are 2-¼ inches high. *Lilly Collection.*

Foreign Costumes

Our natural curiosity with foreign countries and their people probably accounts for the popularity of this genre, which has been exploited by American and foreign manufacturers alike. As with many other subjects, Goebel and Ceramic Arts Studio account for many of the more exquisite examples, but there is also a sprinkling of Japanese shakers that rival or exceed them.

At first I titled this section "People of Other Lands," figuring that the word costume describes a style of dress different than is usually worn. Then I changed my mind after watching several television news shows and realizing that most people in foreign lands today dress very similar to the way we do. In fact, it is probable a collector of these shakers would see more foreign costumes by walking past his china cabinet than by taking an around-the-world trip. And the cost would be so much less.

Both pair are Ceramic Arts Studio. The man at far left is 3-¼ inches high. The man at far right is 4-¼ inches. The sets are called Wee Chinese and Large Chinese. Marks, or lack of them on these two sets were not recorded. Sometimes Ceramic Arts salt and pepper shakers are marked, quite often they are not. *Carson Collection.*

Welcome to the Orient. The couple at left rear, 3 inches high, is *really* Oriental, having a mark written in Chinese on the bottom. Middle rear is 3-¼ inches high with a "Japan" inkstamp, right rear is 2-⅞ inches high, also with a "Japan" inkstamp. The seated couple at left front is unmarked, measures 2-¾ inches, while the pair at front right is 2 inches high and carries a "Japan" inkstamp. *Carson Collection.*

This may be a bastardized set, but it is still rather pleasing. The center shaker, 2-½ inches high, is ceramic. All the rest is plastic. The unmarked base is 4-¼ inches long. The cups have inserts that hold the salt and pepper and lift out. *Lilly Collection.*

The lid is missing on the compote. The heads of the people are the shakers; they lift off. Height of the male is 3-¾ inches. He has a "Japan" inkstamp. *Carson Collection.*

Height of this pair is 3-¼ inches, Each shaker has a "Japan" inkstamp. *Carson Collection.*

The taller member of this ceramic pair is 5-½ inches high. The shakers have "Japan" inkstamps. *Oravitz Collection.*

Another Ceramic Arts set, the sultan and the harem girl. They are 4-¾ inches and 4-½ inches high, respectively. Mark information was not recorded. *Thornburg Collection.*

The heads of this Oriental pair bob up and down when handled. They are 5 inches high, marked "A Commodore Product -3074S Japan." *Carson Collection.*

From the Far East to the Mideast. The shakers of the set on the outside are 3-½ inches high, inkstamped, "Germany." The inside set has "6386" impressed. *Thornburg Collection.*

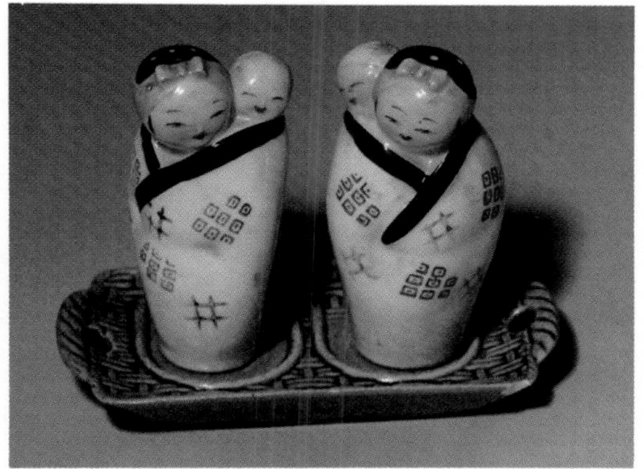

The shakers may have come with the underplate, but the raised outlines on the underplate appear to indicate differently. At this late date, who knows for sure? Shakers are 4-¼ inches high, marked "Japan." *Carson Collection.*

In her pioneer work, *Harrington Figurines,* author Sabra Olson Laumbach refers to these Ceramic Arts Studio shakers as Hindu boys. Others call them Ethiopian palace guards. They are 5 inches high, have Ceramic Arts Studio marks. Note the S and P on their turbans. *Thornburg Collection.*

Nodding snake charmer and cobra. The cobra and basket combine for a height of 4 inches. The set is unmarked. Remember that nodders with figural bases, such as these, are generally more valuable than nodders with non-figural bases. *Thornburg Collection.*

For those who want a break from people shakers, we will allow a little architecture here. The pair on the outside is 3-⅛ inches high, inkstamped "Japan." The pair on the inside is slightly smaller at 3 inches, and is unmarked. *Carson Collection.*

The shaker on the left is 3-⅜ inches high. Inkstamp mark is "Made in Occupied Japan." *Lilly Collection.*

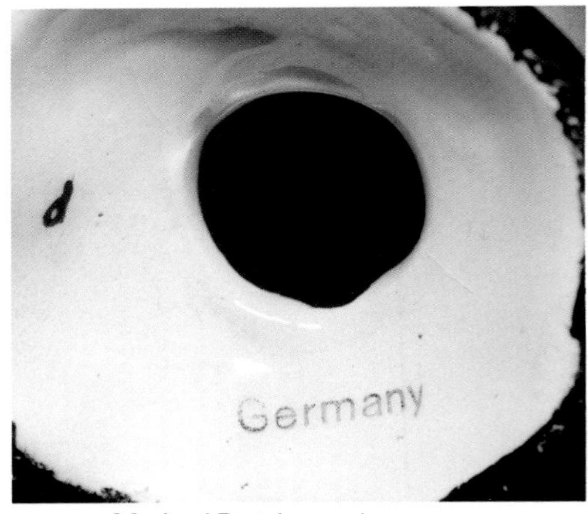

Mark of Dutch couple.

Sitting Dutch couple holding tulips, also with tulips as decoration on their clothes. The boy is 4-⅜ inches high. *Carson Collection.*

Now it's off to Holland, another popular ethnic theme—perhaps the most popular—with yesteryear's salt and pepper shaker manufacturers. The boys in these Shawnee sets are 4-¾ inches high. Each shaker has "USA 323P" impressed. Note the gold trim on the set on the right. *Robert Smith Collection.*

Nicely done unmarked ceramic Dutch couple, 4-⅜ inches high. The set appears to be American judging by style, weight and decoration. *Carson Collection.*

A napkin and shaker set made of plastic. The windmill, split to hold the napkins, is 4 inches high. The Dutch boy is 2-¾ inches high. The mark is "Regaline" in raised letters, and "Made in U.S.A." *Carson Collection.*

Another Dutch couple by Shawnee, but unmarked. Height is 5-¼ inches. Shawneee made an accompanying cookie jar to go with each shaker. *Robert Smith Collection.*

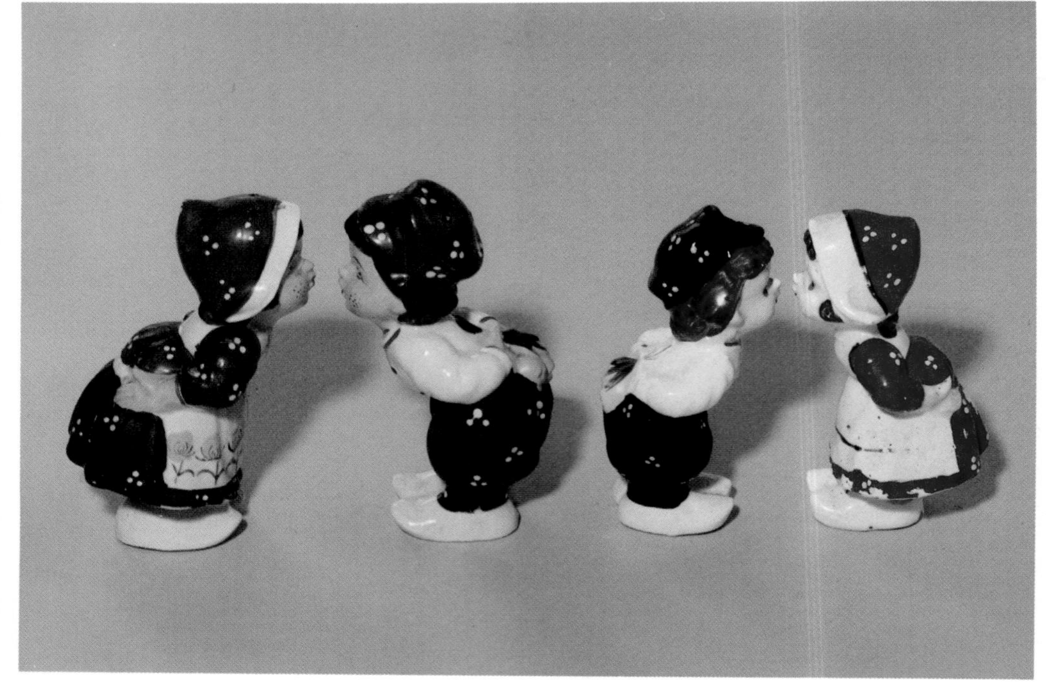

Height of the boy is 2-⅞ inches. The mark is shown below. Although they are not marked as such, they appear to be Goebel shakers P20A and P20B, which were introduced in 1936. *Private Collection.*

Dutch kissers in two sizes. The girl on the left is 5-⅜ inches high. None of the shakers are marked. *Carson Collection.*

The base of this set is 9 inches long, the salt and pepper shakers are 3-¼ inches high. Extensions, similar to tubes, hold the shakers in place. Note the tulip motif of the creamer and sugar. The paper label reads, "A Quality Product." *Thornburg Collection.*

The boat is 5-½ inches long, the boy 3-½ inches high. Neither shaker is marked. Mark of the boat is shown below. *Thornburg Collection.*

Dutch, or something else—you call it. Each ceramic bust is 2-⅝ inches high, each is inkstamped "Made in / Occupied / Japan" on three lines. *Carson Collection.*

The small pair on the inside is the Ceramic Arts Studio Wee Dutch set. He is 3 inches high, she is 2-¾ inches high. Mark of the girl is shown below. Now to the outside set. While it looks like Ceramic Arts, it is actually a Japanese ripoff. Each of the shakers is 3-¼ inches high, inkstamped "Japan." *Thornburg Collection.*

Mark of above boat. While the name Delft, and the color Delft blue, is generally associated with Holland, Delft marks have been used on pottery made in many other places, Germany and New Jersey being two examples.

Three Goebel sets. The girl of the outside set in the rear is 3 inches high; her and her mate have full bee marks. The inside set in the rear has the bee mark along with a "Bavaria" inkstamp and "P449" and "P450" impressed. The front set is inkstamped "Germany." The McHugh's show each of these sets with silver trim. *Thornburg Collection.*

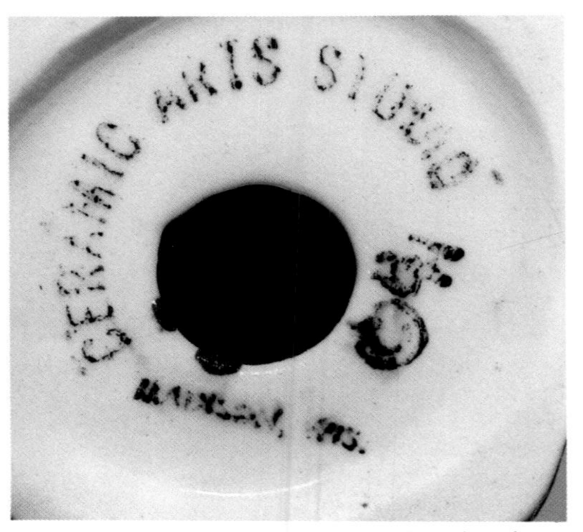

Mark of Ceramic Arts Wee Dutch girl.

The sugar is 4-½ inches high, the girl is 5-⅜ inches. None of the pieces are marked. *Oravitz Collection.*

All these shakers are 3 inches high except the boy on the right, which is somewhat shorter. The set on the left has a "Japan" inkstamp, the set on the right is not marked. *Courtesy of Joyce Hollen.*

Now it's on to Scotland, land of kilts and bagpipes. Each of the shakers in the center is 4 inches high; each is marked with an impressed, "© OC." The pair on the left is unmarked, the pair on the right bears an "H412" inkstamp and a Norcrest paper label, which is shown below. *Thornburg Collection.*

Ceramic Arts Studio Wee Scotch on the right, Wee Swedish on the left. The Wee Swedish boy is 3-⅛ inches high, the Wee Scotch boy is 3-½ inches. All pieces are marked. The mark of the Swedish girl is shown below. *Thornburg Collection.*

Norcrest paper label of above shakers.

These girl bagpipers are 3-⅝ inches high, and unmarked. *Carson Collection.*

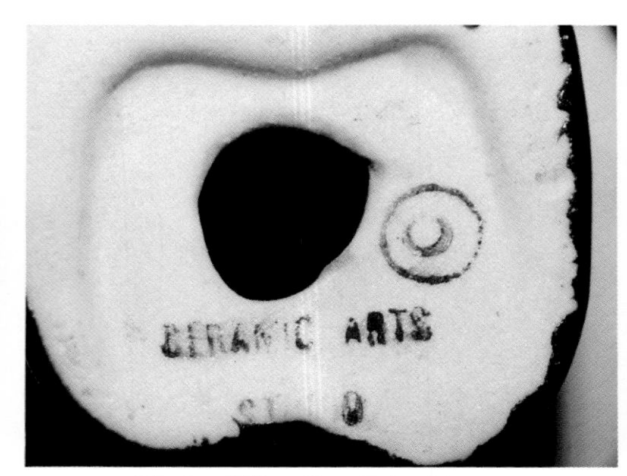

Mark of Wee Swedish girl.

The set on the left is 3-⅛ inches high, has a "Japan" paper label. The set on the right has a Miyad inkstamp, shown below. *Thornburg Collection.*

Ceramic Arts Wee French, 3-½ inches high. As evident in this chapter, Ceramic Arts Studio used many variations of marks. In some cases, even members of a pair were marked differently, as shown below. *Thornburg Collection.*

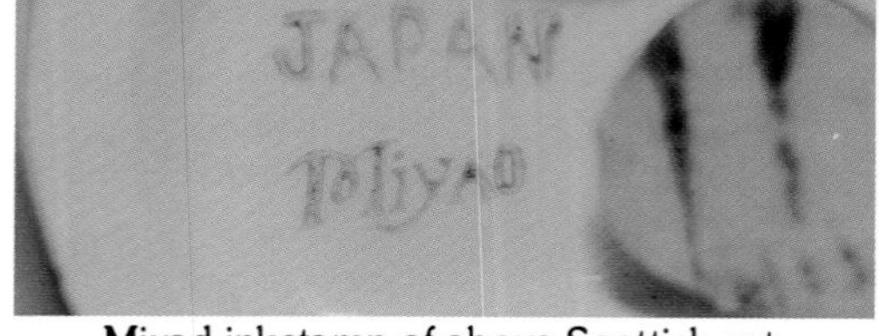

Miyad inkstamp of above Scottish set.

Plastic Scots, one with its tam removed to show how they are used. With the tams on, they stand 3-½ inches high. Mark is "Made in Hong Kong." *Carson Collection.*

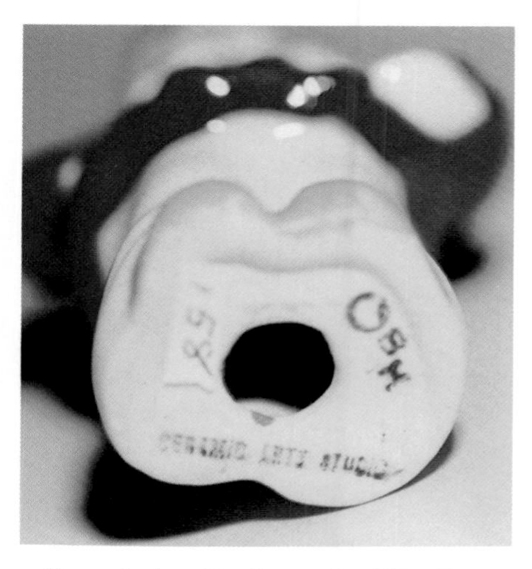

Ceramic Arts Studio mark of Wee French boy.

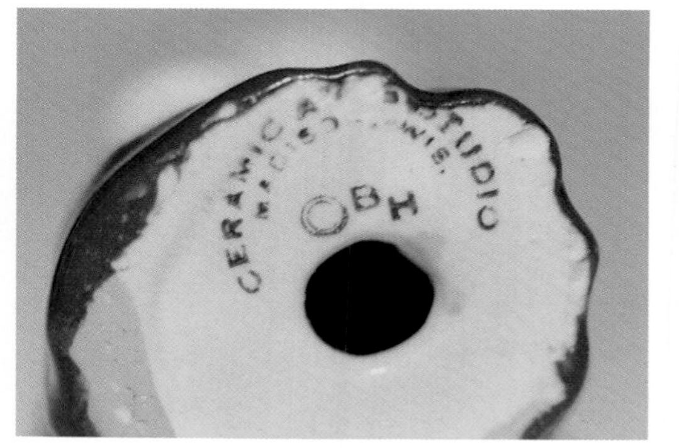

Ceramic Arts Studio mark of Wee French girl. The BH on these marks stands for Betty Harrington, the firm's designer.

Another set of black heads, 4-1/4 inches, unmarked. These are somewhat unusual in that their eyes are flashers, appearing open or closed depending upon the angle from which they are viewed. *Carson Collection.*

A Carribean islander holding a set of drums. Frame is metal wire, shakers are earthenware. With the hat, the unmarked figure is 7 inches high. *Courtesy of Trina Cloud.*

Shawnee Swiss children, with and without gold trim. Height is 5 inches. They are unmarked. *Robert Smith Collection.*

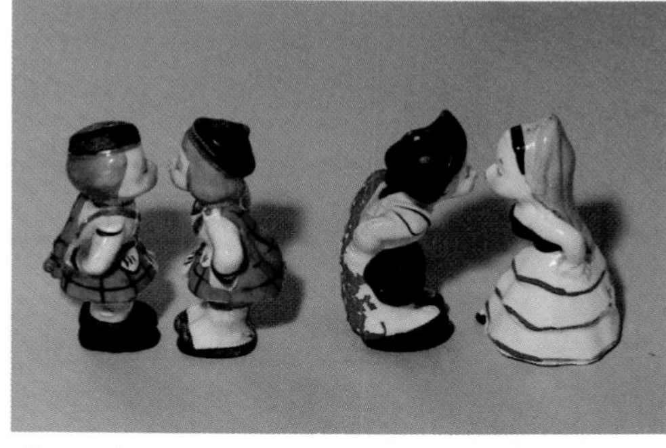

Spanish couple, and also another Scottish couple, both from a Napco series. The Spanish couple is 3-7/8 inches high, marked, "Napco / S1286/T," on two lines, with the slant between 6 and T being part of the mark. The Scottish couple is marked "Scotland / 512 86/2 / Napco / 1955" on four lines. As with the Spanish couple, the slant between 86 and 2 is part of the mark. *Carson Collection.*

African heads, 4-1/8 inches high. The ceramic set is not marked. *Courtesy of Joyce Hollen.*

Another Spanish couple. He is 4 inches high. Both shakers sport "Japan" paper labels. Additionally, there is a penciled date, 1959. *Thornburg Collection.*

This unmarked plastic pair of Mexicans is 4-¾ inches high. *Carson Collection.*

An Eskimo and his igloo, he being 2-¾ inches high. Both pieces have a "Japan" inkstamp. *Carson Collection.*

An unmarked Eskimo and his dog, made by Kathy Wolfe, of K. Wolfe Studios, West Bloomfield, Michigan. The dog is 4-¾ inches high. The address of K. Wolfe Studios is listed in Appendix I Sources. *Thornburg Collection.*

The shakers on the left and right, 3-¼ and 3-⅛ inches, respectively, have "Mexico" written in ink on their bottoms. The set in the middle is 2-⅜ inches high, carries a Japan inkstamp. The

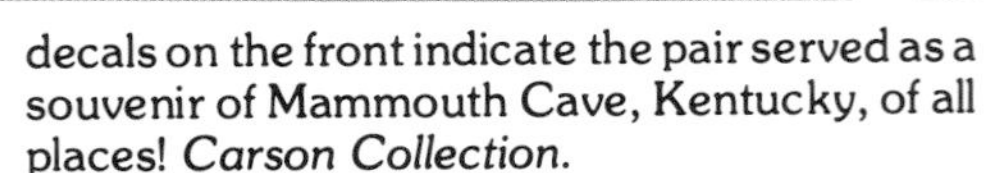

decals on the front indicate the pair served as a souvenir of Mammouth Cave, Kentucky, of all places! *Carson Collection.*

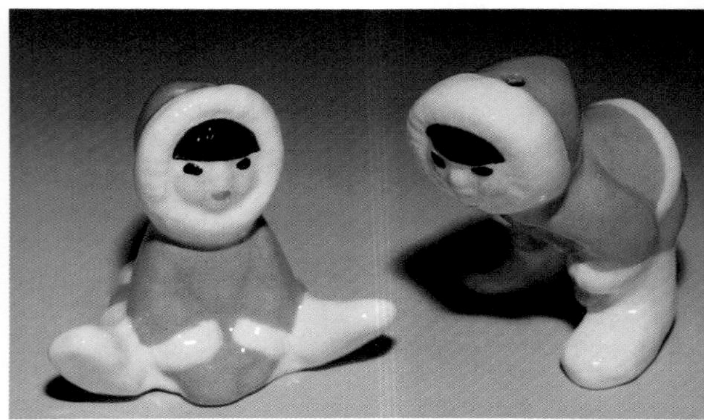

Sandy Srp miniatures, the shaker on the right being 1-¼ inches high. Sandy Srp mark. *Carson Collection.*

The pair on the left, 3 and 3-⅛ inches high, appear to be by the same maker as the set above. They are marked "Mexico." The pair on the right, 2-⅞ and 2-½ inches, are not marked. *Carson Collection.*

The Eskimos on the right are unmarked, 2-¾ inches high. I'm not sure whether the pair on the left are eskimos, but they are defintely from cold country. Height is 3-⅝ inches, they have a Victoria Ceramics paper label. *Carson Collection.*

Another Sandy Srp set, in the original plastic case in which they are often sold. The eskimo is just 1 inch high. Because the shakers are attached to the case, it can only be said that they are *assumed* to have a Sandy Srp mark. *Carson Collection.*

A nicely done but unmarked set of Eskimos, 4-½ inches high. *Carson Collection.*

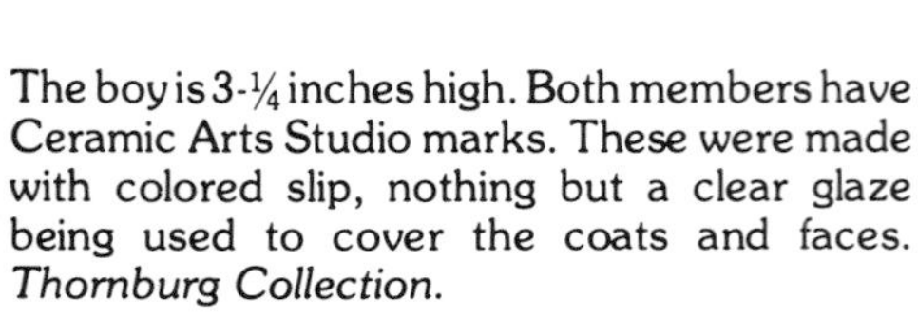

The boy is 3-¼ inches high. Both members have Ceramic Arts Studio marks. These were made with colored slip, nothing but a clear glaze being used to cover the coats and faces. *Thornburg Collection.*

The pair in the center is 4-¼ inches high. It has a Norcrest paper label. The pair on the right carries a "Japan" inkstamp. The ones on the left have an IAAC paper label, which is shown below. *Thornburg Collection.*

The pilgrim man on the left is 4 inches high, the one on the right 4-½ inches. The left pair has an Elbee Art paper label, shown below. The right pair has a "Japan" inkstamp. *Thornburg Collection.*

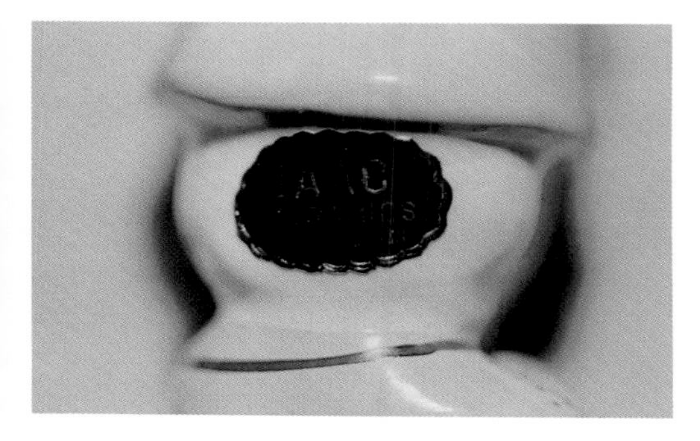

IAAC paper label of above set on the left. A bit dark, it reads, "IAAC / Ceramics / Made in Japan," on three lines.

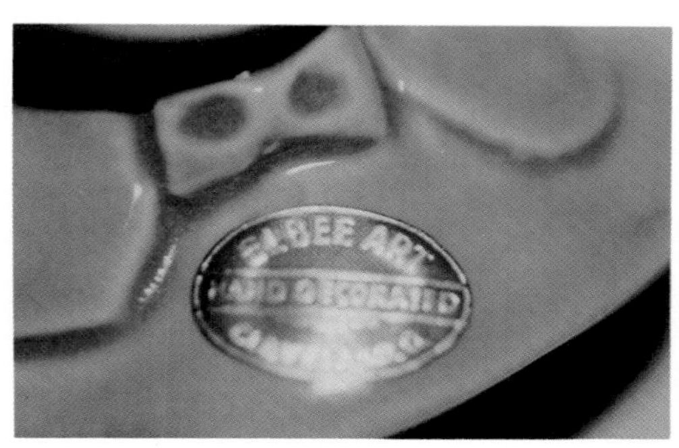

Elbee Art paper label of above pilgrims.

Hula girls, 4-¼ inches high. Paper label says Victoria Ceramics. *Thornburg Collection.*

Looks like some modeller couldn't decide whether to put this pair in a canoe, kayak or rowboat, so he made a contraption that has features of each. The vessel is 5-½ inches long, the shakers are 2-⅞ inches high. *Thornburg Collection.*

The hula girl is 4-½ inches high, the palm tree is 3-¾ inches. Both have "Japan" inkstamps. *Carson Collection.*

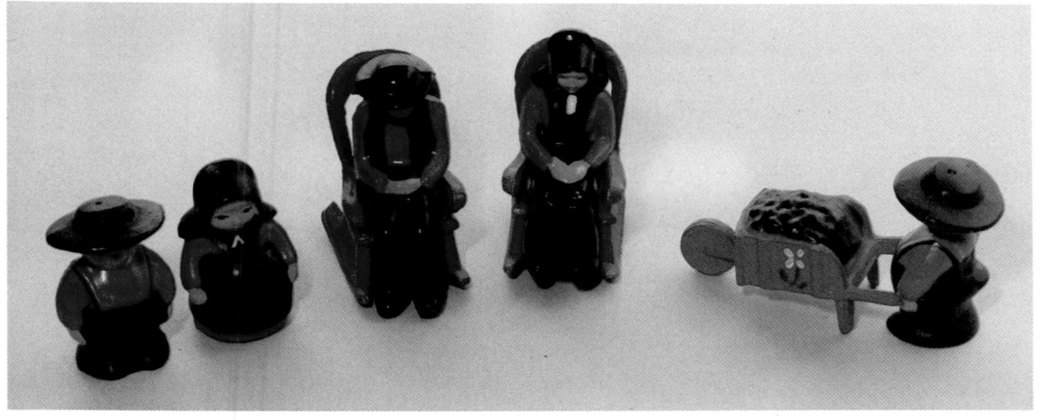

Amish folks, all metal. The two sets in the center are exactly that, two sets. The chairs and their occupants are both shakers. Chairs are 2-¾ inches high. None of these pieces are marked. *Courtesy of Allen and Michelle Naylor.*

Brides and Grooms, Married Couples, and Seniors

I wasn't around in the thirties, was too young to remember the forties, and didn't pay much attention in the fifties or sixties, but my guess is that bride and groom shakers were popular wedding gifts in cases where only token gifts were required. Nearly all of them I have seen were executed with skill and precision, as would be appropriate to present to a couple on their wedding day.

Shakers of other couples run from humorous to highbrow, while in shakers of senior citizens you will typically find the characteristics we came to love so much in our grandparents.

This bride and groom is 4-⅛ inches high. Inkstamp mark of the pair is "Made in Occupied Japan". *Carson Collection.*

Another charming couple, 5 inches high with a "Japan" paper label. *Carson Collection.*

Before-and-after brides and grooms appear to have been favorites with both importers and consumers because you see quite a few of them. This couple, 5 and 4-¼ inches high, have "Japan" paper labels along with an inkstamp, "4501 / 8209" on two lines. The "after" side is shown below. *Carson Collection.*

"After" side of above bride and groom, witty, but at the same time probably quite close to real life for many of us.

Quite a bit younger couple here. He is 3-⅞ inches high, has a pencil date of 10-2-62. She has a "Japan" paper label along with an inkstamp that reads, "Dan Brechner & Co. L-14." *Thornburg Collection.*

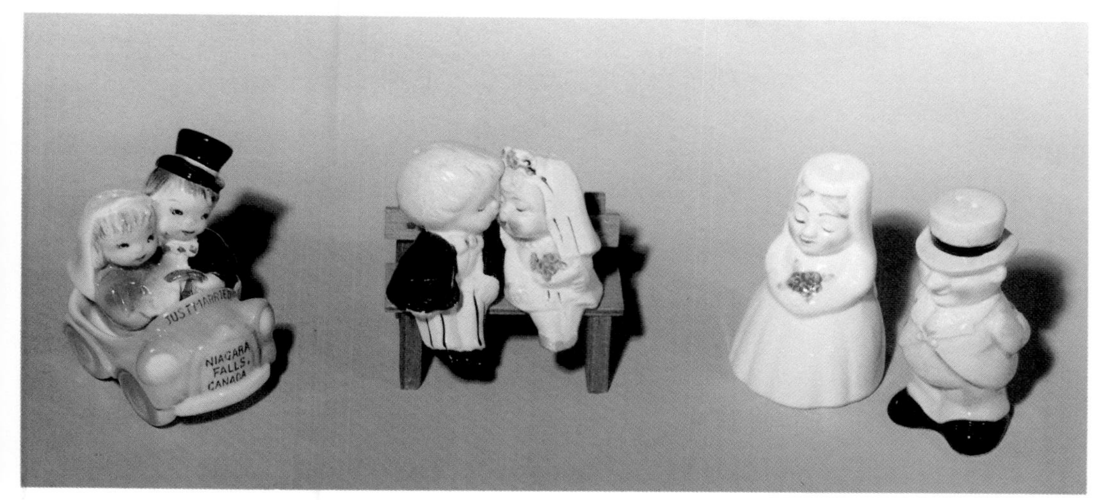

For size reference, the couple on the right is 3-⅛ inches high. It is marked, "© (diamond with flower in center) S-8." The other two sets have "Japan" inkstamps. *Thornburg Collection.*

The lady of this shopping couple is 6 inches high. The shakers carry "Japan" paper labels. *Thornburg Collection.*

I failed to record the height of this temporarily happy couple. I think it is between 4 and 5 inches. The mark is an inkstamp, "Niko China / EW (in circle) / Japan." *Carson Collection.*

The man of this unmarked pair is 5-⅛ inches high. The woman's tag reads, "Made for each other." *Oravitz Collection.*

Reverse of above shakers, not so happy now.

These shakers are 5 inches high, have a "Japan" paper label. *Thornburg Collection.*

Each of these shakers is 5 inches high. Mark is "© 1950 / Reebs / Cour O." on three lines. They were finished in both matte and gloss glazes. Matte glaze is shown here. *Carson Collection.*

Four-eyed nodders. None of the shakers are marked. The base on the left, 4 inches long, is marked "Pat TT." The other base is unmarked. Note that one set nods side to side, the other front to back. *Thornburg Collection.*

A Rick Wisecarver set, miniature versions of a pair of cookie jars the artist made several years ago. Each shaker is 3-¾ inches high. The mark is shown below. *Carson Collection.*

These Artmark shakers are 2-½ inches high, have "Made in Taiwan" paper labels. The box says, "© 1990." *Lilly Collection.*

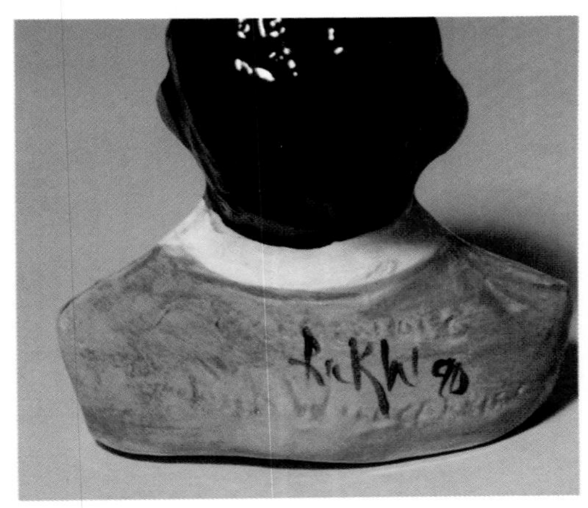

Mark of above Wisecarver salt shaker.

Not a condiment set, just a pair of shakers on a stand which measures 3-½ inches high, 5 inches long. None of the three pieces are marked. *Thornburg Collection.*

This appears to be a picture of old and young, perhaps a grandmother and granddaughter. Height on the left is 4-½ inches; a "Japan" inkstamp is present. On the right there is no mark save for a pencil date of 1962. *Thornburg Collection.*

A couple ceramic bench sets, both marked "Japan." The lady on the left is 4-¼ inches high. *Carson Collection.*

The guy on the left is 2-¾ inches high. All four shakers have "Japan" inkstamps. *Thornburg Collection.*

A modern day Abraham and Sarah. The man is 4-¼ inches high, "Made in Japan" paper label. *Thornburg Collection.*

The lady on the left is 4 inches high, her gentleman partner, 4-⅛ inches. She is marked with a Norcrest paper label, he with a "H3A" inkstamp. The lady on the far right, apparently part of the same series, has a "H12" inkstamp. The picture shows five pair, as each chair is a shaker and so is each person. *Thornburg Collection.*

Babies and Children

This section illustrates how collectibles reflect the times in which they were made. As you will notice, shaker manufacturers presented children in the traditional sense, from newborn babies through graduation from high school or college. If salt and pepper shakers were being made on the same scale now that they were during their heyday, this section would probably include several humorous sets that would carry the process one step further—son or daughter and a couple grandchildren moving back home after the split. If any manufacturers or importers happen to be reading, there is an idea you might try.

An unusual pair in that one's a nodder, one's not. The nodder is 4 inches high. The ceramic set has Sarsaparilla paper labels that indicate it was made in Japan, and copyrighted in 1984. They also have incised marks. *Thornburg Collection.*

Regal China/Ruth Van Tellingen Peek-A-Boo babies, one of the most popular American made sets. Unfortunately, also one of the rarest. The shakers were made to compliment a cookie jar of the same design. The range size set on the outside is 5-½ inches high, while the table size pair on the inside is 3-⅞ inches high. Impressed marks are shown below. *Private Collection.*

The pair of babies at rear center are 3-⅜ inches high. Inkstamp reads, "Inarco 1961 E-184." The two remaining sets have Lego paper labels. *Thornburg Collection.*

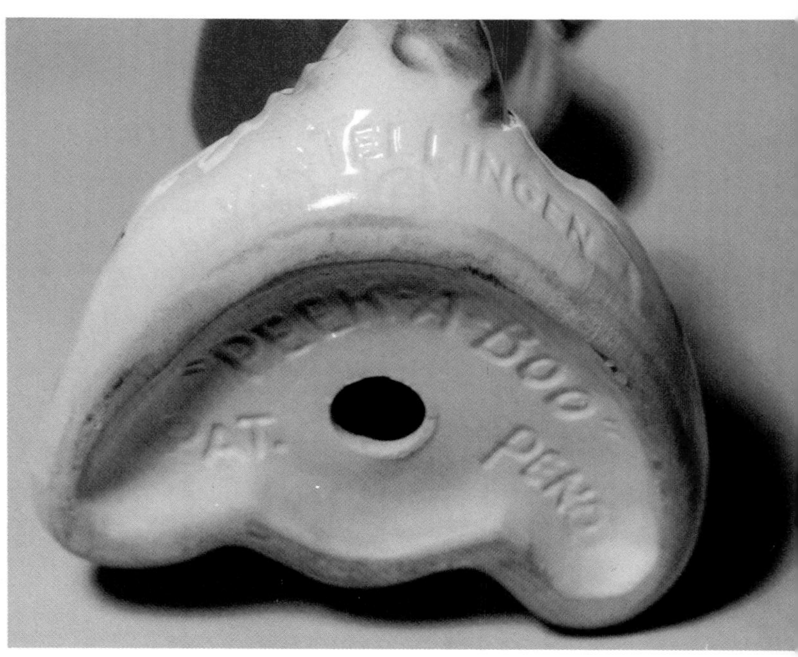

Marks on bottom and back of large Peek-A-Boo baby shakers.

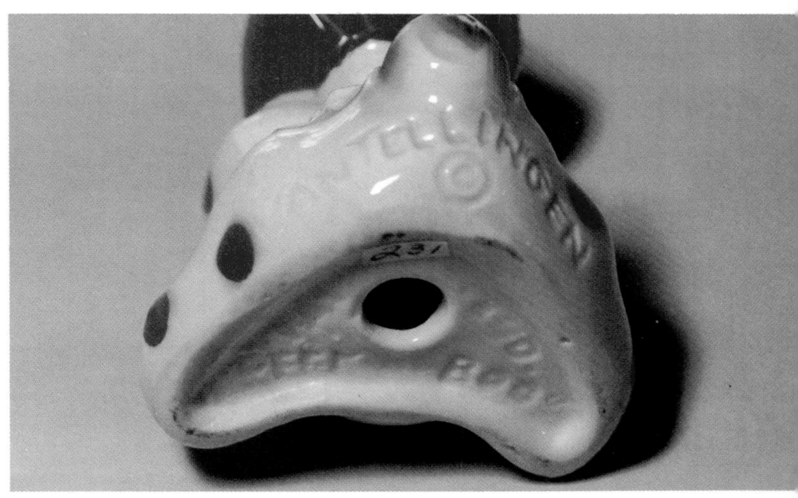

Marks of the small Peek-A-Boo babies, different than those of the large ones.

Undecorated Peek-A-Boo babies, small size. I have heard that undecorated examples were all seconds. Don't know if that is true or not, but I've seen three undecorated sets and each one had a stress crack. *Oravitz Collection.*

A pair of babies in buggies. Height is 2-½ inches, inkstamp reads, "Japan." *Carson Collection.*

Candy cane babies, each 3-⅜ inches high. The stand is 4 inches high. All three members of the set have "Japan" inkstamps. *Thornburg Collection.*

Miniature snow babies by Sandy Srp. Height is 1-¼ inches. *Carson Collection.*

This set has a Clay Art paper label. Height is 3-⅛ inches. *Thornburg Collection.*

More of Sandy Srp's excellent work. The shakers are 2 inches high. *Carson Collection.*

Looks to me like the girl is standing, the boy is sitting. The shaker on the left is 4-⅛ inches high. Mark indicates Japanese origin. *Carson Collection.*

All four are 2-⅜ inches high, all four have Enesco paper labels that indicate Japanese origin. *Thornburg Collection.*

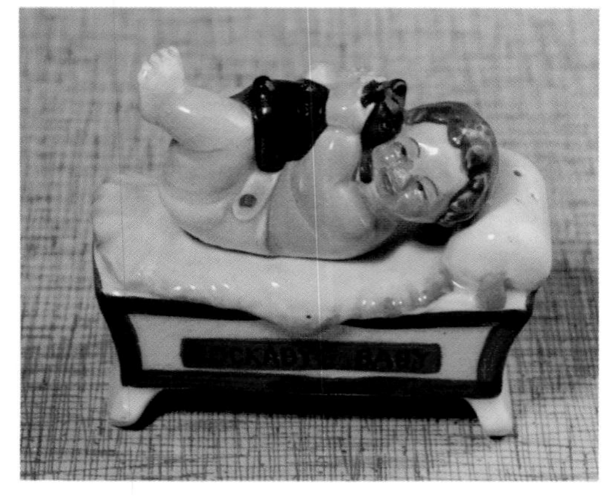

Appears to be a cat this little tyke is holding. The baby is 3 inches long, the cradle is 2-¼ inches high. The set is inkstamped, "Made in Japan." *Oravitz Collection.*

Hard not to fall in love with these little ones saying their prayers. The set on the left is 4 inches high, has a "Japan" paper label. The pair on the right also has a Japan paper label, plus a "#2087" inkstamp. *Thornburg Collection.*

Here the pair on the left is 3-½ inches high, has a "Japan" inkstamp. The pair on the right is inkstamped, "C014 / Japan." The pair in the center, which at first glance appears to be the same as the pair immediately above in the center but isn't, has a Lego paper label. *Thornburg Collection.*

Kissing kids by Goebel. The girl is 3-⅛ inches high. Impressed mark is unreadable, inkstamp says "Germany." *Thornburg Collection.*

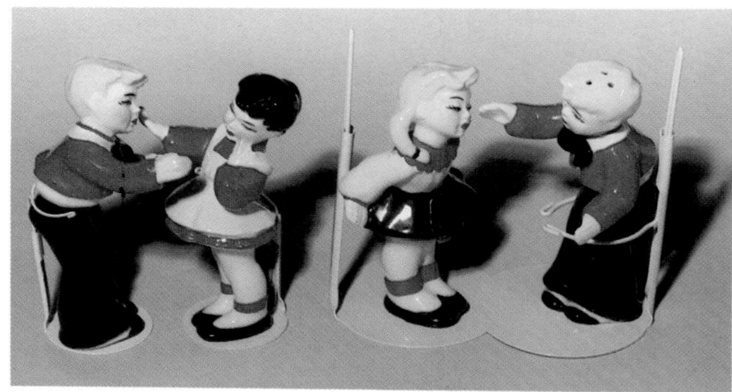

The title of the set on the right is Willing and Valentine Boy Lover, according to Laumbach. On the left must be Bashful and Valentine Boy Lover. Or perhaps we could call the whole picture "Blondes Have More Fun." Willing is 5 inches high. Both sets are products of Ceramic Arts Studio, but mark information was not recorded. *Thornburg Collection.*

Similar to those above, but a big difference for the 3-⅜ inch high pair in the middle. It's plastic. The mark indicates it was made in Hong Kong. The other two pairs are ceramic, neither is marked. *Thornburg Collection.*

These are also Ceramic Arts Studio, and are marked as such. The chairs are 2 inches high, the boy and girl 2-¼ inches high. As you can see from the holes in the shakers, this is two pair, not one. *Thornburg Collection.*

These cuties, Dutch, I believe, are 3-⅛ inches high. They have Napco paper labels with inkstamps that read, "NA4983." *Thornburg Collection.*

Another pair of ceramic heads, also 3-⅛ inches high. Mark indicates they were imported by Lefton. *Thornburg Collection.*

Flower children, each 3-¾ inches high. Inkstamp mark is "© Napco / ICX2248 / 1956" on three lines. There is also a Napco paper label. *Carson Collection.*

Hummel-like shakers, 3-¼ inches high. CNC paper label is shown below. *Thornburg Collection.*

CNC paper label of Hummel-like shakers.

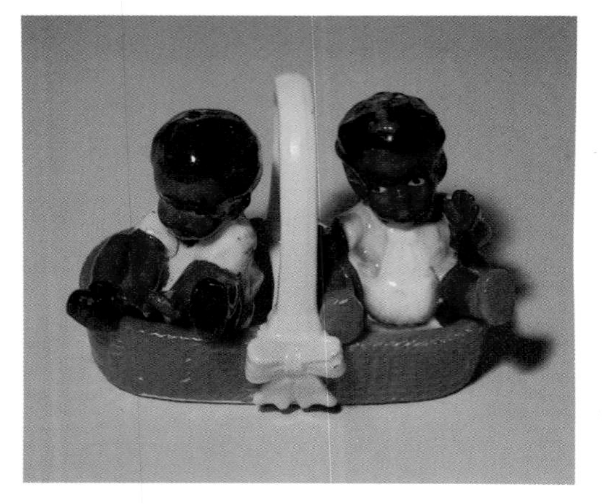

Whenever you see a prospective addition to your collection that has three or more pieces, such as this set, make sure they belong together. These obviously do, the confirming evidence being the matching red-orange on the basket and the girl's shoes, and identical yellows on the bow of the basket and the boy's shirt. The basket is 3-¾ high by 4-¾ inches across the front. It is inkstamped "Japan." *Carson Collection.*

Here is a set that does double duty as an illusion—look at the heavy metal ring at the bottom of the shaker on the left, and after awhile you can't tell whether the shaker is standing up or lying down. (It is standing up.) These shakers are 7 inches high. While a close up of the hang tag appears below, there are also two marks that do not show in the picture. One is a paper label that says, "Made in Japan," and, "Handpainted." Stamped in a circle on the bottom is "Holt Howard © 1959." *Oravitz Collection.*

5062
Rock'nRoll
salt and pepper
Shake and watch the fun
another HOLT HOWARD original

Close up of hang tag on above set of shakers.

Each of these shakers is 4 inches high. Seems like they should have a stamp that says "Germany," but they are unmarked. *Carson Collection.*

This boy and girl are marked "Occupied Japan." The girl is 2-⅜ inches high. *Carson Collection.*

This set has an Enesco paper label, and is 3-¾ inches high. *Carson Collection.*

Amish girl and boy, ceramic instead of metal for a change. The unmarked shakers are 4-3/8 inches high. *Carson Collection.*

Four inches high, unmarked. But there is more; see below. *Carson Collection.*

A sitter, the head lifting off the shoulders. Height as shown is 3-1/2 inches. The set is unmarked. *Carson Collection.*

Farmer boy, farmer girl. Height of each shaker is 3-1/8 inches. They are not marked. *Carson Collection.*

Reverse of above boy and girl.

Nice pair but not a set. Davern, in Book II, pictures this gal with a boy holding a valentine, and there is no mistaking the similar facial features. The shakers are 4-1/4 high, marked, "Shafford / Japan" on two lines. *Carson Collection.*

Souvenir of the 1964 New York World Fair, in patriotic colors. The unmarked shakers are 3-5/8 inches high. *Carson Collection.*

These youngsters are 2-1/2 inches high. To no one's surprise, they have a "Japan" inkstamp. *Carson Collection.*

Goebel flower girls, part of a series. The pair on the left, Pansy, is 2-5/8 inches high, inkstamped "Germany," with "P-52" impressed. The set on the right is Sunflower, same inkstamp with "P-55" impressed. According to the McHugh's this series first appeared in 1945. Other members include P-54 Dahlia, P-56 Rose, P-57 Strawberry, P-58 Raspberry, P-59 Apple and P-60 Plum. Each shaker in the series was finished in several color schemes. *Thornburg Collection.*

These young ladies are 3-½ inches high, have Enesco paper labels. One also has a paper label on the side that says, "Golden Girls." Look complete, don't they? They aren't. Guarnaccia, in Book III, shows the set on a tray, a similarly shaped and decorated dinner bell standing between the salt and pepper. *Thornburg Collection.*

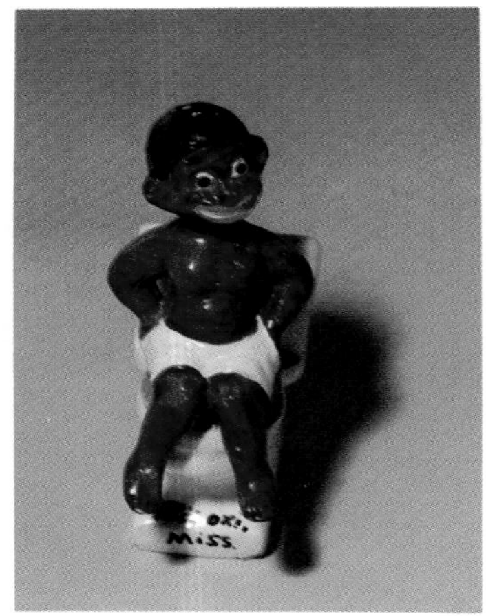

A sitter in more than one sense of the word. As shown this pair is 4-¼ inches high. Mark indicates Japanese origin. *Carson Collection.*

Like the Golden Girls above, this pair also has an Enesco paper label, along with an auxiliary paper label that reads, "Mother's Little Helper." The shakers are 3-¾ inches high. *Private Collection.*

Here's Miss Cutie Pie, who, like the girl above, also came in the form of a cookie jar. The shakers are 3-¼ inches high, carry Napco paper labels, and inkstamps that read, "A3510 / BL," on two lines. In addition to blue, these were finished in pink, yellow, and possibly other colors. This pair of shakers, and the Lefton pair above, illustrate the high quality Japanese ceramics Lefton and Napco demanded and received during the early years of the post World War II era, while many others clearly accepted junk. Looking at these beautiful shakers from out 1993 vantage point, it is easy to understand how the Japanese got the drop on several of our industries, including the pottery industry, once they put their mind to it. *Thornburg Collection.*

Aside from a "439" inkstamp, this pair is not marked but it is definitely Lefton, since that company marketed an identical cookie jar, and other accessory pieces. The cookie jar shows a copyright date of 1957. The shakers are 2-¾ inches high. *Courtesy of Joyce Hollen.*

This is called the Munich condiment set. It was made by Goebel. The mustard was a 1970 addition to the youngsters, which were modelled in 1938, according to the McHugh's. The tray is the same one that is used with the Friar Tuck condiment set. The mustard says "Munchen Welstadt mit Hers," which translates to "Munich the City with a Heart." The children on the tray are 2-¼ inches high on the left, 2-⅜ inches high on the right. The children off to each side are a different pair, measuring 2-½ inches high on the left, 2-⅝ inches high on the right. Marks of all pieces include the words "Goebel," and "West Germany," or "W. Germany." *Thornburg Collection.*

Let's play a game of *What's Missing from This Picture?* It is obvious from the turned up, open palms that these ceramic nodders, instead of being a pair, are each supposed to be holding the second component of the set. I cannot win our *What's Missing* game because I do not have the answer. But I know a likely place for you to look for it—elsewhere in this book, represented as a bonifide set of salt and pepper shakers! Each shaker shown is 4-1/8 inches high, has a "Made in Japan" inkstamp. *Carson Collection.*

Another boy, more dogs. The base is 3-3/4 inches long, the shakers are 2-1/8 inches high. Inkstamp says "Made in Japan." *Thornburg Collection.*

The alligator is 4-1/4 inches long, the elephant 5-1/4 inches high. Both boys, though different, measure 2-5/8 inches. Both sets were made by Ceramic Arts Studio. The animals are marked, the riders are not. An exact replica of the elephant set was made by the Japanese. *Thornburg Collection.*

Although shot from a different angle and hard to say for sure, I believe these shakers are the same as those shown by Guarnaccia (Book II page 58) and Davern (Book I page 129) decorated as blacks. They are 3-1/8 inches high, with "Japan" inkstamps. *Courtesy of Joyce Hollen.*

Here again is a go-with that is unquestionably a pair. The pink on the boy's hands and the dog's feet match, as does the green of the grass on both shakers, and the texture of the grass. The boy is 3-1/2 inches high, has a "Made in Japan" inkstamp. *Thornburg Collection.*

The future astronaut in this set is 3-1/2 inches high, has an Enesco paper label. *Carson Collection.*

Van Tellingen huggers, made by Regal China Company. The boy in each set is 3-1/2 inches high, marked with an impressed "Pat Pend" on the back, plus an impressed "Van Tellingen ©" in a semi-circle on the bottom. The dog is 3-5/8 inches high, with the same impressed wording but the positions of the phrases reversed. *Courtesy of Carol Sparacio.*

The unmarked set on the left is made of chalkware. The shakers are 2-5/8 inches high. The set on the right is ceramic. The head, with a "Japan" inkstamp, is 2-3/4 inches high, the unmarked watermelon is 3-1/4 inches long. *Carson Collection.*

Each of these unmarked shakers is 2-½ inches high. *Courtesy of Jim and Betsy Coughenour.*

The pair of graduates on the inside is 4-⅛ inches high, has Lego paper labels. The outside pair is ⅛ inch shorter, has "Japan" inkstamps. *Thornburg Collection.*

Bums and Hillbillies

Here is another reflection of the times in which we live. Couple the strife of today's large number of homeless people with an economy that always seems to get worse instead of better for many working people, and bums and hobos would probably not be met with much enthusiasm by buyers in gift shops, discount and department stores. Same goes for hillbillies. The subject would not win many friends in the modern industrialized South which has reshaped its image dramatically over the past several decades.

The pair in the center is 6-¾ inches high, has a "Japan" inkstamp. The ones on the outside are 3-¾ inches high, carry a "Japan" paper label. *Carson Collection.*

Sometimes with Japanese salt and pepper shakers, you are not exactly sure what the designers had in mind. Are these bums, or just cute kids? The shaker on the left is 3-⅛ inches high, has a "Japan" paper label. *Carson Collection.*

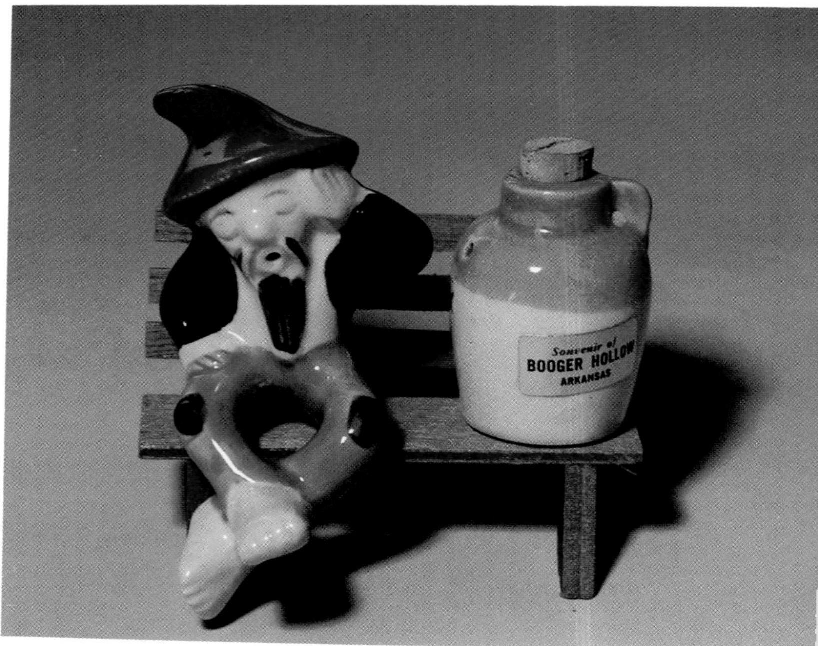

This guy seems to fall somewhere between a bum and a hillbilly. Whichever he is, the jug goes very well. The unmarked man is 4-¼ inches high. *Carson Collection.*

Here's another set that makes choosing a catagory a challenge. The guy on the left is 6-½ inches high, has a Holt Howard inkstamp and "Japan" paper label. *Carson Collection.*

The reclining hillbilly is 4-¾ inches long, has a "Japan" paper label. *Carson Collection.*

This rather common plastic set is 6-½ inches high. Mark says "Made in Hong Kong." *Private Collection.*

Stereotypical hillbilly, his wife probably nagging him to do some work. Length of the man and his headrest is 4-¼ inches. The set is not marked. *Carson Collection.*

Look closely and you will see a pig wrapped around the top of this drunk's head. Length is 4-¼ inches. The set carries a "Japan" paper label, and a G.C. Murphy original price tag showing 49-cents. *Carson Collection.*

The lady (if we may call her that) on the left is 3-¾ inches long, while the tall dude on the right is 6 inches high. The left and center pairs in the rear have "Japan" paper labels, the right pair is unmarked. The pair in front has a paper label that reads "Fairway Japan." *Carson Collection.*

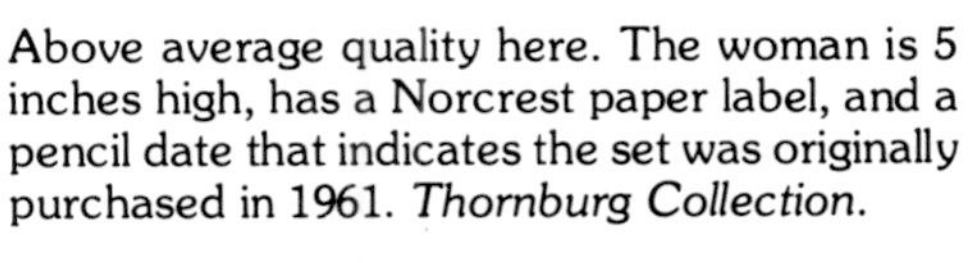

Above average quality here. The woman is 5 inches high, has a Norcrest paper label, and a pencil date that indicates the set was originally purchased in 1961. *Thornburg Collection.*

R- and X-Rated

Some folks have entire collections made up of these kinds of shakers. Others find them offensive. If you belong to the latter group, or if you do not want the little ones to see them, a sharp razor blade and a gentle touch will permanently remove them without damaging the rest of the book.

The woman is 6 inches high. Neither piece is marked. While the couple appears rather benign, the writing on the bases coupled with the position of the man's right arm give clues to why they appear in this section. *Carson Collection.*

This is called a socket couple. The woman is 5 inches high, carries a "Japan" paper label, a "7895" inkstamp, and a pencil date of August 12, 1963. *Thornburg Collection.*

Unmarked boy and thunder mug, the boy being 3-⅝ inches high. Pencil date is 1953. *Thornburg Collection.*

Height here is 3-¾ inches. The Japanese set has an Empress paper label. *Thornburg Collection.*

Reverse of the lady and dirty old man.

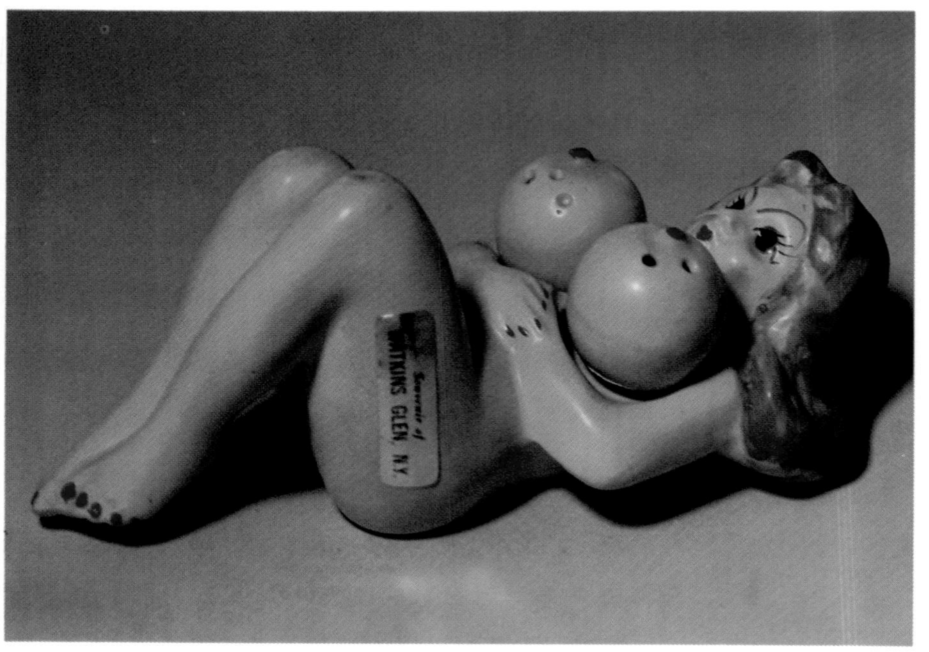

Plenty of room to hold some salt and pepper here. The girl is 6 inches long, bra size is 50-DDDD. None of the pieces are marked. *Carson Collection.*

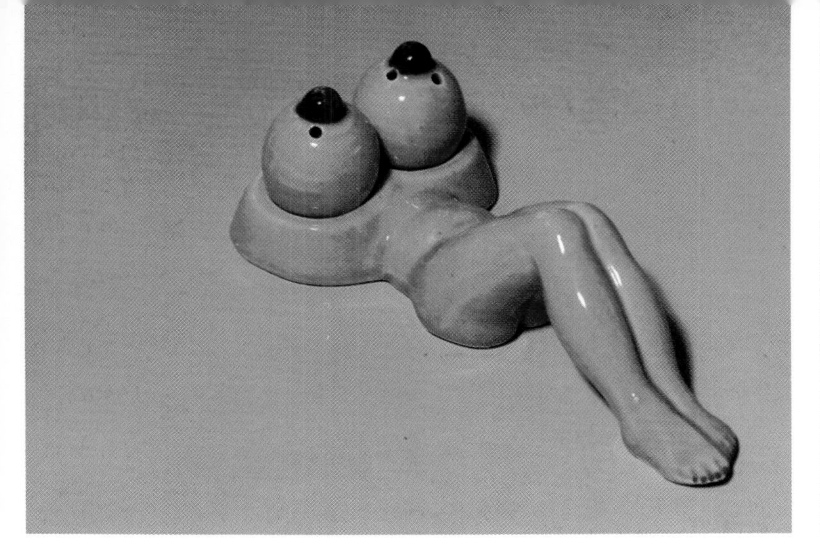

Length here is 6 inches. There are no marks. *Carson Collection.*

I guess even salt shakers enjoy their privacy. The set is unmarked, 3-½ inches long. *Carson Collection.*

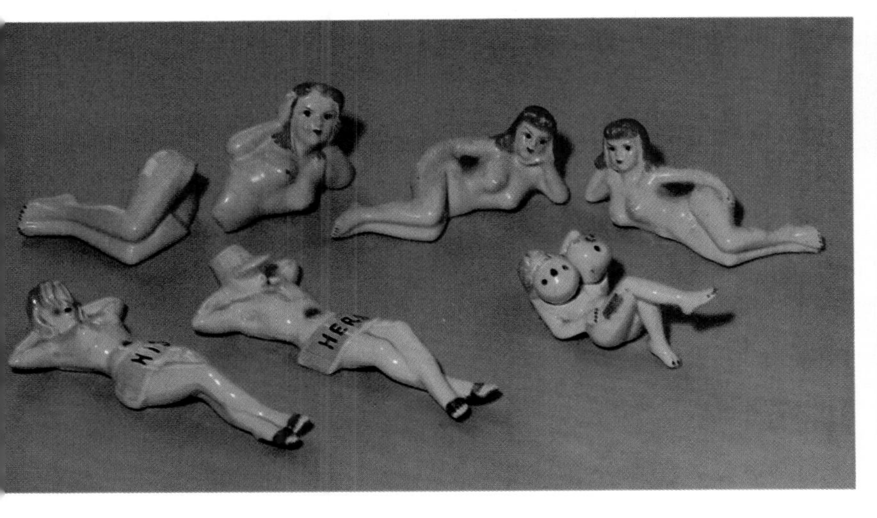

Several more sunbathers. All have marks indicating Japanese origin. The woman at front left is 6 inches long. *Carson Collection.*

This set has a "Japan" paper label. The girl is 4-⅛ inches long, the sombrero 3-⅛ inches high. What do you think she is hiding behind it? *Carson Collection.*

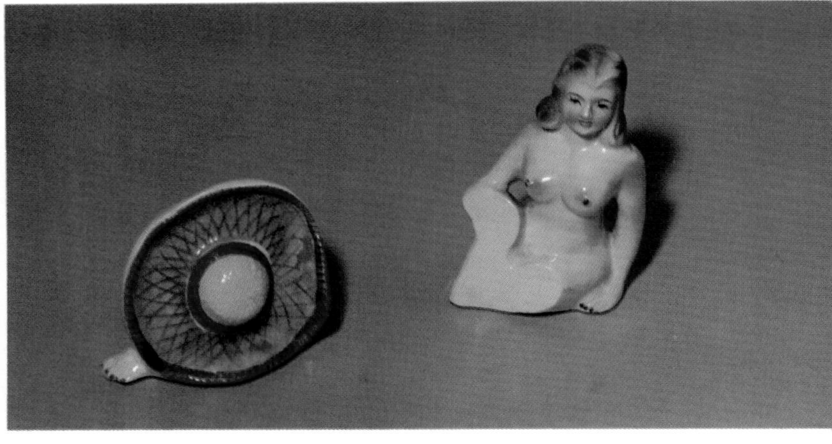

Nothing.

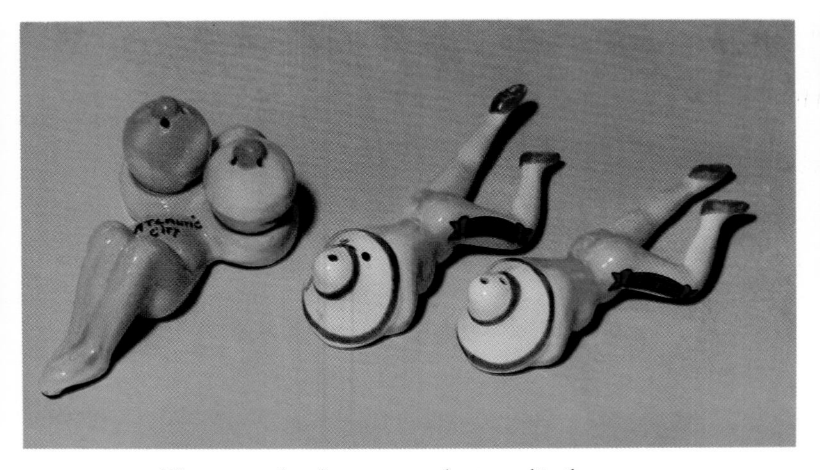

Three people who apparently stayed in the sun longer than they should have. All three of the sets are 4-¾ inches long. None are marked. *Carson Collection.*

My own feeling is that if Adam and Eve were this unattractive, the human race might have never gotten off the ground. She is 4-½ inches high, has a "Japan" inkstamp. *Carson Collection.*

Back of Adam and Eve.

Miscellaneous People and Pieces Parts

The 10, or so, people shown did not seem to fit well into any other category. As far as pieces parts, more feet were made than anything else, but some legs and hands were turned out, too.

These shakers are 4 inches high. They are inkstamped "53 / 424" on two lines. My wife has a matching head vase that is marked, "Holt Howard 1959." *Thornburg Collection.*

Christmas girls decked in holly. Height is 4 inches with a Holt Howard paper label and a "6216" inkstamp. *Oravitz Collection.*

The shaker at far left is 3-⅜ inches high, has a "Japan" paper label. *Thornburg Collection.*

This woman and the refrigerator are obviously well acquainted. Both shakers are 3-¼ inches high. Both have Clay Art paper labels. *Courtesy of Lesia Dobriansky.*

Here is another person who likes to eat. The cannibal is 1-½ inches high, his dinner is 1-⅝ inches. The set is not marked. *Oravitz Collection.*

A second guy about to stir the pot. The cook is 3-¼ inches high. Neither piece is marked. *Carson Collection.*

Seems like this could be parts of two sets, each lacking a cannibal. But the pots are very clearly designated S and P. Height is 5-¼ inches. The set has a "Made in Japan" paper label, along with an inkstamp, "3784/T" on one line. *Lilly Collection.*

These plastic soup cups seem like they should be an advertising set but for whom I don't know. The lids are the shakers. Height is 3-⅝ inches. Each has an F & F mark. *Thornburg Collection.*

The printing on this pair says, "I walked my feet off in Washington D.C." These are quite common, you could probably collect them from every state and most cities. Length is 3-¼ inches, no marks. The suitcases have "Japan" inkstamps. *Carson Collection.*

More feet, more imagination. The pair in the center is 3-⅜ inches high, has a "Japan" paper label, as does the pair on the right. The set on the left is unmarked. *Carson Collection.*

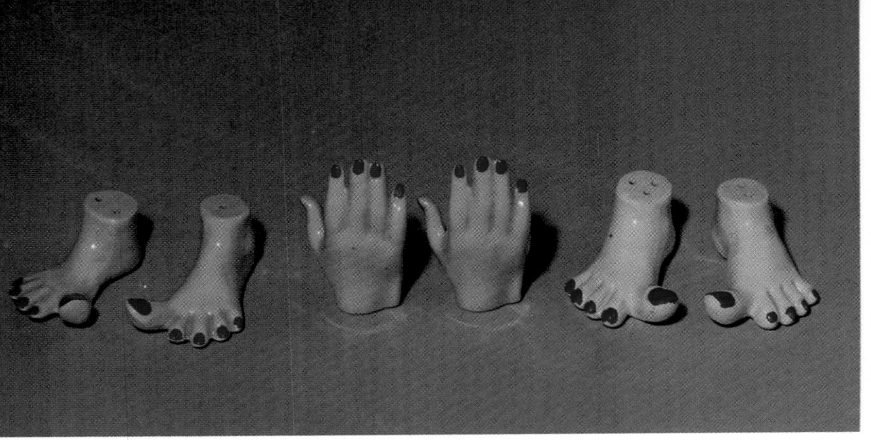

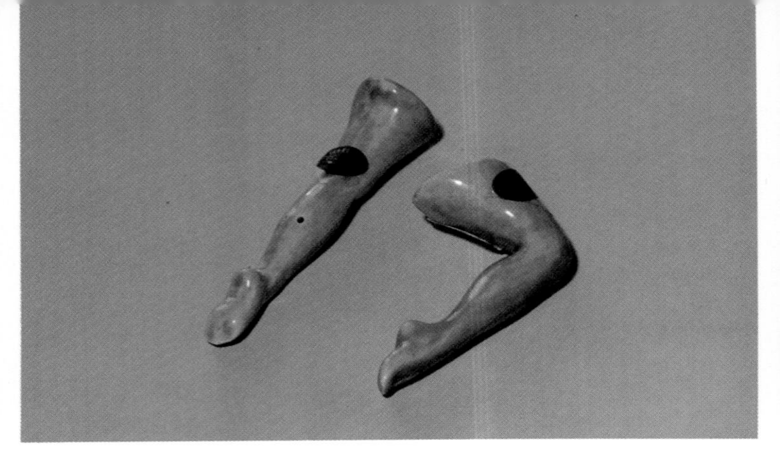

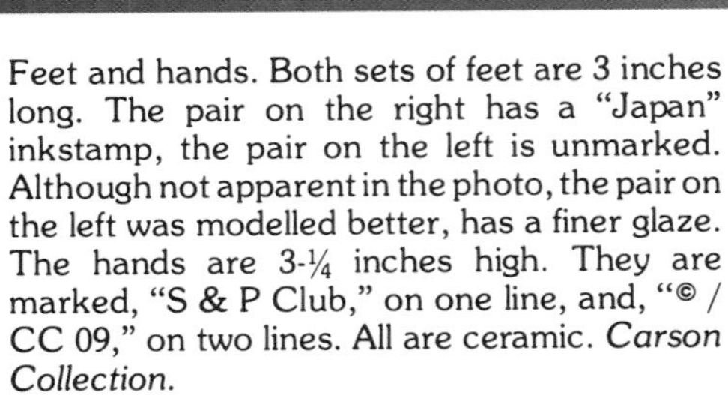

Feet and hands. Both sets of feet are 3 inches long. The pair on the right has a "Japan" inkstamp, the pair on the left is unmarked. Although not apparent in the photo, the pair on the left was modelled better, has a finer glaze. The hands are 3-¼ inches high. They are marked, "S & P Club," on one line, and, "© / CC 09," on two lines. All are ceramic. *Carson Collection.*

Legs only, and I forgot to measure them. The exposed paper label reads, "Davenport, Iowa." One underneath reads, "Elbee Art / Originals / Cleveland, Ohio," on three lines. *Carson Collection.*

Characters and Personalities—Fact and Fiction

This section includes real people such as President Kennedy in his rocking chair, legendary people such as Paul Bunyan, and commercially created people such as the Jetsons. Because there are so many Santa Claus shakers, he and Mrs. Claus appear immediately following.

The Jetsons are exceptionally rare, quite possibly the rarest set of shakers in the book. The story I get, with which most of the salt and pepper shaker collecting community seems to agree, is that Vandor made 12 sets as prototypes, then failed to secure a license from Hanna-Barbera. Meanwhile, the prototypes ended up outside the plant instead of being destroyed. The saucer is 7 inches long, George is 3-⅛ inches high. *Private Collection.*

Popeye and Olive Oyl. Popeye is 7 inches high. While the spinich-eating sailor himself is more than 60 years old—created by Elzie Crisler Segar (American cartoonist 1894-1938) for the comic strip *Thimble Theatre* in 1929—this is a fairly recent set, marketed by Vandor. Its paper label is shown below. *Thornburg Collection.*

Paper label of Vandor Popeye and Olive Oyl.

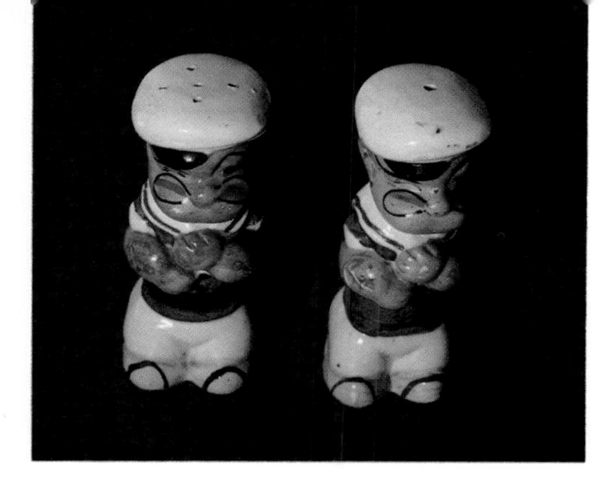

This popeye is 4 inches high, has a "Japan" inkstamp, and was originally a souvenir of Niagara Falls, Canada. *Carson Collection.*

More cartoon characters, Wally and Windy. Height was not recorded but several marks were—"© 1958 / Walter Lantz / Productions Inc." on three lines, "IC3635 / Japan" on two lines, and a Napco paper label. This set is part of a series. Davern, in Book II, shows a similar Oswald and Willy in the same format, and Woody Wookpecker appears later in this book. *Thornburg Collection.*

Paul Bunyan and Babe his big blue ox. Paul is 2-¼ inches high. The unmarked pair was made by Rosemeade. *Thornburg Collection.*

This Paul is 4-½ inches high, has a "Japan" inkstamp. The unmarked Babe is 3-⅞ inches long, "Babe" appearing in raised letters on the opposite side. *Thornburg Collection.*

Three more Pauls. The one on the left is 5-¼ inches high, unmarked. Height on the right is 4-⅛ inches with a "Japan" paper label. The shaker in front is shortest, 4 inches. Its paper label is shown below. *Thornburg Collection.*

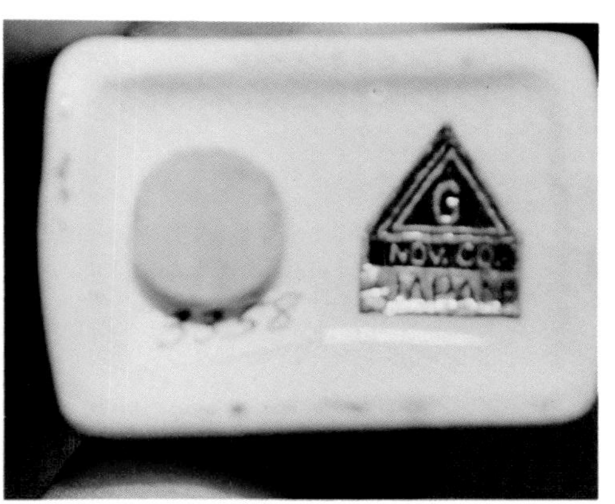

Paper label of front Paul Bunyan in above photo, very likely a firm whose name begins with the letter G, and includes the words "Novelty Company."

Ceramic Arts Studio Paul Bunyan, 4 inches high. Both pieces have Ceramic Arts marks. *Thornburg Collection.*

Bo Peep and sailor boy by Shawnee. The sailor boy is 3-⅛ inches high. None of the pieces are marked. Note that the pair on the left is trimmed with gold. *Smith Collection.*

Hull's Little Red Riding Hood which matches the company's cookie jar and other accessory pieces. Hull made three sets. Shown here are the smallest set, 3-¼ inches high, and the largest set, 5-½ inches high. Hardest to find, and most valuable, is the middle size set which stands 4-½ inches. A fourth set, Little Red Riding Hood kneeling, is shown undecorated in *Collecting Hull Pottery's Little Red Riding Hood,* by Mark E. Supnick (L-W Book Sales, 1989.) *Thornburg Collection.*

Ceramic Arts Studio Little Black Sambo and the tiger. The tiger is 2-½ x 5-¼ inches. Like most Ceramic Arts pieces, some of these sets are marked, others aren't. *Thornburg Collection.*

Snow White and Dopey. Dopey is 3-¾ inches high. Snow White is inkstamped "©Walt Disney Productions," while Dopey's inkstamp is "©Walt Disney." *Thornburg Collection.*

Mowagli and the elephant from *The Jungle Book,* by Rudyard Kipling. Mowagli is 4-⅞ inches high, marked "©MCMLXIV / Walt Disney / Prod." on three lines. There is also an Enesco paper label. *Thornburg Collection.*

Another Little Red Riding Hood set, this time with each piece having a "Japan" inkstamp. The wolf is 3 inches high. *Courtesy of Allen and Michelle Naylor.*

The old woman who lived in a shoe, by Sandy Srp. The miniature woman is 1-¼ inches high. *Carson Collection.*

Two renditions of Goldilocks by Regal China Company. The two-piece shaker on the right is 3-½ inches high as shown. *Carson Collection.*

Rub a dub dub, three men in a tub condiment set. Height to the top of the center figure is 4-¾ inches. He is attached to the lid, the tub is the mustard. Inkstamp reads "E3120." *Carson Collection.*

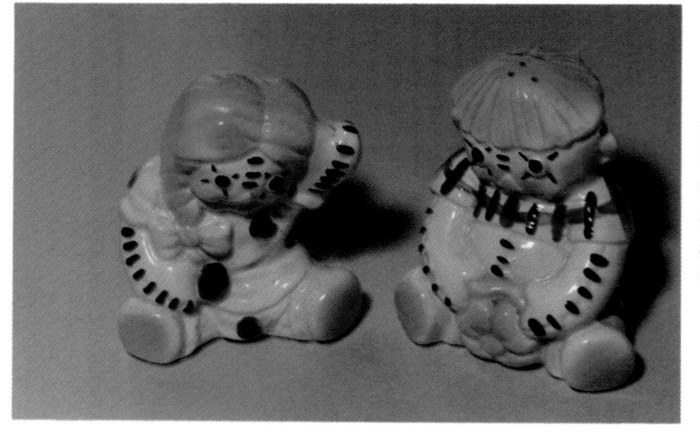

Raggedy Ann and Andy, 4-½ inches high, unmarked. Whatever pottery produced these it also made Raggedy Ann and Andy figurines, a cookie jar, and gingham dog and calico cat shakers and figures in the same style. *Carson Collection.*

The paper tag across the front of the shaker set on the left reads, "There was an old woman." Height is 4-¾ inches as shown. Little Jack Horner is also 4-¾ inches high. Both sets have "Japan" inkstamps. Based on format, color and finishing techniques it appears these two shakers, plus Robinson Crusoe, Bobby Shaftoe and the Magician, shown in the next two pictures, belong to the same series.

Robinson Crusoe and Bobby Shaftoe, both with "Japan" inkstamps. Bobby Shaftoe is 3-⅛ inches high. *Carson Collection.*

The Magician is 4-¾ inches high, unmarked. *Carson Collection.*

Marilyn Monroe made by Clay Art during the 1980s. The shaker on the right is 3-½ inches high. A decal on the back reads "Marilyn Monroe / © 1988 Clay Art / San Francisco / Made in Japan" on four lines. *Thornburg Collection.*

Old King Cole, possibly an extension of the above series. The chair is 2-¼ inches high, has a "Made in Japan" inkstamp. The king is 2-⅞ inches high, unmarked. *Oravitz Collection.*

Jonah and the whale. Jonah is 1-½ inches high, the whale is 4 inches long. Neither piece is marked. *Carson Collection.*

Richard Nixon and Spiro Agnew, currently in production by K. Wolfe Studios. Nixon is 3 inches high, Agnew 3-⅛ inches. Neither is marked. *Courtesy of Kathy Wolfe.*

President John F. Kennedy. This set is 3-½ inches high as shown. Its inkstamp mark appears below. *Thornburg Collection.*

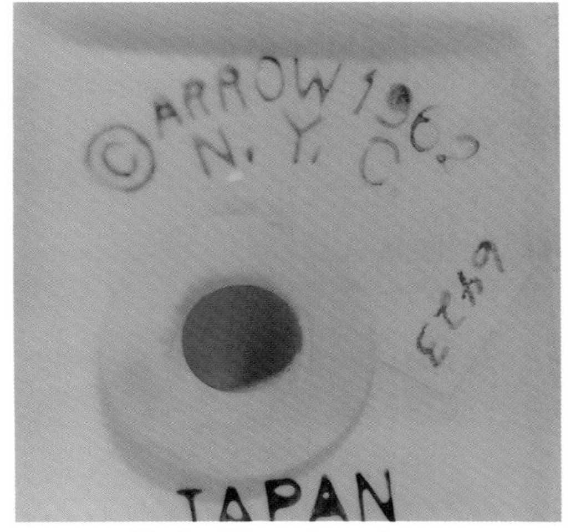

Mark of Kennedy rocker.

This is the Marilyn Monroe cake. Together, as shown, the shakers are 4-½ inches high. There is a Clay Art paper label. *Thornburg Collection.*

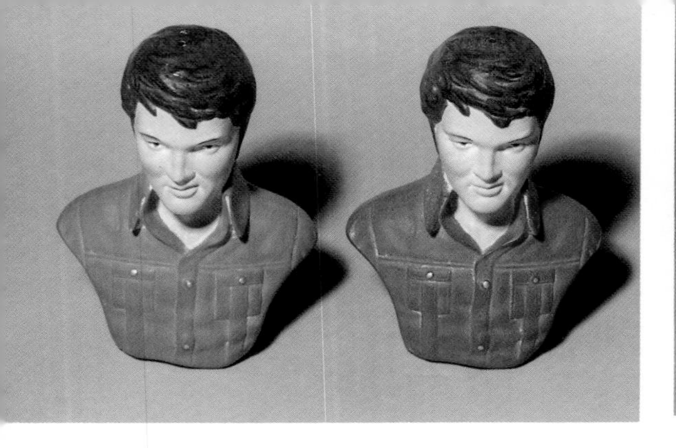

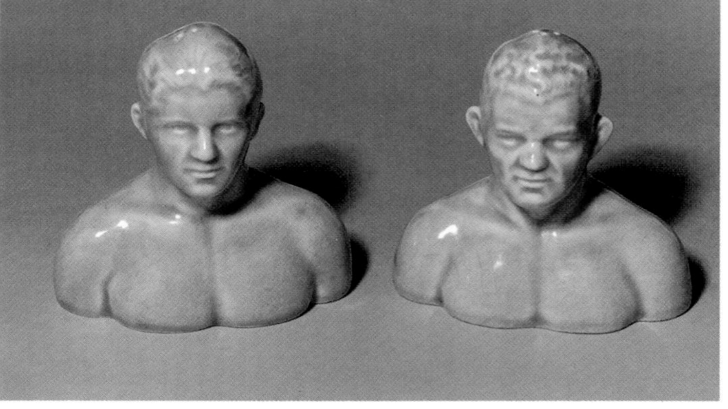

Collectors tell me this set represents Jack Dempsey and Gene Tunney. The shaker on the left is 3-¼ inches high. Neither is marked. *Thornburg Collection.*

Elvis Presley, 4-½ inches high, no mark. *Thornburg Collection.*

Tray set Laurel and Hardy, made by Beswick. The shaker on the left is 3-¾ inches high. A "Beswick, England" inkstamp appears on all three pieces, an impressed "Beswick England 575" on the tray only. Also made was a cruet, 4-½ inches high. The pieces were modelled by a Mr. Watkin in 1938, were out of production no later than 1969, according to Harvey May in *The Beswick Collector's Handbook. Thornburg Collection.*

Blue Bonnet Sue, 5 inches high with a "Made in Taiwan" paper label. The shakers are identical to a Blue Bonnet Sue cookie jar (11-¾ inches high) that also has a "Made in Taiwan" paper label along with a more informative impressed mark of "© 1989 Nabisco." *Thornburg Collection.*

Sairey Gamp and Mr. Micawber, another Beswick set. Mr. Micawber is 3-½ inches high. Inkstamp mark on both pieces is "Beswick England." ***Thornburg Collection.***

Buffalo Bill Cody and his namesake. Buffalo Bill is 2 inches high. Both pieces are unmarked. *Carson Collection.*

This is Tee and Eff, advertising symbols for Tastee Freeze ice cream drive-ins. They are 3-½ inches high and unmarked. *Courtesy of Jazz'e Junque Shop, Chicago, Illinois.*

Sandeman Brandy decanter with matching salt and pepper shakers. The shakers are 5-½ inches high, the decanter is 10-¾ inches. The decanter has a Wedgwood mark, shown below. The shakers are unmarked but of the same quality, and, thus, *assumed to be Wedgwood. Thornburg Collection.*

Mark of Sandeman brandy decanter.

Two Big Boys, a 1992 introduction of K. Wolfe Studios. The shaker on the left is 5 inches high, the one on the right 2-½ inches high. Neither is marked. *Courtesy of Kathy Wolfe.*

Japanese Big Boys, 4-¼ inches high with "Japan" paper labels. *Oravitz Collection.*

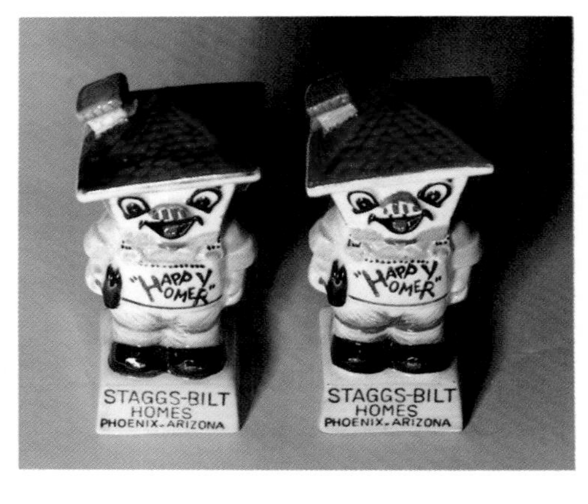

I'm not sure if Happy Homer is supposed to be a person or not, he looks a little like a pig. The shakers are 4 inches high, carry OMC paper labels. A spokesman for Phoenix's Staggs Realty—unafiliated with Staggs-Bilt Homes—said as best as he could remember, Staggs-Bilt Homes was operated by a man named Ralph Staggs during the late 1950s and throughout the 1960s. Attempts to reach Ralph Staggs were unsuccessful. *Thornburg Collection.*

Here is Kellogg's Rice Krispies Snap and Pop. Only Crackle is missing. The shakers are 2-⅝ inches high, have "Japan" inkstamps. *Thornburg Collection.*

Maggie and Jigs with Jigs being 3-¼ inches high. The mark is "Cop / Ps" in a circle, along with "MP822." on the back of Jigs. These are said to date from the 1920s, and are quite expensive. *Private Collection.*

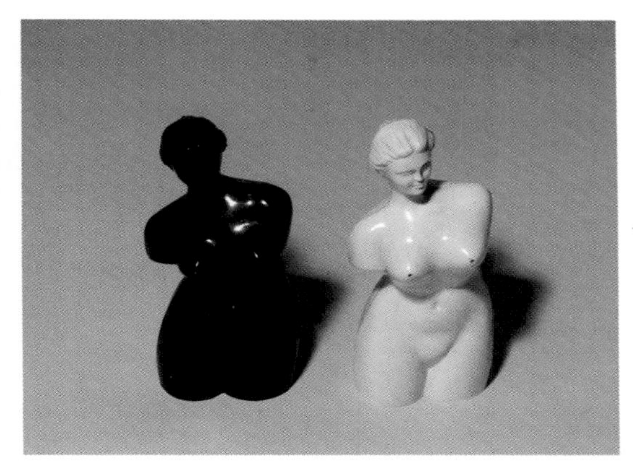

Plastic Venus de Milo, 4 inches high and unmarked. When I put a pair of these on display at an antique show once, a woman criticized me for selling lewd merchandise. She was obviously not a fan of classics! *Carson Collection.*

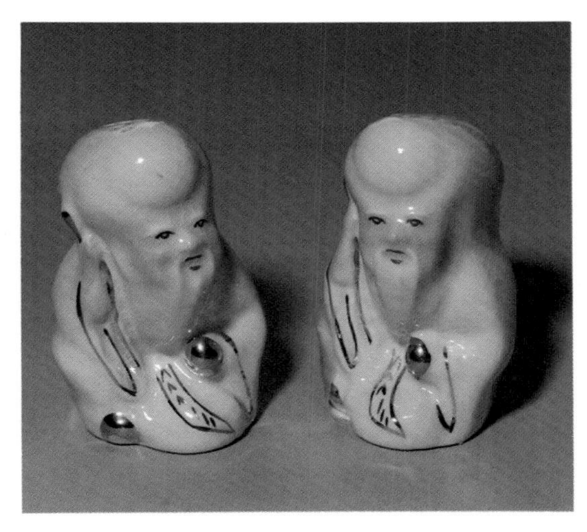

Father Time, 3-¼ inches high. Unmarked, these shakers are believed to have been made by Coventry Ware, Inc., of Barberton, Ohio. *Carson Collection.*

Father Time and New Years Baby. The baby is 3-½ inches high. Father Time is marked with an inkstamp, "ICX4706," and a Napco paper label. *Thornburg Collection.*

Santa and Mrs. Claus

Santa and Mrs. Claus are perhaps the most enduring figural salt and pepper shakers that ever graced a table. They were made very early in the figural movement and are still being made today. Walk into your favorite discount store or gift shop during the Christmas season and you will spot the Claus's waiting to be purchased and taken home to provide flavorings for your Christmas dinner.

Kelvin paper label of Mrs. Claus on the left above.

The Santa on the left is 3-¾ inches high. Santa has a Kelvin inkstamp, Mrs. Claus a Kelvin paper label. Both marks are shown below. An identical set was photographed that had Napco paper labels. Santa in the middle is 4-¼ inches high, has a Brinn's paper label. In the set on the right, Santa is 3-⅞ inches. Each shaker has a Lefton inkstamp and paper label, which appear below. I find it interesting that with these three pair, as with several other pair shown that include Mrs. Claus, Mrs. Claus is slightly taller than Santa. *Carson Collection.*

Lefton inkstamp and paper label of shakers on the right above, dating the set at 1956 or later.

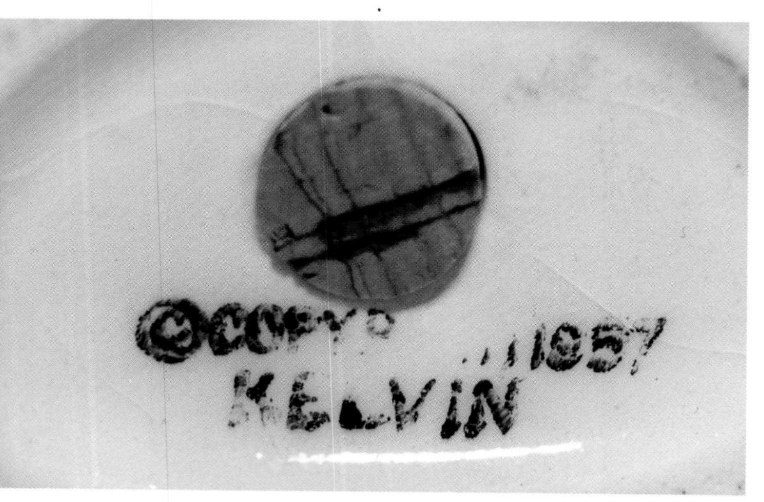

Kelvin inkstamp of Santa on the left above, showing a copyright date of 1957.

The head shakers are 3 inches high, marked "Japan." The unmarked pair on the inside is 3-½ inches high. *Carson Collection.*

December 26. The shakers are 2-⅝ inches long, have "Made in Japan" paper labels. *Thornburg Collection.*

Heights of the Santas from left to right are 4-½, 4-⅛ and 5-¼ inches. The outside pairs are marked "Japan." The pair in the middle is unmarked. At first I was suspect of that pair but upon comparing the texture of Mrs. Claus's hair with the texture of Santa's beard, it became obvious they are truly a pair. *Carson Collection.*

The bell ringing Mrs. Claus is 4 inches high, carries a Napco paper label. The rocking chair couple is 3-¼ inches high. Its mark indicates it was a product of Taiwan. *Carson Collection.*

Hard to believe these Sandy Srp miniatures are only 1-⅜ inches high. *Carson Collection.*

The rocking chair shakers are each 4-¼ inches high, marked Lefton. The bench set is marked "Japan." *Carson Collection.*

These shakers are 5-½ inches high, have a Lipper & Mann paper label which is shown below. *Thornburg Collection.*

Apparently the sleigh must have broken down. Height is 3-½ inches, "Made in Japan" paper label. *Lilly Collection.*

This is a napkin holder, the salt and pepper shakers' heads being removable. Shakers themselves are 1-⅞ inches high, have "Japan" inkstamps in addition to the Commodore paper label shown below. *Thornburg Collection.*

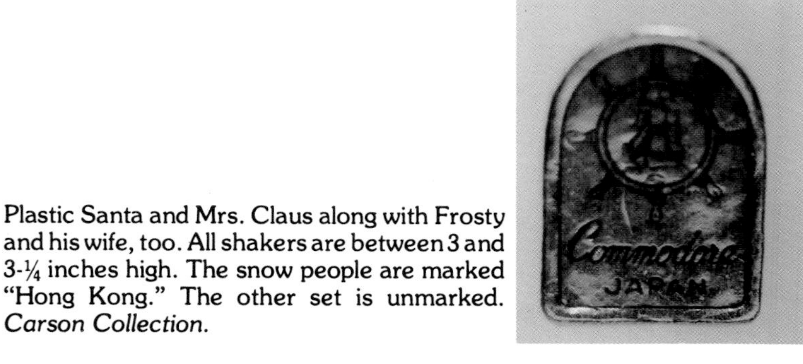

Commodore paper label of above napkin set.

Lipper & Mann paper label of Santa and Mrs. Claus with Noel gifts.

Plastic Santa and Mrs. Claus along with Frosty and his wife, too. All shakers are between 3 and 3-¼ inches high. The snow people are marked "Hong Kong." The other set is unmarked. *Carson Collection.*

These shakers are 3 inches high, unmarked. *Zera Collection.*

Santa and Rudolph. Santa is 3-¼ inches high. Neither piece is marked. *Carson Collection.*

Ceramic Arts Studio Santa and tree. Santa is 2 inches high; the tree is the same shown earlier with Paul Bunyan. *Thornburg Collection.*

The unmarked heads are 3-¾ inches high. Santa and his bag of presents are marked Holt Howard. The rocking horse just sort of crept in. It's Lefton. *Carson Collection.*

The Santa heads are 3-½ inches high, are inkstamped "© 1959 / Holt Howard," in addition to having Holt Howard paper labels. Inkstamps of the chimney Santas indicate Japanese origin. *Thornburg Collection.*

This set was purchased at a discount store in 1990 for $1.99. The shakers are 3-¼ inches high, have "Made in Taiwan" paper labels. Printed on the box is, "Distributed by Midwestern Home Products, Inc., Wilmington, DE, 19803." *Oravitz Collection.*

The set on the left is chalkware, 3 inches high. Like most chalk shakers, it is not marked. The set on the right stands 5 inches, has a "Japan" mark. *Carson Collection.*

Rudolph is 3-¼ inches high, Santa is 3-½ inches. Marks indicate Japanese origin. *Carson Collection.*

Almost Human

This section covers those animate creatures that do not quite meet the standard to be called people, but cannot be relegated to the animal chapter, either—angels, devils, ghosts, mermaids, pixies, brownies, witches, etc. To some they are imagined. To others they are real. To salt and pepper shaker buffs they are charming additions to their collections.

Angel and bow miniatures by Sandy Srp in original case. The angel is 1-¼ inches high, the bow 1-⅛ inches. *Carson Collection.*

The elves are 3-¾ inches high, have Napco paper labels. The angels are marked "Japan." *Carson Collection.*

These angels are 3-¾ inches high, unmarked. *Oravitz Collection.*

This devil is 4-¼ inches high. Both pieces are inkstamped "© Kreiss 1955." They also have "Japan" paper labels. *Carson Collection.*

A pair of cherubs, one having an Armita paper label, shown on the next page. Height of the member on the left is 3 inches. *Thornburg Collection.*

Angel and fallen angel, better known as the devil. Each is 4-⅛ inches high, neither is marked. *Carson Collection.*

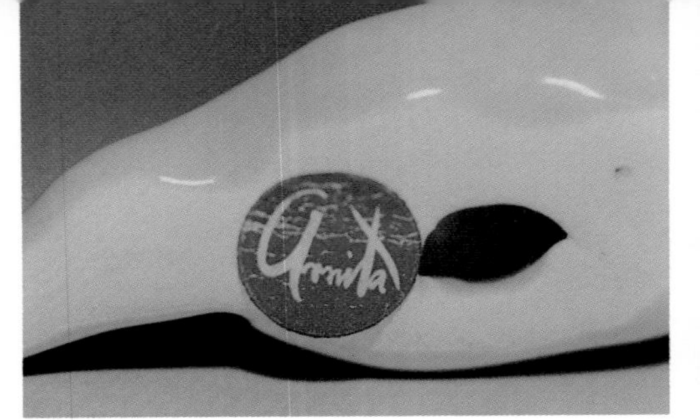

Plastic witches, 3-¼ inches high. Mark is "© J.S.N.Y. / Hong Kong" on two lines. *Carson Collection.*

Armita paper label of above cherubs.

The ghosts in the center are 3-½ inches high, have a "Japan" paper label. The set on the left is inkstamped "Japan," while the pair on the right carries a paper label that reads "JSNY / Taiwan" on two lines. *Thornburg Collection.*

Ghost and pumpkin miniatures, 1-¼ and ⅞ inches high, respectively. They have Sandy Srp marks. *Carson Collection.*

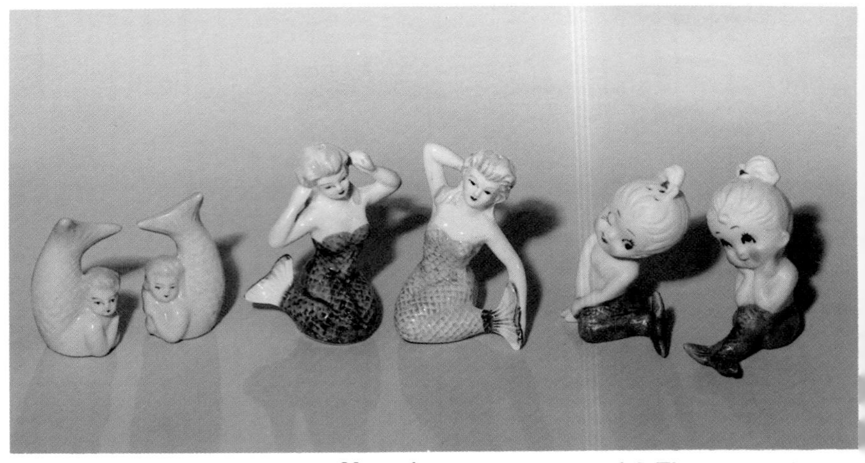

How about some mermaids? The pair in the center measures 3-½ inches high, has a "Japan" paper label. The pair on the left carries a "Japan" inkstamp, while the set on the right is unmarked. *Thornburg Collection.*

A witch making her brew. The witch is 3-¾ inches high. The set has a Fitz & Floyd mark. *Carson Collection.*

More mermaids. As you can see, the mermaid at right rear separates at the waist. It is 4-⅞ inches high. Both members have "Japan" inkstamps. Those at left rear are 3-½ inches high, have a Norcrest paper label and an inkstamp that reads "H-422." The two sets in the center of the picture are unmarked. *Thornburg Collection.*

Pinocchio on the left is 4-¾ inches high. Its inkstamp mark reads "Hand / Painted / Japan" on three lines. The unopened set in the center carries an Enesco paper label. The pair in front was made by Goebel. The cat is Cleo, who belonged to Geppeto. *Thornburg Collection.*

Height on the left is 3-½ inches. The shakers have a "Japan" inkstamp. Height on the right is 3 inches. This pair has a "Japan" paper label, plus a "3963" inkstamp. *Thornburg Collection.*

PIXIES

These pixie heads are 3-¾ inches high. They are not marked. *Oravitz Collection.*

These are kind of different as they have movable eyes. The shakers are 2-⅞ inches high. Mark information was not recorded. *Thornburg Collection.*

These shakers are 2-¼ inches high. They have Enesco paper labels, but don't seem to be on a par with that company's generally fine work. *Carson Collection.*

This is believed to be an advertising set but not Kellogg's Rice Krispies symbols Snap, Krackle or Pop, in the opinion of owner Irene Thornburg who is a retired Kellogg's employee. That's good enough for me. The shakers are 3-½ inches high, have "Japan" inkstamps. *Thornburg Collection.*

A Holt Howard set, marked "© Holt Howard." The shakers are 2-¾ inches high. A pencil date indicates they were originally purchased in 1960. *Oravitz Collection.*

Pixies and vegetables. Heights are 2-½ and 2 inches. Neither shaker is marked. *Carson Collection.*

Play ball! Each shaker is 3-¼ inches high. They are inkstamped "Japan." *Oravitz Collection.*

Well executed teapots, 4-¼ inches high. Marks, or lack of them, were not recorded. *Carson Collection.*

Note how the bases of these kissers fit together. Unfortunately, all information about size and marks was misplaced. *Carson Collection.*

From Hot Stove to the kitchen stove. The shaker on the left is 4-⅝ inches high. While both of these are unmarked, one has an unreadable remnant of a paper label. *Oravitz Collection.*

These are 3-¼ inches in length, have an impressed "22," and a Relco paper label. *Oravitz Collection.*

A nicely finished pair, one having a "Japan" paper label. The mushroom shaker is 3-¾ inches high. *Oravitz Collection.*

Hard to tell what they are supposed to be holding, perhaps cue balls. Height is 2-¾ inches, inkstamp reads "© / Japan" on two lines. *Oravitz Collection.*

Each shaker is 3-⅛ inches high. They each have a "Japan" inkstamp. *Oravitz Collection.*

The pipes have faces on both sides, mouth open on one side, closed on the other. Height is 2 inches, "Japan" inkstamp. *Lilly Collection.*

The shakers of this condiment set are 1-¾ inches high, have "Japan" paper labels. The mustard is 3 inches high, unmarked. *Oravitz Collection.*

On these, the bodies and caps were glazed, but the faces were left in the bisque state. The unmarked shakers are 3-⅜ inches high. *Oravitz Collection.*

More Ceramic Arts Studio, the pieces having been cast with yellow slip. The mushroom is 2-⅜ inches high. Both pieces are marked. *Thornburg Collection.*

The pair on the left is 4-⅛ inches high, have Enesco paper labels. Paper labels of those on the right read "An / Enterprise / Exclusive / Toronto / Canada." *Carson Collection.*

Gnome in shoe sitter, 4-½ inches high as shown. Mark information was not recorded. *Thornburg Collection.*

Ceramic Arts Studio, which made these shakers, called them oak sprites on leaves. The sprite on the left is 3-¾ inches high. Both shakers have Ceramic Arts Studio marks. *Thornburg Collection.*

Unmarked, 4-¾ inches high. *Courtesy of Joyce Hollen.*

Ernie the Keebler elf. Height is 4-¼ inches, and there are two paper labels. One says "© 1989 / KC" on two lines. The other says "Made in Taiwan." *Thornburg Collection.*

Here is a plastic set, advertising pieces for a LaCrosse, Wisconsin brewery. Height is 5 inches. Neither piece is marked. *Thornburg Collection.*

Pixies at home. Both shakers have "Japan" inkstamps, the one on the right is 2-⅜ inches high. *Oravitz Collection.*

The feather in the cap of the pixie on the left doesn't show up very well due to it being damaged over the years. Height of that shaker is 3-⅛ inches. Both pieces are unmarked. *Oravitz Collection.*

This set is a mixture of materials. The shakers are plastic, the base is chalkware, and the elf is rubber. The shakers are 1-¾ inches high, "Emmel" is marked on their bottoms. *Carson Collection.*

These little guys are 2-⅞ inches high. They are not marked. *Oravitz Collection.*

Chapter 6

Animal Life

Name an animal and chances are that somewhere there is a salt and pepper shaker to represent it. Name certain popular animals such as dogs and cats, and chances are that somewhere there are collectors who have china cabinets overflowing with them.

This chapter is divided into five sections. In order of appearance they are Mammals, Birds, Amphibians and Reptiles, Insects and Other Little Creatures, and Creatures of the Deep.

Included along with those animals nature bestowed upon us are character animals, Mickey Mouse in Mammals and Woody Woodpecker in Birds, for instance.

Mammals

There are more salt and pepper shakers made to depict mammals than any other kind of animal. The reason is possibly twofold. First, from earliest times to modern, there is our long dependence upon mammals for various necessities such as labor, food, fabric, security and companionship. Second, there is our general fascination with them, which many of us exhibit on a daily basis by visiting zoos, walking in the woods, or just sitting in the living room in the evening and watching our pet cats tear through the house with the nighttime crazies.

Mammals appear here in roughly alphabetical order, although you may find discrepancies as some mammals are referred to by more than one name. For instance, in some areas of the country skunks are called polecats and woodchucks are called ground hogs. Therefore, if you cannot find what you are looking for, be sure to check the index which is cross-referenced.

A pair of apes, 5 inches high. They have "Japan" paper labels. *Courtesy of Jazz'e Junque Shop, Chicago, Illinois.*

These monkeys are 2-⅝ inches high, have a Goebel mark. According to the McHugh's, they were first released in 1961, and were also used for a condiment set featuring a chimpanze mustard. *Thornburg Collection.*

King Kong and the Empire State Building. Kong is 4 inches high, the building 5-½ inches. Kong is marked "447" on one foot, "© SDD 1988" on the other. The building has a Five and Dime/Sarsaparilla paper label, shown below. *Lilly Collection.*

Five and Dime/Sarsaparilla paper label on Empire State Building.

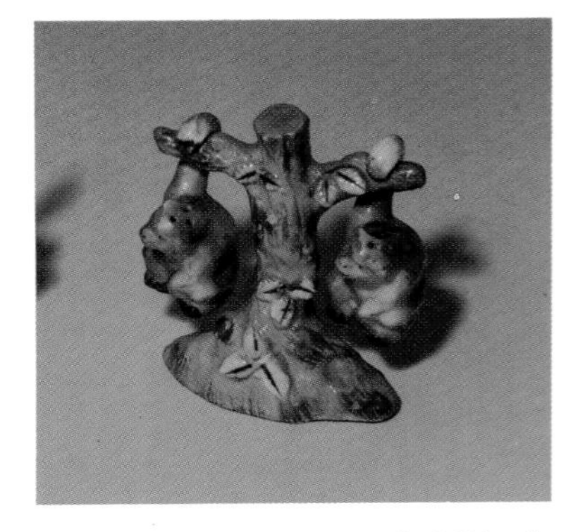

The base of this set is approximately 4-¾ inches high, has a "Japan" paper label. *Carson Collection.*

Redware bears, 2-½ inches high with a "Japan" inkstamp. *Oravitz Collection.*

This pair is made of redware, has an inkstamp that reads "H509" along with a "Made in Japan" paper label. One of the shakers is 3-⅝ inches high. *Carson Collection.*

This monkey is 2-⅝ inches high. It is not marked. *Carson Collection.*

Unmarked armadillos, 2-¼ inches high. *Thornburg Collection.*

Some monkey business going on here with a sitter and a hanger. The members of the sitter are 5-¼ and 3 inches high. They have a paper label, "Atlas / Japan" on two lines. Heights of the hanger pieces are 3-⅝ inches and 3 inches. Its paper label says "Victoria Ceramics," and indicates the set was made in Japan. *Carson Collection.*

These bears are 3 inches high, have a paper label that reads "Victoria Ceramics / Made in / Japan" on three lines. *Carson Collection.*

Nodder monkeys and nodder bears; no other information. *Carson Collection.*

These are unmarked Rosemeade bears, 3-¼ inches high. *Thornburg Collection.*

More bear nodders, each set marked a different way. The pair on the left has "PAT" impressed. The one on the right, "Patent TT," also impressed. The middle set has a paper label that reads, "A Quality Product." *Thornburg Collection.*

Rosemeade bears in a different color, both bears the same. According to an advertising sheet put out by Rosemeade, this is the way the bears were sold, each set composed of two like bears. *Carson Collection.*

Bear nodders, with a "Made in Japan" inkstamp and "Patent TT" impressed in the base. *Carson Collection.*

Here is another case where something is missing, and probably only the most observant buyers would recognize that the set is incomplete. The missing piece is a bicycle built for two made out of brass coated wire. The bears are each 3-¼ inches high. *Carson Collection.*

When Rosemeade shakers are absent paper labels, with a little familiarity you can identify them by their overall design features and the beach sand color of the bisque, as shown here.

The bear is 5 inches high, the fish, which are the shakers, are 3-⅛ inches high. The set has an Artmark paper label that indicates Japanese origin. *Carson Collection.*

Close up of Rosemeade paper label on above bears.

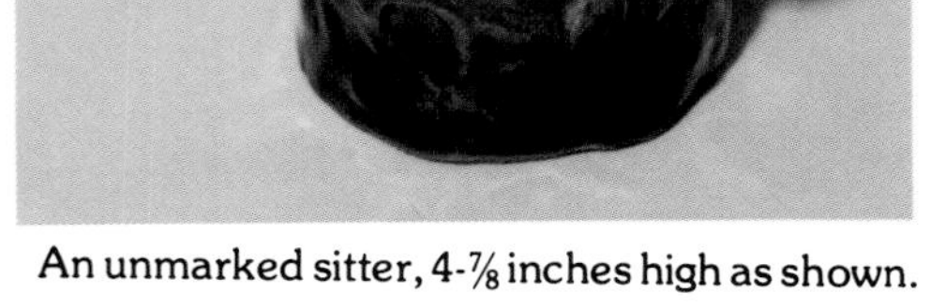

An unmarked sitter, 4-⅞ inches high as shown.

This is part of a series that includes a racoon, and possibly other animals. The bear is 4-¾ inches high. The set has "Japan" inkstamps. *Carson Collection.*

These bears are 2-⅝ inches high, have a "Japan" inkstamp. *Carson Collection.*

Each panda is 2-⅝ inches high, has a "Japan" inkstamp. *Lilly Collection.*

The bears are 3 inches high, have "China" impressed. The box is labeled, "JTC-103 / Made in China" on two lines. *Lilly Collection.*

Unmarked, 2-¾ inches high, and rather cute. *Lilly Collection.*

Christmas bears, also by Otagiri. Santa bear is 4 inches high, marks are the same as above, paper label is shown below. Some folks feel these two sets are mice, so you may choose for yourself. *Thornburg Collection.*

Winnie-the-Pooh and Rabbit. Pooh is 4 inches high. There is a "Walt Disney Productions" inkstamp on the bottom of each. *Carson Collection.*

Bears with bibs and cookies, 3-¼ inches high. They were made by American Bisque in the 1940s to match a cookie jar of the same design. They are not marked, but the cookie jars were marked "Royalware." In addition to white, the cookie jars were finished in gray. Possibly the shakers were, too. *Zera Collection.*

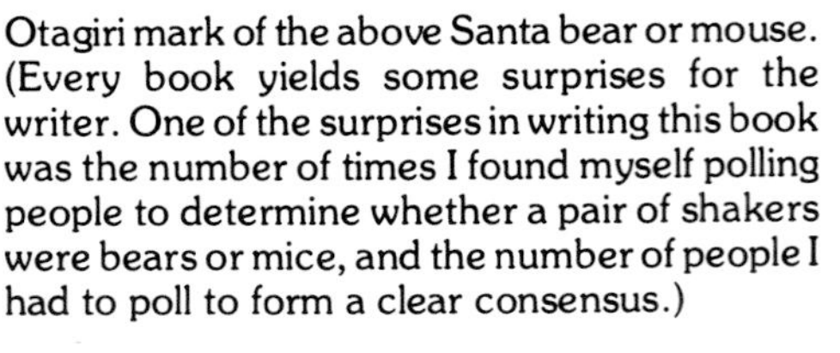

Otagiri mark of the above Santa bear or mouse. (Every book yields some surprises for the writer. One of the surprises in writing this book was the number of times I found myself polling people to determine whether a pair of shakers were bears or mice, and the number of people I had to poll to form a clear consensus.)

The bear in front and the tricycle contraption form the salt, the mother and cub in the rear is the pepper. Height is 4 inches. There is no mark. *Courtesy of Allen and Michelle Naylor.*

Thanksgiving bears, 4-¼ inches high. Impressed mark is "© Otagiri" on one line. They also have an Otagiri paper label. *Thornburg Collection.*

Here's Smokey, 4-¼ inches high, with an Empress paper label. *Carson Collection.*

Three more sets of the famed forest fire fighter. Set on the left is 3-½ inches high, has a Norcrest paper label, along with an "H-415" inkstamp. The middle set is 3 inches high, carries a "Made in Japan" paper label. The set on the right stands 3-⅝ inches, has a "Japan" paper label. *Thornburg Collection.*

Same size as above Smokey, with a "Made in Japan" paper label. *Carson Collection.*

Polar bears by Sandy Srp in original plastic case. Height is 1-⅜ inches. Note the detail on the extremely small fish. *Carson Collection.*

I shot this set from four different angles, never did get a satisfactory picture. But this one shows its most important characteristic, that it's a one-piece that rocks back and forth on its base. Overall height is 3-¾ inches; its mark is shown below. *Thornburg Collection.*

Van Tellingen huggers by Regal China Company, 3-½ inches high. Impressed marks are "Van Tellingen" in a semi-circle with "©" below on back, "Pat. Pend." on bottom. *Oravitz Collection.*

The unmarked purple bears are 2-¼ x 3 inches. *Carson Collection.*

Inkstamp mark of one-piece rocker.

The bear at far left is 4-¼ inches long. Its mate is believed to be at far right. Both have "Japan" paper labels. Sitters, or leapfrog set, in the middle stand 4-¾ inches high as shown. The brown Smokeys are Twin Winton; they match a cookie jar. The chefs have marks that indicate Taiwanese origin. *Courtesy of Millie Gisler and Joyce Hollen.*

The printing on the bases of the pair on the left says, "No hugging allowed." Each unmarked shaker is 3-½ inches high. The set at center has a metal foil label that reads, "Handpainted / PAC / Japan" on three lines. The set on the right carries a "Japan" inkstamp. *Carson Collection.*

Camel figural nodder, the base being 3 inches high. There is a "Made in Japan" inkstamp. *Thornburg Collection.*

Nesting bears, obviously, but the face of the mother resembles a lion. The larger piece is 3-½ inches high, has a "Japan" inkstamp. *Carson Collection.*

Unmarked Rosemeade bison, 2-⅝ inches high. **Rosemeade made many of its smaller figurals as both salt and peppers and figurines. The bison** came in three finishes—gloss glaze, as shown here, matte glaze, and what is often referred to a a bronze-like finish, which is actually a very dark luster glaze. *Thornburg Collection.*

Ceramic Arts Studio camels, 5-⅝ inches high. Each has a Ceramic Arts Studio mark. *Thornburg Collection.*

The beavers on the outside are 3-¼ inches high. They are not marked. The beaver of the inside set is 2-⅝ inches high, the log is 3 inches long. Both have Victoria paper labels. *Thornburg Collection.*

Like beavers, you don't see an overabundance of bison salt and pepper shakers. Each of these is 4-⅝ inches long. Each is unmarked. *Carson Collection.*

The set on the left is 3-⅜ inches high, has a Victoria Ceramics paper label. The set on the right is 3 inches high, carries a Norcrest paper label. *Thornburg Collection.*

Intertwined necks on these camels. The tallest stands 4 inches. One has a "Made in Japan" paper label. *Oravitz Collection.*

Sometimes collectors of advertising salt and peppers have to look beyond what readily shows on a seller's table. This plastic camel illustrates why, as shown below. The unmarked camel is 2-½ x 3-¾ inches. *Thornburg Collection.*

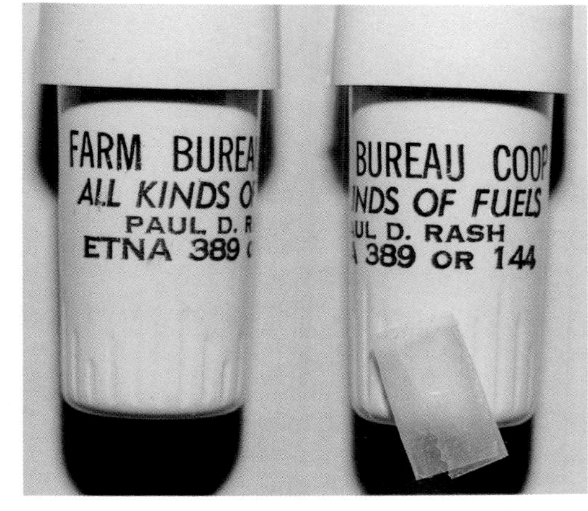

Individual shakers of the above camel.

Same plastic camel as above but in two different colors. Handpainted examples, such as that on the right are considered more desirable. *Carson Collection.*

Cats or dogs, your call. (Owner Susan Oravitz calls it a "Cat-a-Lac," if that's any help.) In any event, it is a condiment set, the trunk opening to reveal the mustard. The car is 7 inches long, the driver is 1-¾ inches high. There is a "Japan" inkstamp. *Oravitz Collection.*

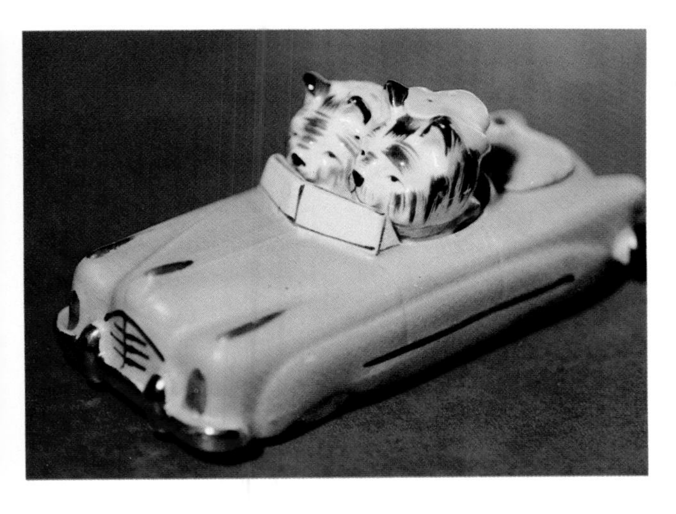

The pair at left rear is 4 inches high, the pair at right front 2-⅜ inches. The others fall inbetween. Left rear has a "Japan" paper label, right rear a "TV-1305" inkstamp and a Brinn's paper label. The front left set is unmarked, while the front right set has a paper label that reads, "Bone / China / Japan" on three lines. *Courtesy of Allen and Michelle Naylor.*

Height here is 2-¾ inches. The set has a "Japan" inkstamp. *Courtesy of Trina Cloud.*

As you can see, this pair can be pushed together so both cats are drinking out of the same bowl. Heights are about 4 inches, mark is "Made in Japan." *Carson Collection.*

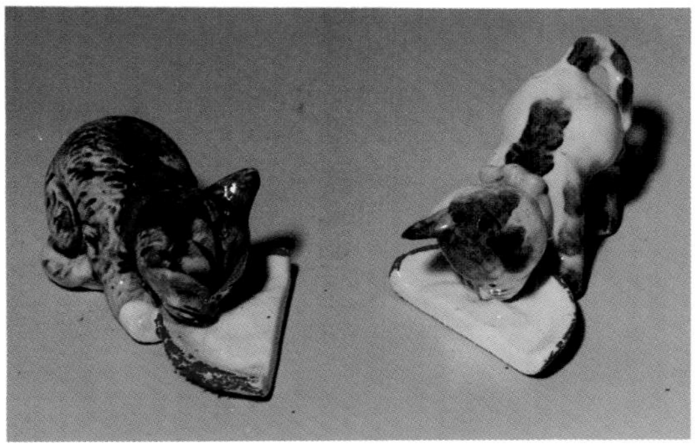

The shakers on the left in the rear are 2-¾ and 3-¼ inches high, have Brinn's paper labels. The colorful pair next to them stand 3-¼ inches with an inkstamp mark that reads "5803 / Handpainted / Japan" on three lines. The pair next to them on the right, reading the "Salty Times" and "Peppery Times," are unmarked, 3-⅝ inches high. The two cats in front are 3 and 2-½ inches long, with "Japan" inkstamps. *Carson Collection.*

Unmarked Shawnee Puss 'N Boots cats, 3-⅜ inches high. They complement an identical cookie jar, and were also made with gold trim. *Smith Collection.*

An assortment of Rosemeade cats. Those in the back row are 3-½ inches high, while those in the middle row are 2-¾ inches. The white cats in front stand 2-½ inches. The pair of blue cats have a Rosemeade mark, which is shown below. *Thornburg Collection.*

Unnamed Ceramic Arts Studio cats, much easier to find than Thai and Thai-Thai, below. Heights are 4-⅜ and 3-¼ inches. Mark of the larger cat is shown below. *Private Collection.*

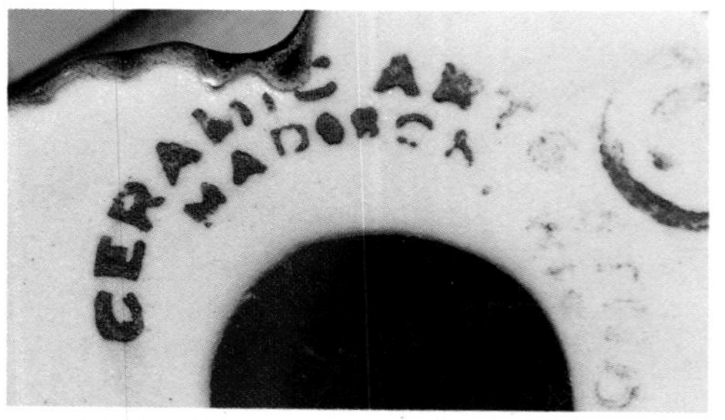

Mark of larger unnamed cat above.

The owl and the pussycat by Fitz and Floyd. The cat is 2-½ inches high, the owl just slightly shorter at 2-⅜ inches. The mark of the boat is shown below. *Oravitz Collection.*

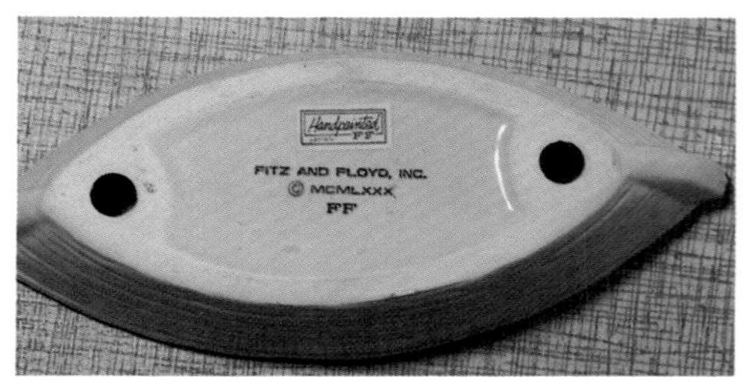

Paper label and inkstamp mark of the boat that holds the owl and the pussycat.

Mark of blue Rosemeade cats above.

This is a figural nodder, 4 inches high. It has a "Made in Japan" inkstamp. *Thornburg Collection.*

Like the white and red bears on page 84, these cats were apparently made by American Bisque Company during the 1940s, as they come pretty close to matching a Royal Ware (American Bisque) cookie jar, and appear to have the same bisque and glaze. The unmarked shakers are 3-¾ inches high. *Zera Collection.*

Here is Thai and Thai-Thai, an often hard to find set of shakers by Ceramic Arts Studio. Thai-Thai, who is being used for a pillow, is 5 inches long. The mark of Thai-Thai is shown below. *Private Collection.*

The shakers on the left are 2-½ inches high, and unmarked. Those on the right are 2-⅝ inches high with a "Japan" inkstamp. *Carson Collection.*

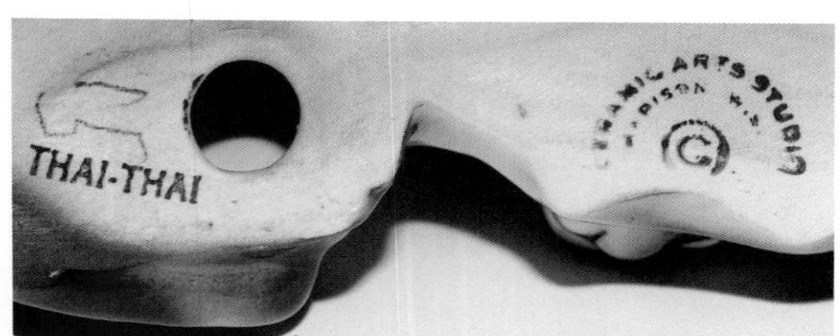

Mark of Thai-Thai.

Three sets of cats, all made by Goebel. The black one at front right is 3-¼ inches high. All have Goebel marks. The baskets are plastic. *Thornburg Collection.*

Wooden black cats with leather ears and plastic whiskers, also leather tails which do not show in the picture. As you can see, they are also magnetic. Height of the larger ones is 4-⅜ inches. *Carson Collection.*

Black cats made of redware. All are unmarked but Shafford would be a good bet. Heights left to right are 4-⅝, 3-½ and 5-½ inches. *Carson Collection.*

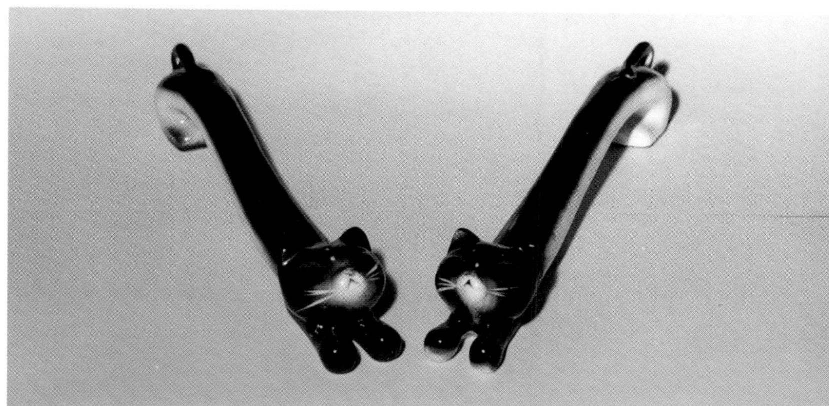

While Shafford sold a lot of black cats, it did not have a corner on the market. These shakers, 8-¾ inches long, possess a Wales paper label which is shown below. *Carson Collection.*

Shafford black cat salt and pepper shakers and other accessories. The teapot is 7 inches high. The piece at far left is a wallpocket. Paper labels read, "Product of / Shafford / Japan" on three lines. *Courtesy of Cloud's Antique Mall, Castalia, Ohio.*

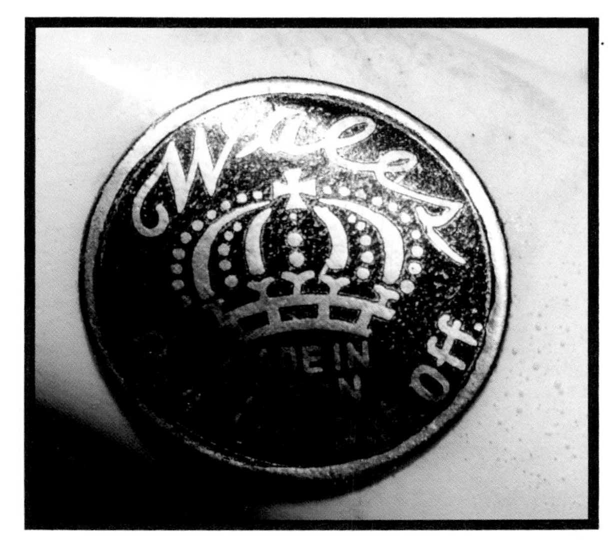

Wales paper label of long black cats.

The white cats in the center are 4-¼ inches high. They have an "HH Japan" paper label, and an inkstamp that reads, "© 1958 / Holt Howard" on two lines. The cat and shoe shakers carry a Lego paper label. The pair on the outside is the same as above. *Carson Collection.*

Height here is 2-⅞ inches. The kitties have paper labels that read, "Giftcorp." *Lilly Collection.*

All wood. Heights are 5, 4 and 2 inches. All have "Japan" paper labels. The tallest pair have bellows type noisemakers in their bottoms that meow when the shakers are used. *Carson Collection.*

Here are two singles that appear to be part of a series. Each is 4-½ inches high, and unmarked. Each originally carried an umbrella. *Carson Collection.*

The cat is 4-⅝ inches high, has a "Japan" paper label. *Lilly Collection.*

More felines with their favorite toys. The cat on the left is 3-¼ inches high, unmarked. The one on the right is 4 inches high. It is not marked, either, but has a dry foot similar in shape to those used by American Pottery and American Bisque. The unmarked pair in the center, 3-⅞ inches high, is rather unusual. They have meowers, and the yarn is real. *Lilly Collection.*

The shakers on the outside are 5-¼ inches high. None of the four are marked. *Carson Collection.*

Same pair with different glaze. Although the difference in size appears to be greater, heights are 4-¼ and 4 inches. Paper labels read "Made in Taiwan." *Lilly Collection.*

Plastic Fifi and Fido, made by F & F of Dayton, Ohio, as an advertisement for the Quaker Oats Company's Ken-L-Ration dog food and Puss 'n Boots' cat food. Height of the dog is 3-¼ inches. A cookie jar was made in the image of Fido, while wallpockets were made to match both shakers. *Zera Collection.*

The black cats are 2-¾ and 4-½ inches high. The set in the center is two pieces, the cat having a height of 2-¾ inches. All have Ceramic Arts Studio marks. *Thornburg Collection.*

Sandy Srp dog and cat miniatures. Each is 1-⅛ inches high. *Carson Collection.*

The cat is 2-½ inches high, has a "Japan" inkstamp. *Carson Collection.*

Ceramic Arts Studio gingham dog and calico cat. Heights are 3 and 2-⅞ inches, respectively, and both are marked. *Thornburg Collection*

Sorry about the birds. But since they are here, we will say they are 3-¼ inches high, have "Japan" inkstamps. The cats are 3-⅛ inches high, include meowers, and have marks indicative of Japanese origin. *Courtesy of Joyce Hollen.*

These probably look like ceramic to you, but they are plastic. The shakers are 4-¾ inches high, and have meowers as stated on the box. While the shakers are not marked, the box is. A close up is shown below. *Lilly Collection.*

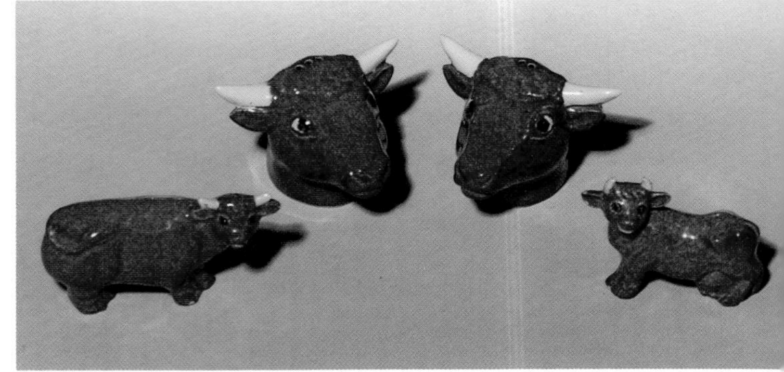

Rosemeade cattle. The heads are 2 inches high, the full figures 3-¼ inches long. None of the four shakers are marked. *Thornburg Collection.*

Close up of box containing white plastic cats.

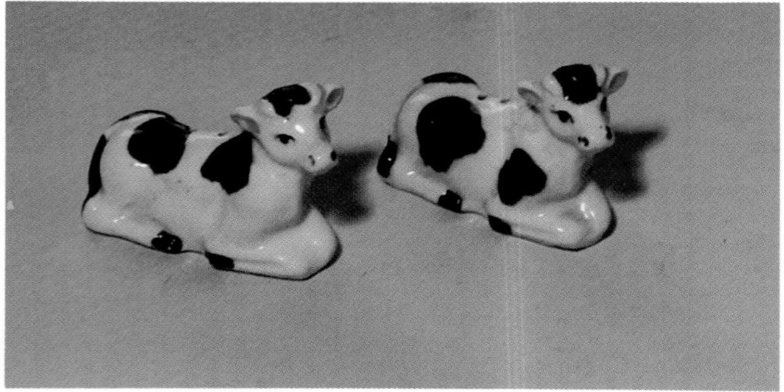

These shakers are 2-⅝ inches long. They have "Japan" inkstamps. *Carson Collection.*

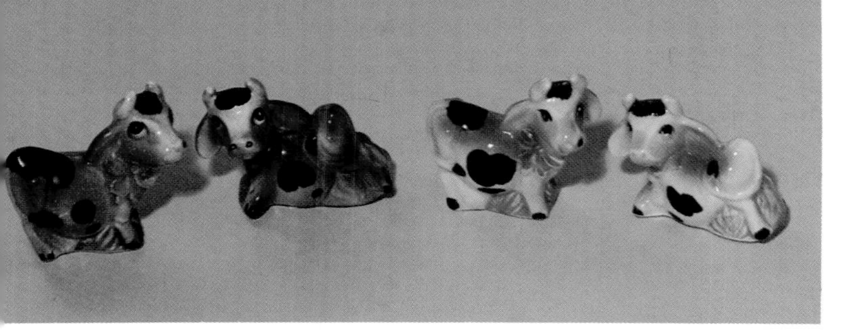

Same pair, different decoration. The shaker on the far left is 2-⅝ inches high, the one on the far right is 3-⅛ inches long. One pair is marked with a "Japan" paper label, the other with a "Japan" inkstamp. *Carson Collection.*

Height is 2-⅝ inches. The shakers are unmarked. *Carson Collection.*

A couple bulls, 3-¼ inches high on the left, 4-⅝ inches long on the right. Each has a Lego paper label. *Carson Collection.*

These ladies are 3 inches high, have Lefton paper labels. *Carson Collection.*

Elsie and Elmer, symbols of Borden's milk. The shaker on the left is 4-¼ inches high. Impressed mark on this pair is "© The Borden Co." *Thornburg Collection.*

Elsie sitters. The pair on the left is 4-¼ inches high. It is marked "© The Borden Company." The unmarked pair on the right, 4 inches high, is a counterfeit, meaning it was produced without licensing by Borden's. Note the more detailed and higher quality decoration on the original, which is usually the case when dealing with counterfeits. *Thornburg Collection.*

Similar to Elsie but certainly not official due to lack of a copyright notice. The mother is 5 inches high, has a "Japan" inkstamp. The calves are 2-½ inches high, unmarked. Another set photographed had calves with "Japan" inkstamps. Ceramic Arts Studio made a similar set, but with just one calf, it being the pepper. *Oravitz Collection.*

The cow is 4-¼ inches long, the wreath 3-⅞ inches in diameter. The set has a Vandor paper label, also a Pelzman Designs paper label. *Thornburg Collection.*

Another Artmark set, each member being 2-⅞ x 3 inches. As above, "Taiwan" is impressed on each. *Lilly Collection.*

Bull with salt and pepper teapots. The bull is 3-⅝ x 5-½ inches. The shakers are 3 inches high. All pieces are unmarked, all are redware. *Lilly Collection.*

The cow jumped over the moon. The cow is 3-¼ inches long, the moon 4 inches high. "Taiwan" is impressed on both pieces, while the paper label of the moon is shown below. *Lilly Collection.*

Cow and pig by Sandy Srp. The cow is 1-⅝ inches high. *Carson Collection.*

Paper label of above moon.

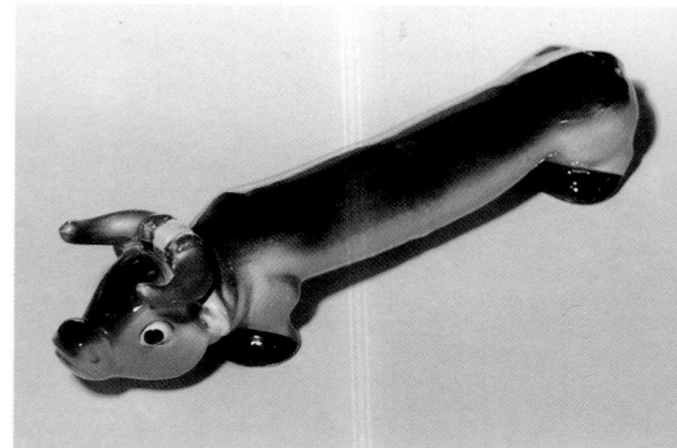

A long one-piece that measures 9-½ inches. It has an Empress paper label. *Thornburg Collection.*

One-piece cows, 3-⅛ inches high. "Made in Japan" paper label, plus an inkstamp that has a "C" superimposed on a "N" alongside the characters "B468." *Thornburg Collection.*

This is a highly prized set made by Regal China Company for its Farmer John/Old MacDonald kitchen artware series. Each shaker is 5-⅛ inches high. Impressed in the bottom of the salt is "Patent Pending 384." Impressed mark on the pepper was unreadable. *Carson Collection.*

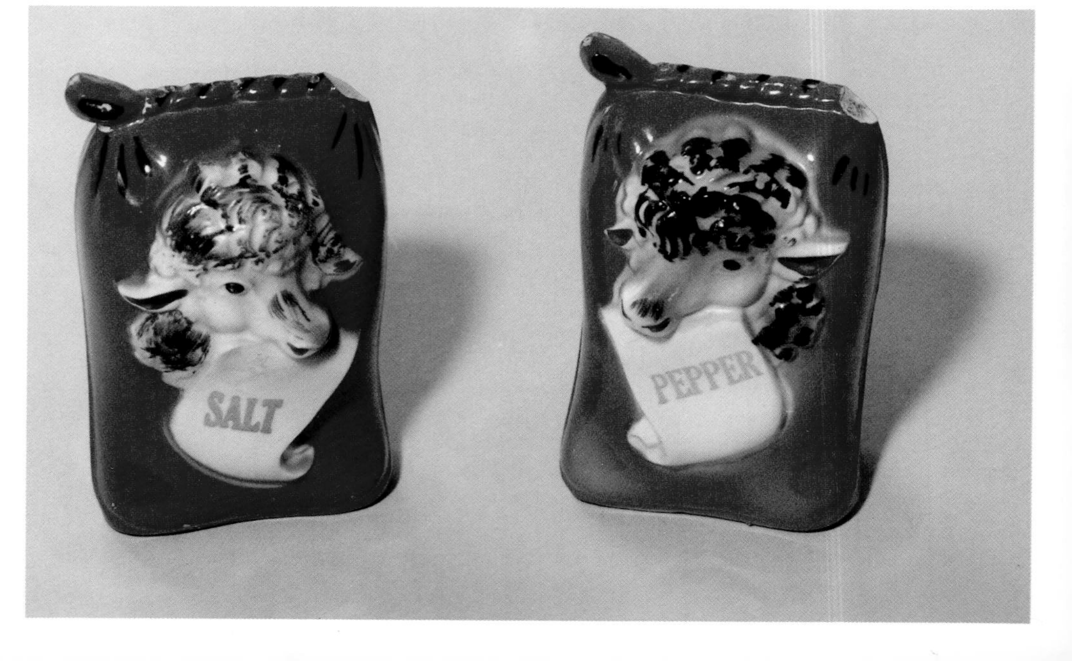

Cows and bears by either American Bisque or American Pottery. The cows are 3-½ inches high. The bears, 3 inches high, were shown above. *Zera Collection.*

Here is a married pair, each member originally having a different partner, but apparently the same shakers with different decoration. Height is 2-⅞ inches. One has a "Japan" inkstamp, the other a "Japan" paper label. *Lilly Collection.*

Plastic, 4-⅜ inches high, with "Made in / Hong Kong" impressed on two lines in very small letters. *Private Collection.*

Rosemeade fawns, 4-¼ inches high. Like most Rosemeade shakers, the paper label is the only mark. *Thornburg Collection.*

Though unmarked, these fawns have styling and decoration traits indicative of the Kass China Company, of East Liverpool, Ohio. A very similar Kass figurine of a standing deer appears in my book, *Animal Figures*. Size of these shakers was not recorded. *Carson Collection.*

Ceramic Arts Studio stylized doe and fawn. The doe is 4-¼ inches high, the fawn 2-¼ inches. Both have Ceramic Arts Studio marks. *Thornburg Collection.*

The shaker at far left is 4 inches high, has a "Japan" paper label. The one at far right is 4-½ inches high, carries a Brinn's paper label. The tallest member of the set in front is 3-½ inches high, has an unreadable paper label. *Carson Collection.*

Rosemeade fawns. The one on the left is 3-½ inches long, while the one on the right is 2-½ inches high. Unmarked except for the paper label. *Thornburg Collection.*

These fawns stand 4-¾ inches high. They have "Japan" inkstamps. *Oravitz Collection.*

These shakers are 2-¼ and 3-⅛ inches high. They have "Japan" inkstamps. *Lilly Collection.*

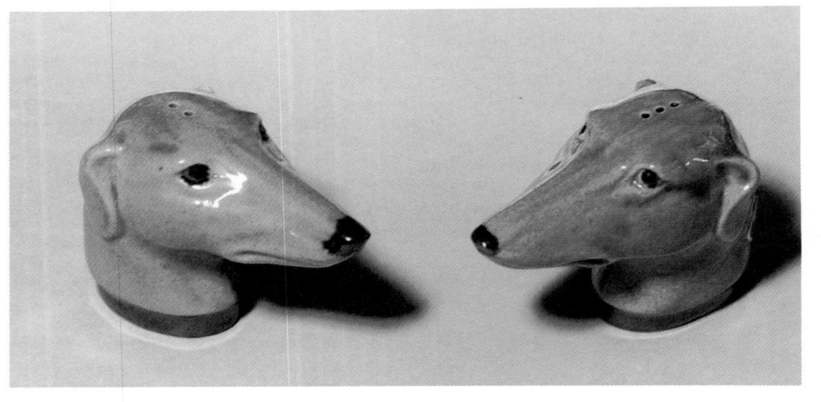

Rosemeade greyhound heads, 2-⅜ inches high, unmarked. *Thornburg Collection.*

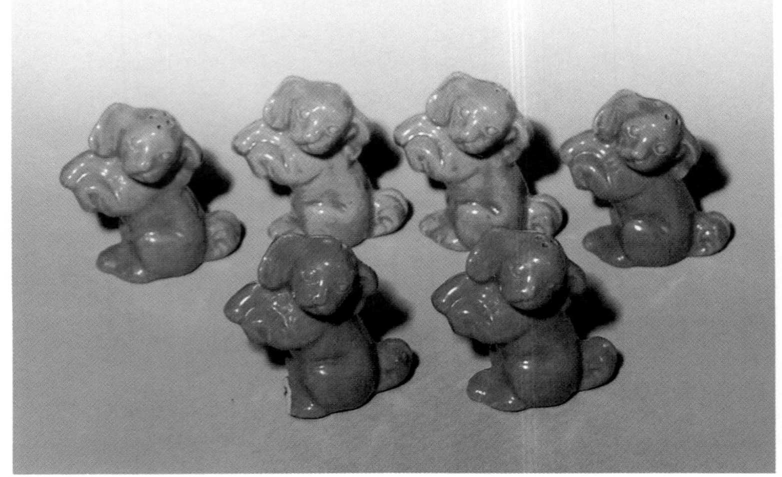

Rosemeade begging dogs, each 2-¾ inches high, and unmarked. *Thornburg Collection.*

More Rosemeade dog heads, also unmarked. The chow heads, on the left, are 2-⅛ inches high. The chows were also made with a black glaze. *Thornburg Collection.*

In this group of unmarked Rosemeade shakers, the English bulldogs, in the center, are 2-⅜ inches high. *Thornburg Collection.*

The dalmations are 2-⅞ inches high. As above, none of these Rosemeade heads are marked. These four pictures show all known Rosemeade dog head shakers with the exception of the chihuahuas, which are quite rare. *Thornburg Collection.*

These dogs are 2-¾ inches high. Their paper labels read "Grindley Ware / Manufactured in / Sebring, Oh." on three lines. *Oravitz Collection.*

Continuing on with American made dog shakers, here is Shawnee's Mugsey in the table size, 3-½ inches high. The shakers are unmarked. *Courtesy of Joyce Hollen.*

Shawnee Mugsey in range size, 5-¼ inches. The example in the middle is a single, shown because of the more detailed decoration and gold trim. Shawnee made a Mugsey cookie jar based on a design similar to the shakers. *Smith Collection.*

Brayton Laguna Pottery gingham dog and calico cat, inspired by American poet Eugene Field's (1850-1895) poem "The Duel." (This set isn't Field's only contribution to the hobby. Other shakers based on this works include Little Boy Blue, and Wynken, Blynken, and Nod.) Heights are 3-⅜ inches and 4-¼ inches, respectively. They are not marked. These were also made as figurines, creamers, sugars, pencil holders and cookie jars. *Carson Collection.*

Same as above but in yellow, brown and green. *Private Collection.*

The dogs on the left are Billy and Butch. Heights are 3 inches and 2 inches. The spaniels on the right are unnamed. They are 2-½ and 1-⅝ inches high. All four shakers are by Ceramic Arts Studio. *Thornburg Collection.*

That's Suzette on the pillow. She is 3-¼ inches high while her pillow is 2-¾ inches across. As for the other pair, the member on the right is 2-¾ inches high. All are Ceramic Arts Studio, with the mark of Suzette's pillow shown below. *Thornburg Collection.*

Pluto on the far left is 3-¾ inches high, has a "Japan" inkstamp. The Pluto next to him is 4-½ inches long. The other set was marketed by the Good Company; Pluto is 3 inches high. *Thornburg Collection.*

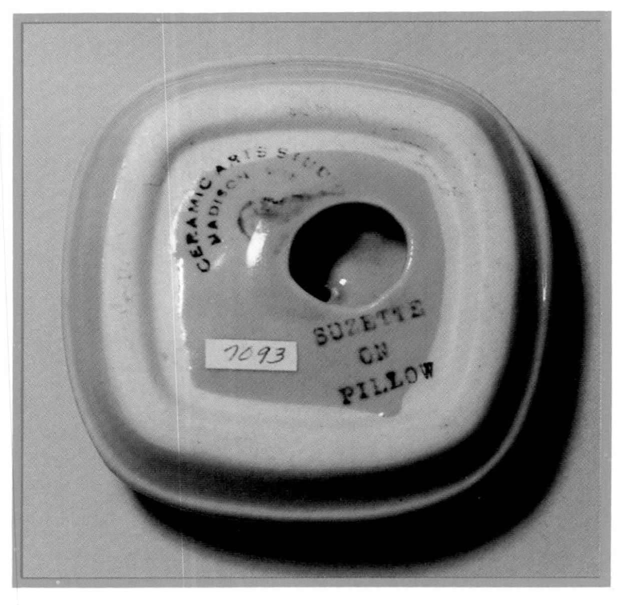

Mark of Suzette's pillow above.

Again, Ceramic Arts Studio. The dog is 1-⅝ inches high and its house is 2-¼ inches long. Both are marked. *Thornburg Collection.*

Unmarked Ceramic Arts scotties, the one on the left is 3 inches high. The pomeranians are also Ceramic Arts Studio shakers, and also unmarked. The pomeranian on the left stands 2-¾ inches. *Thornburg Collection*

Unmarked Pluto, similar to the work of American Bisque and American Pottery, and probably marketed through the Leeds China Company, a Chicago distributor licensed by Walt Disney from 1944-1954. The shakers are 3-⅛ inches high, and are unmarked. The bottom of one is shown below. *Oravitz Collection.*

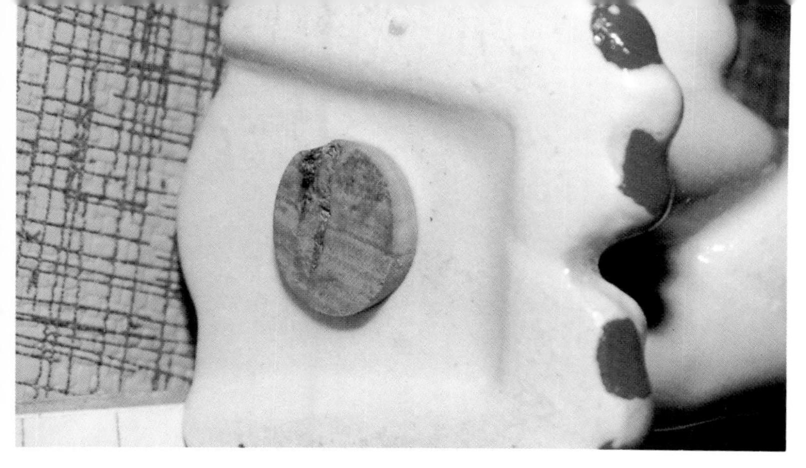

Bottom of Pluto. The u-shaped dryfoot was often used on these white and red shakers by American Pottery and/or American Bisque.

The pair on the right is a recent Artmark set with paper label reading "©1990 Artmark." The hydrant is 3 inches high, the dog 2-⅞ inches. The set on the left has a "Japan" inkstamp. Height of the dog is 3-¼ inches, that of the hydrant 2-¾ inches. *Lilly Collection.*

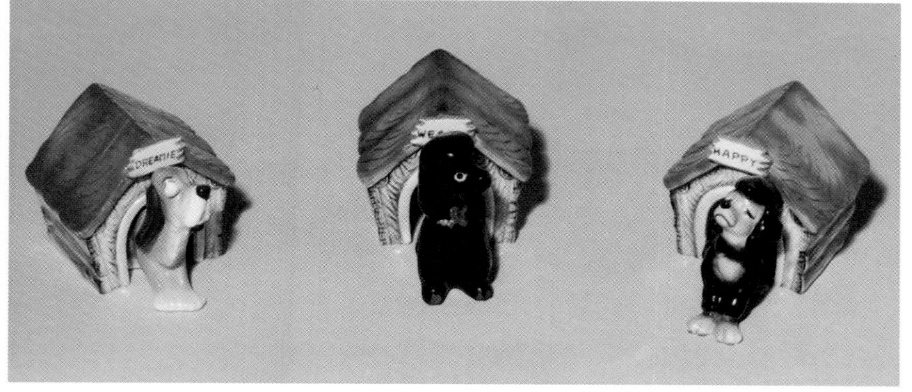

Dreamie, Weary and Happy, all unmarked. All of the dogs are approximately 2-¾ inches high, and their houses are approximately one-quarter inch taller. In addition to these three, Guarnaccia, in Book III, shows four others in the series—Choosie, Sad Sack, Grouchie and Snootie. *Lilly Collection.*

What goes together as well as a dog and a fire hydrant? Looks like a cow and a milk can. The only measurement taken was that of the cow, which is 3 inches high. None of the pieces are marked. *Carson Collection.*

Posh poodles. The male—perhaps "gentleman" would be more appropriate—is 3 inches high. Marks are shown below. *Courtesy of Allen and Michelle Naylor.*

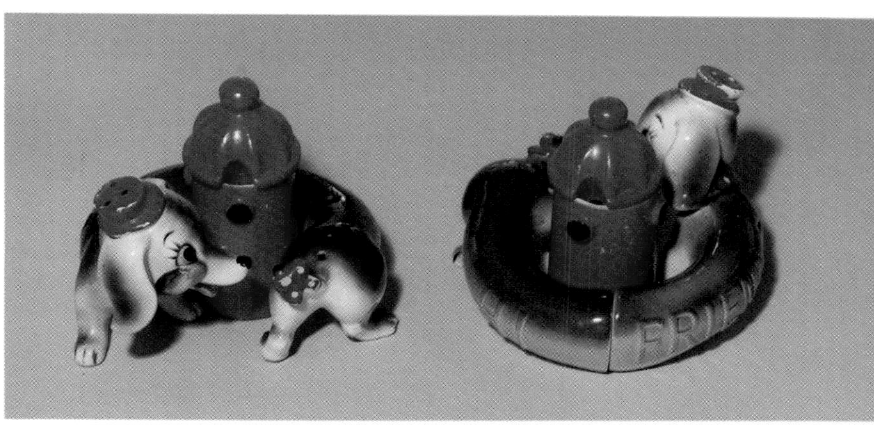

Back to just dogs—same set, but different decoration. It is a condiment set, with the hydrant as the mustard. The front half of the dog is 5 inches long, the rear half 4-½ inches long. The hydrant is 4-¼ inches high. As you can possibly see on the set on the right, the dogs have lettering—"Hi Friend"—on them. An Artmark paper label points to Japanese origin. *Carson Collection.*

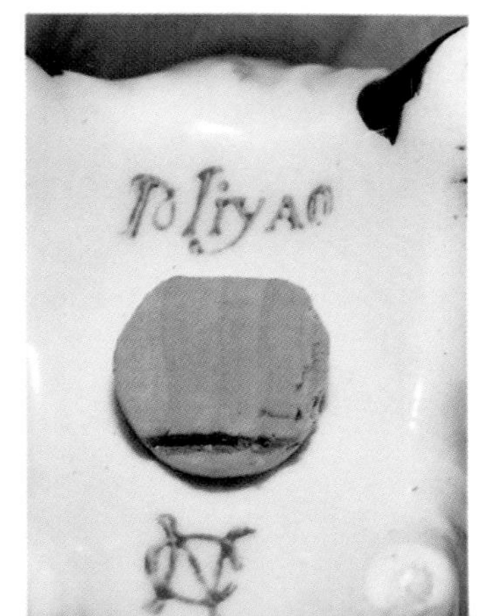

Mark of the posh poodles.

No shyness from this guy. The dog is 3-½ inches high and has a Vandor paper label. *Courtesy of Mark Taylor.*

A plastic set. The dogs are 3-¼ inches high and the holder is 4-⅝ inches long. The holder is marked, "©JSNY / Made in Hong Kong," on two lines. *Lilly Collection.*

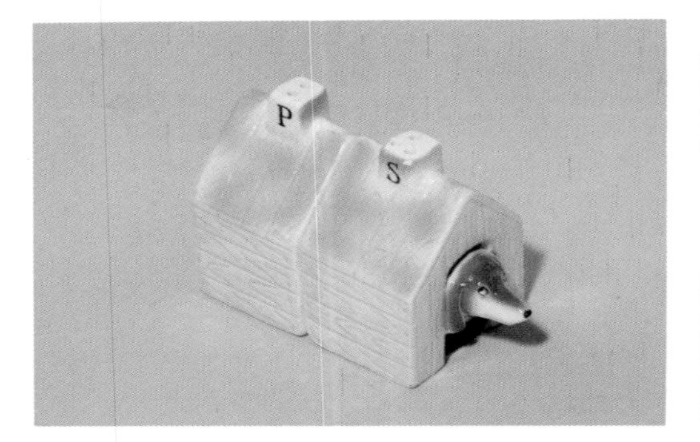

This set is 2-¾ inches high and has a Brinn's paper label. *Carson Collection.*

The basket is 5-½ inches long and the dog on the right is 3 inches high. None of the pieces are marked. *Lilly Collection.*

This is a one-piece, 2-⅞ inches high. It has a "Made in Japan" paper label. *Thornburg Collection.*

The standing pair on the right are 3 inches high. Both pair are Goebel, but types of marks were not recorded. *Carson Collection.*

Height is 4-⅛ inches with an Artmark paper label. *Lilly Collection.*

The dalmations are 3-½ inches high, unmarked. *Lilly Collection.*

When turned toward each other and set close together, these two become kissers. Each is 3-¼ inches high, and each has a "Made in Japan" inkstamp. *Lilly Collection.*

Christmas pooches, 5-⅛ inches high. They have "Japan" paper labels. *Oravitz Collection.*

Here is Nipper, the trademark of RCA Victor. The base of the set on the left is plastic, the shakers are ceramic. Nipper is 2-⅞ inches high. Both pieces of the set on the right are plastic. The dog is 2-⅞ inches high, and is unmarked. The Victrola is marked "(RCA Symbol) / Trademark / R (in circle)." *Thornburg Collection.*

The shaker on the left is 2-½ inches high while the one on the right is 3 inches. Both are marked "Japan." A paper label indicates they were originally a souvenir of The Blue Hole, in Castalia, Ohio. *Carson Collection.*

Nipper is 3-⅛ inches high. None of the pieces are marked. *Carson Collection.*

Each of these dogs is 2-¾ inches high. Inkstamp mark is "S1518A," plus there is a Napco paper label. *Carson Collection.*

Spuds McKenzie, 3-¼ inches high. "SDD © 1987" is impressed, in addition to a Five and Dime paper label. *Thornburg Collection.*

Seems like sanitation would be a problem if these shakers were actually used. Height is 3-¾ inches. There is a "H713" inkstamp. *Carson Collection.*

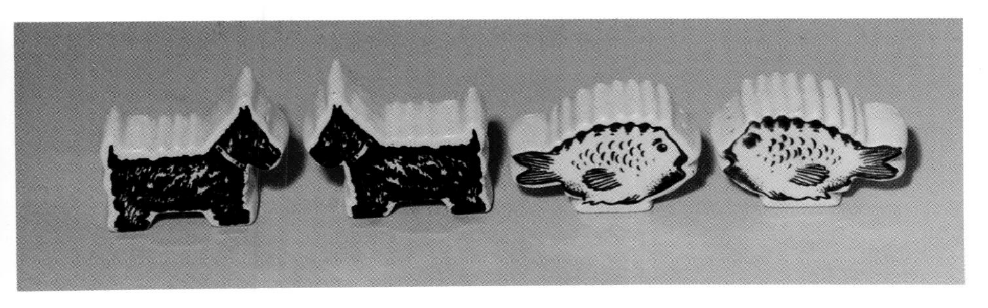

The scotties are 2-½ inches high. The fish, obviously part of the same series, are 3-⅛ inches long. Neither set is marked. *Carson Collection.*

Two sad-looking fellows, each 4 inches high. They are made of redware and have paper labels that read "Imports / Enesco / Japan" on three lines. *Coutesy of Virginia Sell.*

Height of the pair on the left is 3-⅞ inches. They have "Japan" inkstamps. The pair at midle rear has an inkstamp which reads "U437," in addition to a "Japan" paper label, while the pair on the right has a "Japan" inkstamp. The set in front is 2-½ inches high and is inkstamped "© JSNY / Taiwan" on two lines. *Carson Collection.*

On the left are Poof and Woof, each 4 inches high and unmarked. The middle set, originally sold as a souvenir at Cedar Point, Ohio, is 3 inches high, also unmarked. The pair on the right is 4 inches high, has a "Japan" inkstamp, along with a paper label that reads, "Commodore / Japan" on two lines. This set also has a bellows in each piece that simulates the sound of a barking dog when the shakers are turned over. *Carson Collection.*

The white dogs on the outside are 4 inches high and have both a Lego paper label and an inkstamp that reads "© Lego." The basket holding the blue poodles is 4-½ inches long and has a GNCO paper label which is shown below. The poodles themselves have "Japan" paper labels. *Carson Collection.*

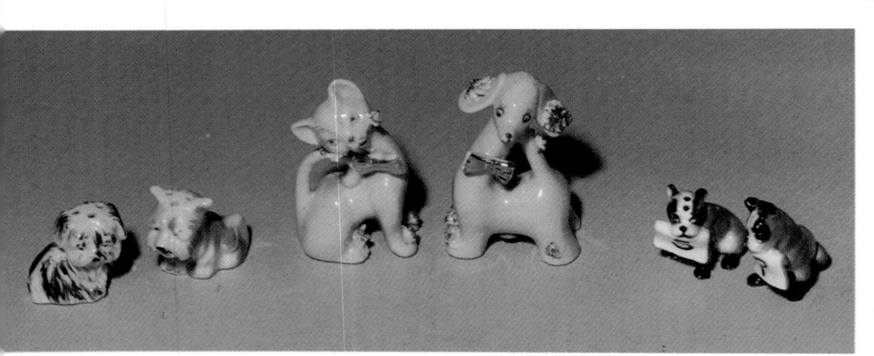

The scotties are 1-½ inches high and are marked "Japan." The pink poodle and cat are 3-½ inches high and have "Japan" paper labels. The boxers are part of a bone china series, more of which is shown in Chapter 2. They are 1-⅞ inches high and have "Japan" paper labels. *Courtesy of Joyce Hollen.*

Paper label of the above blue basket.

The puppy is 2-⅝ inches high. Neither piece is marked. *Carson Collection.*

Heights of these donkeys were not recorded, but the one on the right is about 3 inches. Both have Ceramic Arts Studio marks. *Thornburg Collection.*

Long beasts of burden for farmers, donkeys were also popular beasts of burden for giftware manufacturers and importers. This one is 7-¼ inches long and holds vinegar, oil, salt and pepper. None of the pieces are marked. *Carson Collection.*

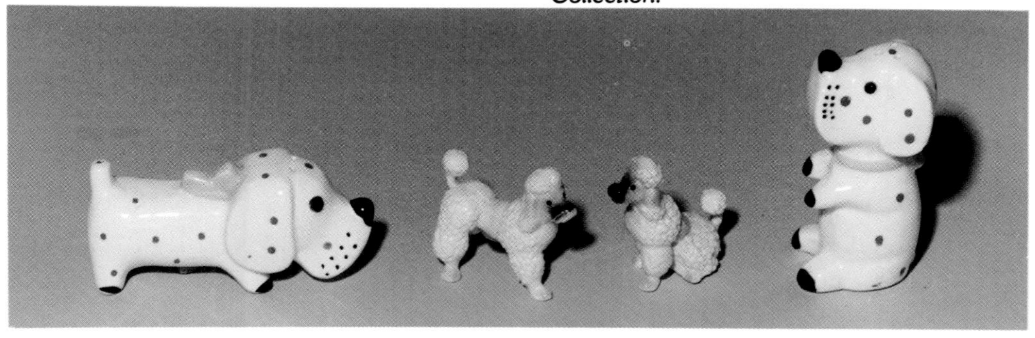

The dog on the left is 4-¼ inches long, his mate on the right is 4 inches high. They have Napco paper labels. The inside pair is 2-⅛ inches high, unmarked. *Courtesy of Joyce Hollen.*

A slightly different set here, with cream and sugar plus salt and pepper. Size is the same as above, however, with the donkey and cart 7-¼ inches in length. The Palomar paper label of the cart, which indicates Japanese origin, is shown below. *Courtesy of Jazz'e Junque Shop, Chicago, Illinois.*

A pair of sitters photographed prior to deciding upon the book format. The dog in the chair is 4-½ inches high as shown. The chicken on the nest is 3-¼ inches high as shown. Neither set is marked.

Miniature set by Sandy Srp. The donkey is 1-½ inches high. *Carson Collection.*

Paper label of donkey cart above.

This poor donkey has to carry a man besides all the other paraphernalia. Overall height is 6-¾ inches. Paper label reads "Made in / Thames / Handpainted / Japan" on four lines. *Carson Collection.*

These donkeys have Relco paper labels. They stand 3-⅛ inches high. *Carson Collection.*

Miner's donkey stands 4-¾ inches high. The tool opposite the pick, which is the second shaker, is a shovel. The set has a "Japan" paper label. *Carson Collection.*

Tambo and Tembino, also by Ceramic Arts Studio, and bearing that company's marks. Tambo is 6-¾ inches high, Tembino 2-½ inches. *Thornburg Collection.*

Circus elephants, 4-¼ and 2-⅞ inches high. Each has a "Japan" inkstamp. *Carson Collection.*

A donkey on the farm. Height is 3-½ inches—there is no mark. *Carson Collection.*

Donkey and elephant to represent Democrats and Republicans. The donkey is 2-½ inches high, and the pair is made of chalkware. *Carson Collection.*

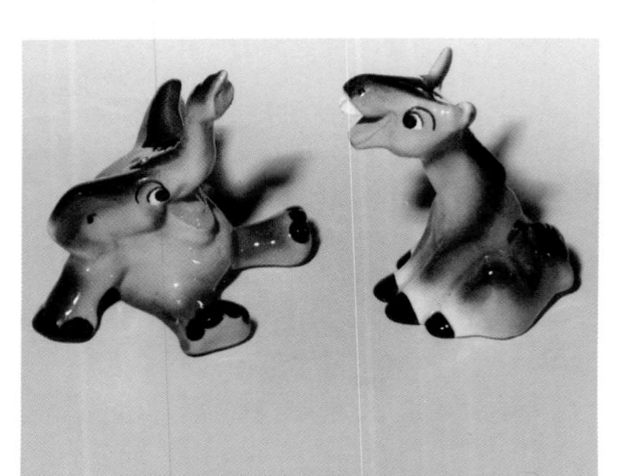

Here are Dem and Rep, Ceramic Arts Studio's entry into political symbolism. Rep is 4 inches high. Both have Ceramic Arts Studio marks. *Thornburg Collection.*

More Ceramic Arts Studio on the right. On the left is a pair of Japanese imposters, complete with a "Japan" inkstamp. Heights of both originals and imposters is 3-¾ inches and 3-¼ inches. Both sets are turned wrong in the picture, but when turned correctly, the trunks form the shapes of S and P. Mark of the Ceramic Arts pair is shown below. *Thornburg Collection.*

Mark of the pink elephants above. Note the pink bisque.

Rosemeade elephant shakers, 2-⅝ inches high. The brown pair on the left has the usual Rosemeade paper label. Paper label of the dark pair on the right reads "Souvenir of / Outlaw Trading Post / Winner, South Dakota" on three lines. *Thornburg Collection.*

These elephants have lace collars. The blue example is 2-3/8 inches high. They have Palomar labels that indicate Japanese origin. *Carson Collection.*

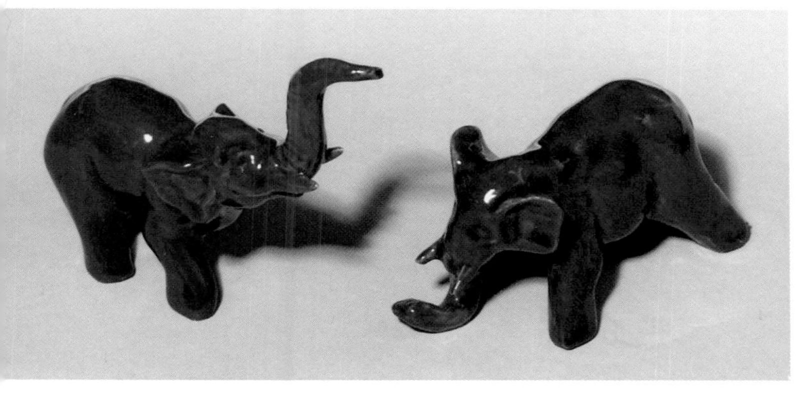

The elephant on the left is 3-1/2 inches high. It has a Meridan paper label. *Thornburg Collection.*

Dumbo made of redware with "Japan" inkstamps. Height of the one on the left is 3-1/8 inches. *Carson Collection.*

Unmarked Dumbos believed to be by American Pottery or American Bisque. Heights are 4-1/4 inches and 3-1/4 inches. *Oravitz Collection.*

The pair of foxes in front is by Enesco, as shown on paper labels. The one on the left is 3 x 5-1/4 inches. In the set to the rear, the left-most member is 2-1/2 x 5 inches and has a Relco paper label. *Thornburg Collection.*

Snazzy rendition of Old Reynard, 4-7/8 inches high. Neither shaker is marked. *Thornburg Collection.*

Heights are approximately 4-3/4 inches. Each has a "Japan" inkstamp. *Thornburg Collection.*

The shaker on the left is 3-⅛ inches high. All have Enesco paper labels. *Thornburg Collection.*

The taller shaker is 6 inches high. Its mark is shown below. *Carson Collection.*

Mark of the foxes shown above. Word at bottom right is "Nissen."

These foxes are 2-⅞ inches high. Paper label, which indicates Japanese origin, is shown below. *Carson Collection.*

Paper label of the above foxes.

A couple foxes, a couple rodents. The foxes are 2-¾ inches high. None of these pieces are marked. *Carson Collection.*

Fox and goose. The fox is 4-½ inches long, the goose 2-½ inches high. Both have Ceramic Arts Studio marks. *Thornburg Collection.*

The taller of the set in the center is 5-½ inches high. The tag reads, "Gi-Gi / Neckin 'n Kissin / Giraffes / Salt and Pepper / Chase Import Japan" on five lines. The set on the left has a Vandor paper label, while the pair on the right has a "Japan" paper label. *Thornburg Collection.*

The giraffes on the left are 5-¼ inches high, have a "Japan" paper label. Those in the middle stand 4 inches. They also have a "Japan" paper label. The ones on the right are 5 inches high, have a "Japan" inkstamp. *Carson Collection.*

Giraffe and zebra, the giraffe being 4-¼ inches high. Both have "Japan" paper labels. *Carson Collection.*

The taller of this handsome couple is 6-¼ inches high, the shorter 5-¾ inches. Neither is marked, but both are Ceramic Arts Studio. *Thornburg Collection.*

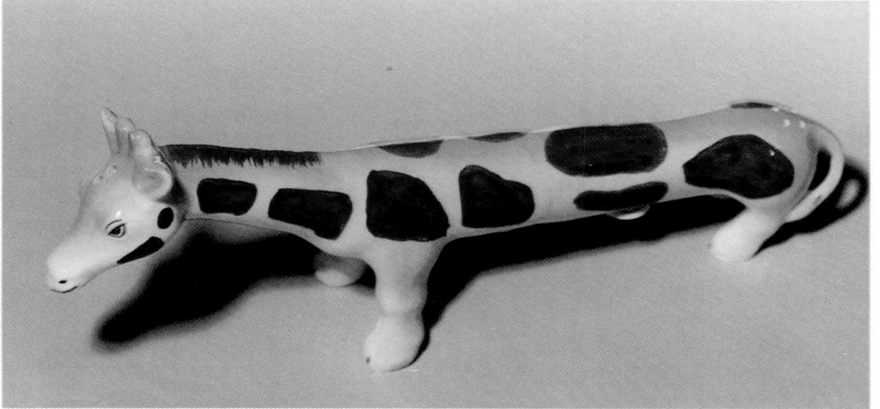

Long one-piece that measures 4 x 9-½ inches. Impressed mark is "SDD / © 1989 / 342" on three lines. It also has a Five and Dime paper label. *Thornburg Collection.*

These realistic goats have a Brinn's paper label. The one on the left is 3-¾ inches high. *Carson Collection.*

Hedgehogs, 3 inches high and unmarked. *Thornburg Collection.*

Unmarked Rosemeade mountain goats, 2-⅛ inches high. *Thornburg Collection.*

These adorable hedgehogs are 3 inches high, have a "Japan" inkstamp. *Carson Collection.*

These are the only gohpers that were found, and all are by Rosemeade. The pair at left rear is 4-¼ inches high, the pair at right rear is one inch shorter. The ones in front are called flickertails. They are 2-½ inches high. None of the shakers in this picture are marked. *Thornburg Collection.*

The pair with the bowties stand 3-⅝ inches high, have Enesco paper labels. The other two pair are not marked. *Thornburg Collection.*

The hippo at far left is 3-⅜ inches long. It has a "Japan" inkstamp. The other pair has "Japan" paper labels. *Thornburg Collection.*

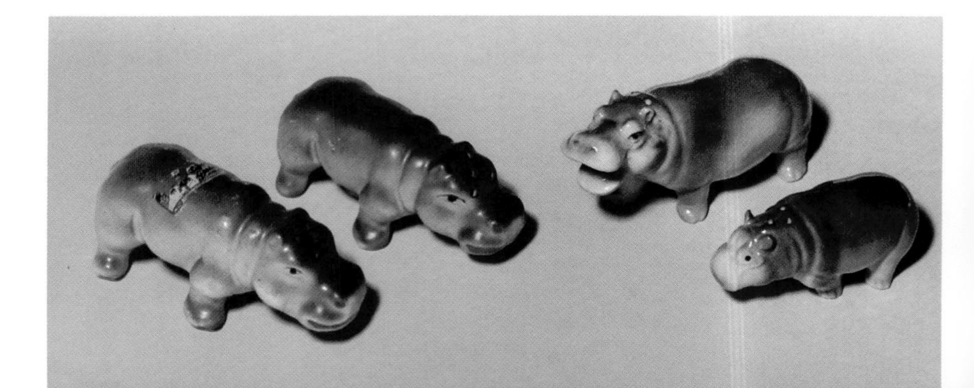

These appear to be stoneware. Length is 3-½ inches. They have a United China and Glass paper label that indicates Japanese origin. *Carson Collection.*

Miniature horse salt and pepper by Ohio ceramist Sandy Srp. Height is 1-¾ inches. *Carson Collection.*

Hippos and frogs, apparently taken when I was contemplating a water-related chapter. The frogs are 4 inches high, have Brinn's paper labels and inkstamps that read, "T-2143." The hippos have "Made in Taiwan" paper labels. *Carson Collection.*

Juanitaware horseheads made by Triangle Pottery, first of Carrolton, Ohio, later Malvern, Ohio. They are 2 inches high, and unmarked. *Oravitz Collection.*

Height of these Rosemeade ponies was not recorded. One is marked "Rosemeade / N.D." on two lines. Irene Harms, in *Beautiful Rosemeade,* shows an identical pair in what appears to be an olive color. *Thornburg Collection.*

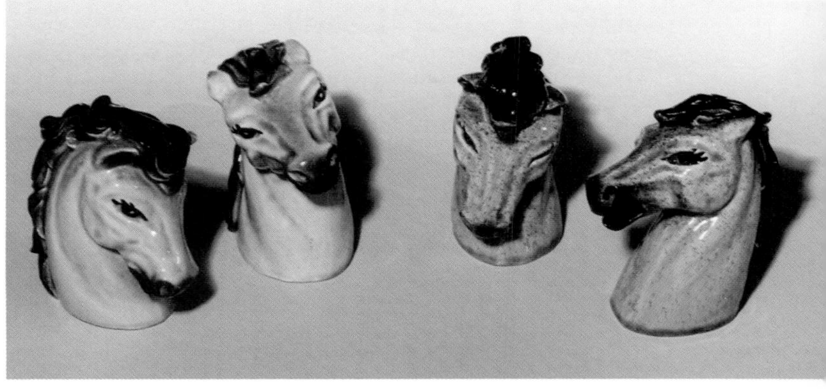

Jumping from Ohio to Wisconsin, here are two pair of Ceramic Arts horsehead salt and peppers. Height is 3-⅜ inches, at least one is marked. *Thornburg Collection.*

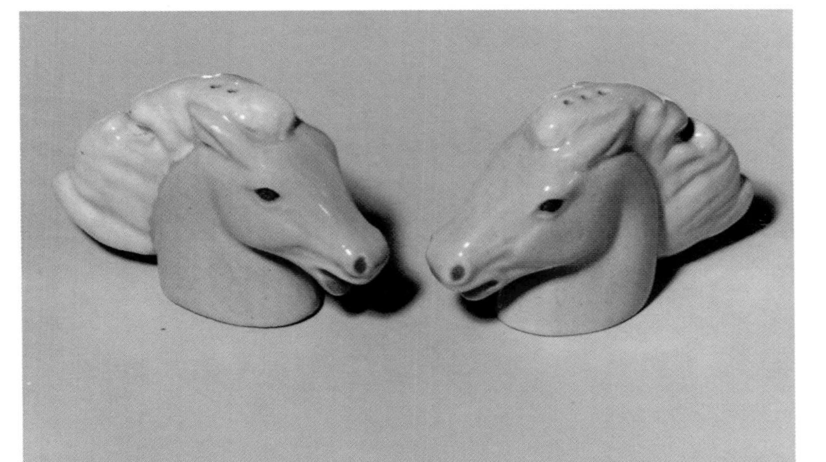

These heads are also Rosemeade, but unmarked. They are 2 inches high. *Thornburg Collection.*

And off to Japan. The rearing pair in the center is 4-½ inches high, has a "Japan" paper label. Both pair of heads have "Japan" inkstamps. *Carson Collection.*

Although size was not recorded, if they were just a little bigger they could serve as bookends. The pair is not marked. *Carson Collection.*

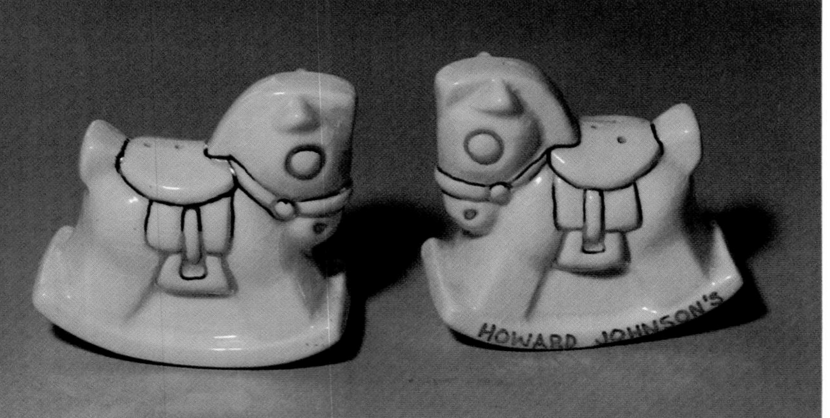

A rocking horse advertising set for Howard Johnson's, the restaurant and motel chain. Height is 2-⅝ inches. Neither shaker is marked. *Carson Collection.*

Stylish stallion and mare, each 4-¾ inches high. Paper label reads "Hand Decorated / RELCO / Japan" on three lines. *Carson Collection.*

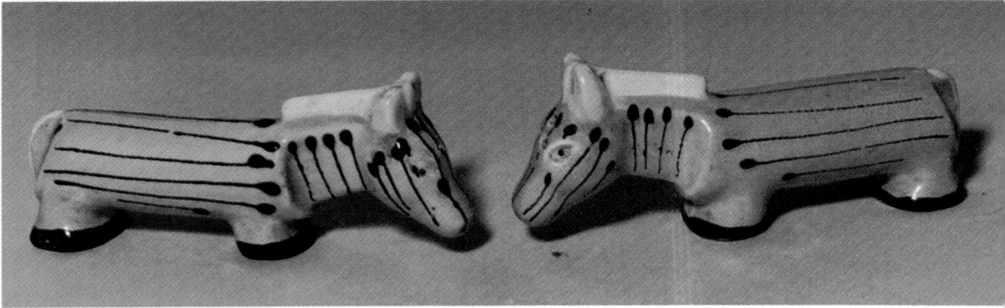

These Zebras are 4 inches long, have a "Japan" paper label. *Carson Collection.*

The saddle acts as a toothpick holder. The base is 4-½ inches long, has "Japan" impressed. *Carson Collection.*

The small set on the left is 4-½ inches high, has a "Japan" inkstamp. The similar set on the right has a Brinn's paper label and an "IV1420" inkstamp. The tallest set in the picture is 7 inches high, unmarked. The paper label of the set to the immediate left of the tallest set reads, "Made in Taiwan." *Thornburg Collection.*

Here is a figural nodder of a kangaroo and joey. The set is 4-¼ inches high as shown, has a Palomar paper label. Note that the mother nods forward and backward, while the joey nods side to side. *Thornburg Collection.*

This is Kanga and Roo from *Winnie The Pooh.* Kanga is 5 inches high, Roo 1-⅞ inches. Kanga is marked, "© MLXIV / Walt Disney / Productions" on three lines. *Thornburg Collection.*

Here the salt is 4-½ inches high, the pepper 2-¼ inches. Incised is "Sutton Ware / Canada" on two lines. The decal on the front shows that the original owner bought it as a souvenir of Niagara Falls, Canada. *Carson Collection.*

Here is a very dapper-looking gentleman, 4-¾ inches tall. (We won't tell the modeller that only the females have pouches.) The set has a "Japan" paper label. *Carson Collection.*

The set at center rear is 3-½ inches high. It, along with the set on the left, has "Japan" inkstamps. The pair in front is 2-¾ inches high, inkstamped "H985." The set at right is unmarked. *Thornburg Collection.*

The mother is 5-⅛ inches high, the joey 2-½ inches. There is a "Made in Taiwan" paper label. *Lilly Collection.*

An apparent relative of the above pair, this set actually lays on its side as shown. The larger member is 6 inches long, the smaller is 3-½ inches high. It has an unreadable remnant of a paper label. *Thornburg Collection.*

The left and middle sets are 5-½ and 3-¾ inches high, respectively. Neither is marked. The set on the right, 5-½ inches high, bears a paper label reading, "Original / Dee Bee Co. / Imports / Handpainted / Japan" on five lines. *Carson Collection.*

Leopards are not very common subjects for salt and pepper shakers for some reason. This pair has a Vandor paper label. The shaker on the left is 3-¾ inches long, the one on the right 3-½ inches high. *Thornburg Collection.*

Kangaroo salt and pepper (4-¼ inches high), cream and sugar, cookie jar (9-¾ inches high). The salt and pepper each have an inkstamp, "© FF." The other pieces all are marked "Fitz and Floyd, Inc. / © MCMLXXVII / FF" (1977) on three lines. One more note, due to yours truly adjusting the camera incorrectly, you have to use your imagination for this picture because every piece is pure white. *Oravitz Collection.*

These Ceramic Arts Studio leopards are figurines, not shakers. But since the only difference between these and the shakers is the pour holes, they should do quite well for the purpose of illustration. The one on the left is 6 inches long, the one on the right 3-¾ inches high. Both have typical Ceramic Arts Studio marks. *Private Collection.*

Ceramic Arts Studio stylized lions. The male is 3-¼ x 7-¼ inches, the female 1-¾ x 5-¼. Both have Ceramic Arts Studio marks. *Thornburg Collection.*

The lion on the right is 2-¾ inches high. This pair is marked by a "Japan" paper label and an inkstamp that reads "H121." *Carson Collection.*

Mice in the corn. The tray is 5-⅜ inches long. Paper label reads "Brinn's / Pittsburgh PA / Made in Japan" on three lines. *Carson Collection.*

The unmarked mouse is 2-⅛ inches high, the cheese wasn't measured. Looking at it now, it appears this set could be chalkware but that wasn't recorded in my notes. *Carson Collection.*

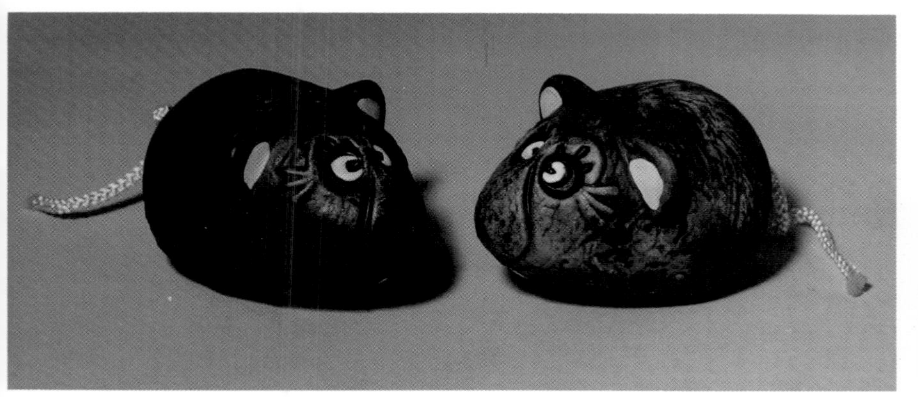

This is Stoney Mouse. The reason I know is that it says so on the pair's Enesco paper label. Each shaker is 1-¾ inches high. *Carson Collection.*

A lady and gentleman of mousedom. He is 3-¼ inches high. Inkstamp mark is "PY" enclosed in an ellipse. There is also a remnant of an Enesco paper label. *Oravitz Collection.*

Rosemeade mice, unmarked by conventional standards but instantly recognizable three booths up the aisle by their sandy color bisque showing through the glaze. The one on the left is 1-¾ inches high. *Thornburg Collection.*

Both of these mice are 3-¼ inches high. They are unmarked with the exception of the "Wyo." on the cowboy's hat, and the plastic hatband that shows the original owner apparently purchased them at Yellowstone National Park. *Carson Collection.*

Two go-withs. The mouse with the cheese is 5 inches high. Neither piece is marked. The animal with the garbage can—you call the species—is 4-¾ inches high, and has a "Japan" inkstamp. *Carson Collection.*

Hickory dickory dock mouse. The mouse is 1-¾ inches high, while the clock, striking one, stands 3-½ inches. The set is unmarked. *Oravitz Collection.*

The mouse is 3-⅛ inches high, the piano 2-⅝ inches long. Inkstamp mark is a "PY" in an ellipse. *Carson Collection.*

A Christmas Mickey and Minnie. Mickey is 4 inches high. His Applause paper label is shown below. *Thornburg Collection.*

Mickey and Minnie, both sets appearing fairly recently on the salt and pepper shaker timeline. Mickey in the car is 2-½ inches high. The car is 2-⅞ inches long, with "© Disney" impressed. Minnie is 3 inches high, the Mickey with her 2-¾ inches. Each has "© Disney" incised. Both sets have Good Company paper labels, the one on the car shown below. *Lilly Collection.*

Applause paper label of the above Mickey Mouse.

Paper label of car holding Mickey Mouse.

Christmas mice made of plastic. Height is 3-¼ inches, mark is "© JASCO 1981 / Hong Kong" on two lines. *Carson Collection.*

Minnie Mouse is 3-⅛ inches high. Both pieces are marked, "Walt Disney Prod. / Japan" on two lines. *Courtesy of Millie Gisler.*

These Mickeys stand 4-⅝ inches high, have a "WD-21 / © Walt Disney / Productions" inkstamp along with a Dee Bee Company paper label. *Thornburg Collection.*

These unmarked moose shakers are 4 inches high. They are unusual in that they have the type of moving eyes that are often found on teddy bears. *Carson Collection.*

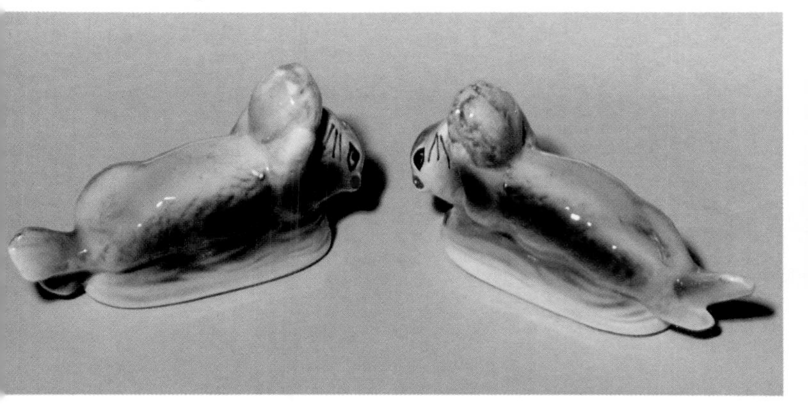

The sea otters are 4-⅛ inches long. They are not marked. *Thornburg Collection.*

Regal China Company made three sizes of huggie pigs. Those shown here are 6-½ and 4-½ inches high. The third size, not shown, is approximately 3-⅝ inches high. The smaller set shown (note the gold trim) is marked "© Van Tellingen" on the back, and "Pat. Pend." on the bottom. Marks of the larger set are what appears to be "Copr / 1958 / R. Bendel" on three lines on the back with only the Copr in question, and "Pat. No. 2560755" on the bottom. The number refers to an invention patent, not a design patent. The patent was issued in 1951. All of the marks referred to above are impressed. Additionally, the 6-½ inch pigs have "His" and "Hers" incised in script near their bases. This was apparently included because they could also be made to serve as banks, as illustrated below. *Thornburg Collection.*

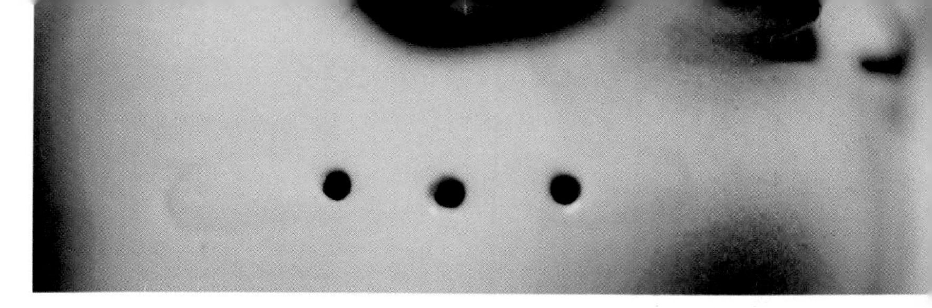

Top of large pig. Barely visible is the outline of the slot that was to be cut out if the piece was to be sold as a bank.

Three Shawnee Smiley pigs and one Winnie pig. Winnie is 3-¼ inches high. None of the shakers are marked. While some debate persists among collectors as to whether two Smileys or a Smiley and a Winnie constitutes a true pair, it seems probable the shakers were sold both ways, if not by Shawnee, then certainly by retailers. *Smith Collection.*

Smileys and a Winnie in different colors, and different sizes, too. These are range size, 5-¼ inches high. As above, they are not marked. Although not shown here some of these shakers were trimmed with gold, which can be said of most Shawnee products. The Smileys and Winnies were designed to complement matching cookie jars. *Smith Collection.*

Shawnee farmer pig shakers, 3-⅜ inches high, unmarked. *Smith Collection.*

Goebel pigs in the original plastic baskets in which they were issued. Height is 1-⅝ inches, marks are shown below. *Thornburg Collection.*

Ceramic Arts Studio pigs, 3-½ inches high. They have typical company marks. *Oravitz Collection.*

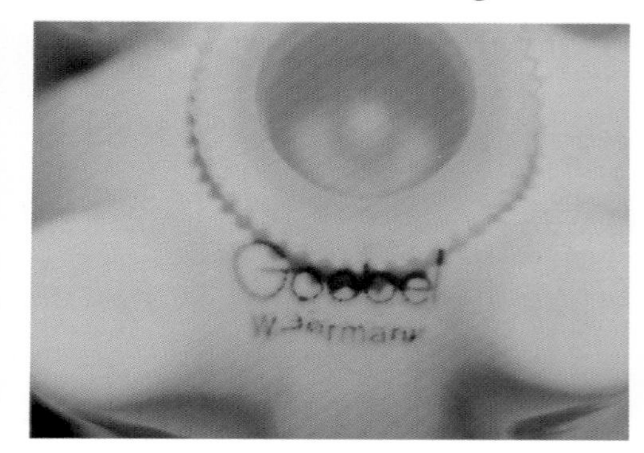

Mark of the pink Goebel pigs above.

These pigs are 3-⅛ inches high, have a "Taiwan" paper label. *Courtesy of Joyce Hollen.*

Paper label of the white Goebel pigs above.

Here is another Regal China Company entry. The one-piece shaker stands 3-⅞ inches high, is marked "Copyright / C. Miller / ©" impressed on three lines. Unrelated to this set but one of the mysteries of shaker collecting is why—with hogs being a staple of North Dakotan agriculture, and Laura Taylor Hughes' affinity for animals—Rosemeade never put out a pair of pig salt and peppers. *Thornburg Collection.*

One-piece plastic pushbutton set of chef pigs. Height is 2-⅝ inches. *Carson Collection.*

Figural nodder of walking pigs, 4 inches high, "PAT TT" impressed with a "Made in Japan" inkstamp. *Thornburg Collection.*

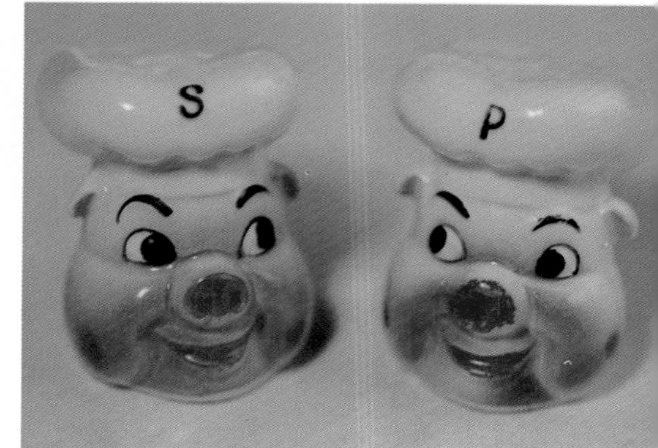

More chef pigs, 2-⅞ inches high, "135" inkstamp with a "Made in Japan" paper label. *Oravitz Collection.*

Lots of gold on these little pigs. The pair is unmarked, 2-½ inches high. *Lilly Collection.*

Unmarked, 2 inches high. *Lilly Collection.*

Pigs on corn holder. Each shaker is about 3 inches long. *Courtesy of Dawn and Jack Glow.*

Another nice pair of American made pig shakers, this time by Imperial Porcelain. Height is 5-¼ inches. Mark is "Imperial Porcelain Corp. / Zanesville, Oh" impressed on two lines. *Carson Collection.*

Miniatures by Sandy Srp, just ⅝ x 13/16 inches. Mark is "SS." *Carson Collection.*

The pigs on the outside are 3 inches high, have a "Japan" inkstamp. Height on the inside is approximately 3-¾ inches. Marks, or lack of them, were not recorded. *Carson Collection.*

Farmer pigs by Enesco, 4-⅞ inches high. Enesco inkstamp and paper label. *Carson Collection.*

These handsome gentlemen stand 3 inches high, and are unmarked. *Oravitz Collection.*

These pigs are 2-½ inches high. They are handpainted but not marked. *Carson Collection.*

Here is a pig of a different color—luscious golden brown. Height is 3-⅜ inches. There is no mark. *Carson Collection.*

The largest pair is 3-¼ x 4-¼ inches, unmarked. The pair on the outside is 3-¾ inches long. They have a "Japan" inkstamp. The pair of "pigs" in front are actually bears, placed there when speed apparently superceded accuracy. They are 2 inches high, have a "Japan" inkstamp. *Carson Collection.*

Pumas, or mountain lions, 4-¼ inches long, by Rosemeade. *Thornburg Collection.*

Kissing bunnies, another unmarked creation of Ceramic Arts Studio. Heights are 4 inches and 2-⅛ inches. *Thornburg Collection.*

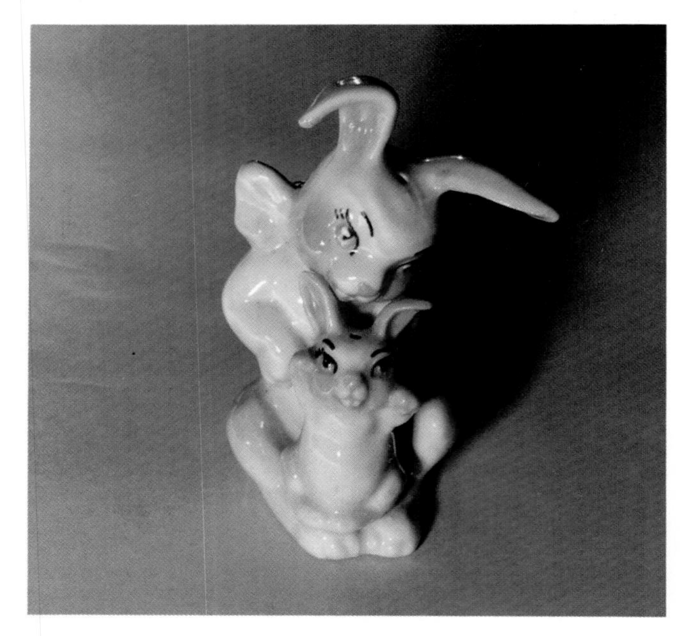

Mother and baby rabbit. The mother is 4-⅜ inches high, the baby 2-¼ inches. Both have Ceramic Arts Studio marks. *Thornburg Collection.*

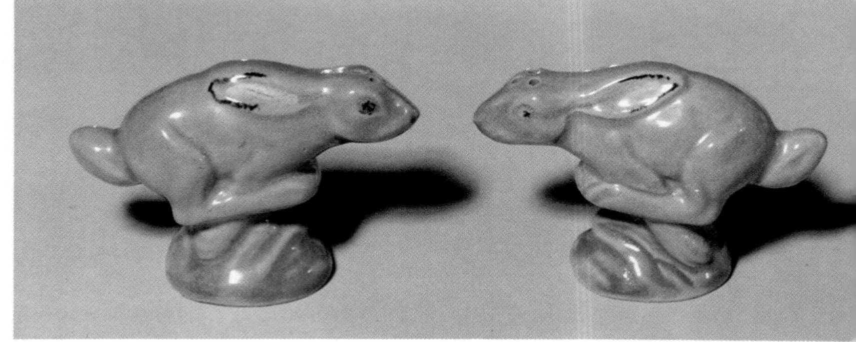

Rosemeade running rabbits, 2 inches high, and unmarked. *Thornburg Collection.*

Running rabbits, also by Ceramic Arts Studio, but unmarked. Height is 2-¼ inches. This is not a true pair. The real mate is about one inch shorter. *Thornburg Collection.*

Regal China rabbits. The ones with the longest ears are 3-¾ inches high. The yellow pair and the green pair are marked "Bunny Hug / Van Tellingen / ©" impressed in the back. The white pair in front is not marked, but appears to be a genuine Regal China item, not a ripoff. *Oravitz Collection.*

Same as above but look at the wonderful handpainted flower decoration, a really rare find. *Thornburg Collection.*

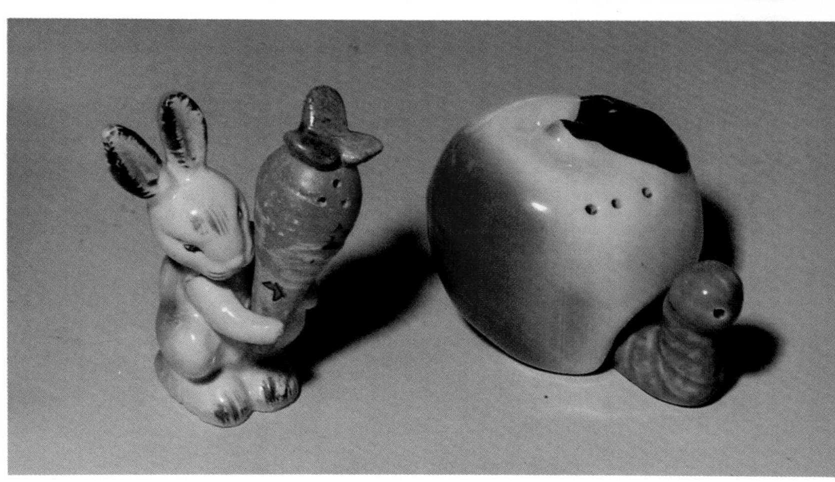

The rabbit is 3-⅜ inches high, has a "Japan" inkstamp. The worm, who I am sure must have dragged himself and his apple into the picture since I took it, is 2-½ inches high. The apple is 2-⅜ inches high. Neither is marked. *Carson Collection.*

All of these Disney character sets except the tray set were made by Goebel. It has an Enesco paper label. Bambi of the Enesco set is 4 inches high while Thumper is 3-½ inches. Flower, in front, is 2-½ inches high. *Thornburg Collection.*

A real bandit here, about to snatch a meal from a garbage can. Height of the racoon is 4-¾ inches. It has a "Japan" inkstamp. *Thornburg Collection.*

Luster, 3 inches high, marked "Japan." *Courtesy of Cloud's Antique Mall, Castalia, Ohio.*

The racoons on the left are 2-¾ inches high. They have a "Made in Japan" paper label. The ones on the right are of similar size, have a "Japan" inkstamp. *Thornburg Collection.*

Rosemeade racoons. The bandit on the left is 1-⅞ inches high, the one on the right is 2-⅞ inches long. Unmarked except for the Rosemeade paper label. *Thornburg Collection.*

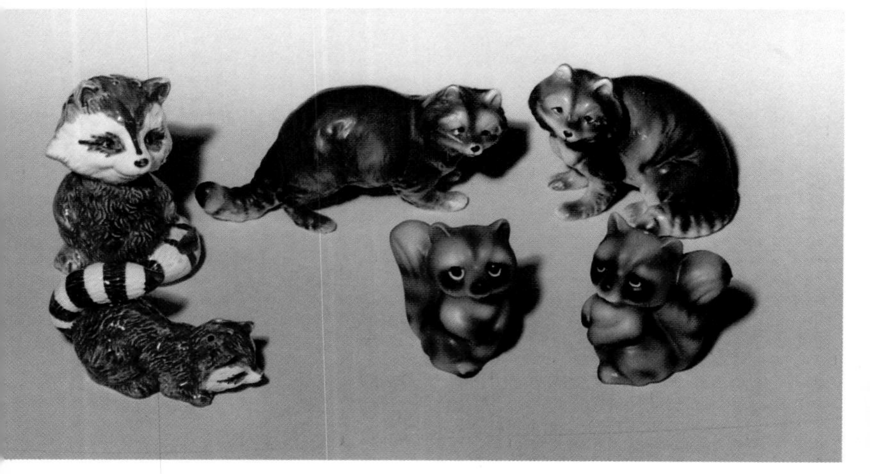

The racoon at right front is 2-½ inches high, and is marked with a "Japan" inkstamp. The pair at right rear have a Victoria paper label in addition to a "Made in Japan" paper label. The seated animal at rear left is 3-¾ inches high, has a Gibson inkstamp, shown below. *Thornburg Collection.*

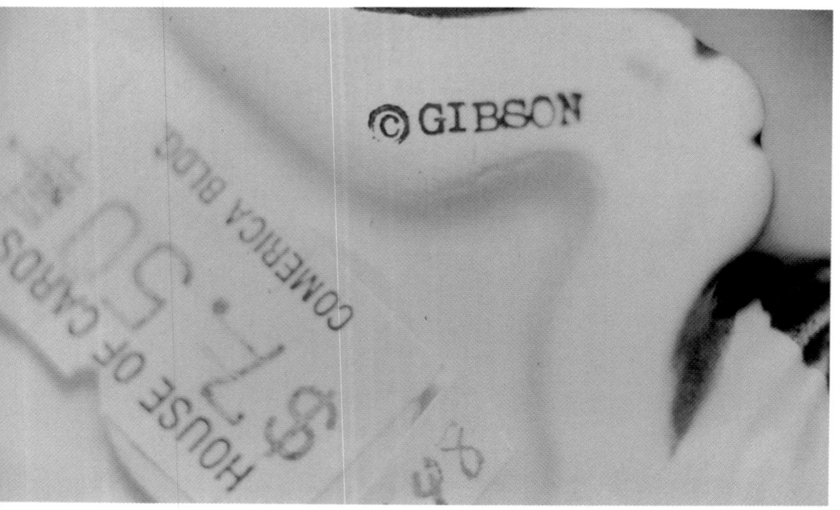

Inkstamp mark of the seated racoon at rear left above. A second pair photographed had an identical mark. Note the original price tag. Some collectors leave these on, others take them off. By now it shouldn't surprise you that I favor the first group, believing that the more information you can preserve about any collectible, the better off you and future collectors will be.

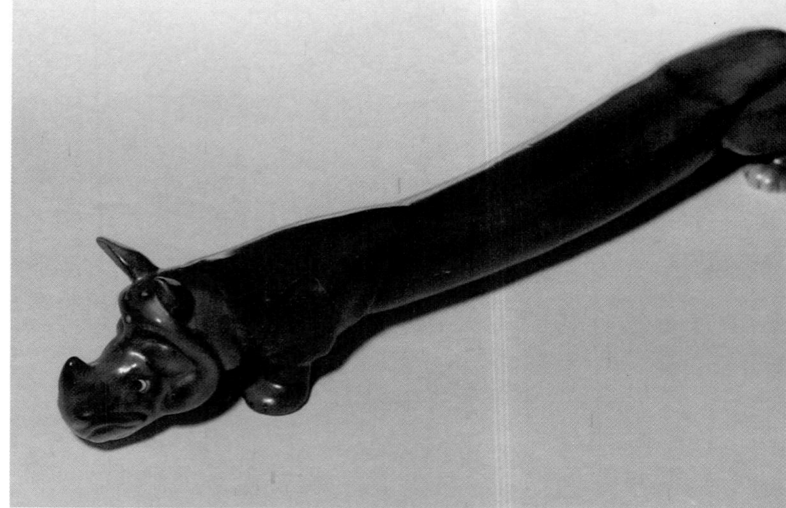

One very long rhinocerous measuring in at 10-¼ inches. A one-piece, it is marked with an inkstamp, "Craftsman China / Japan / 266" on three lines. The same mark appears, and is shown, on a rather lengthy zebra later in this chapter. *Thornburg Collection.*

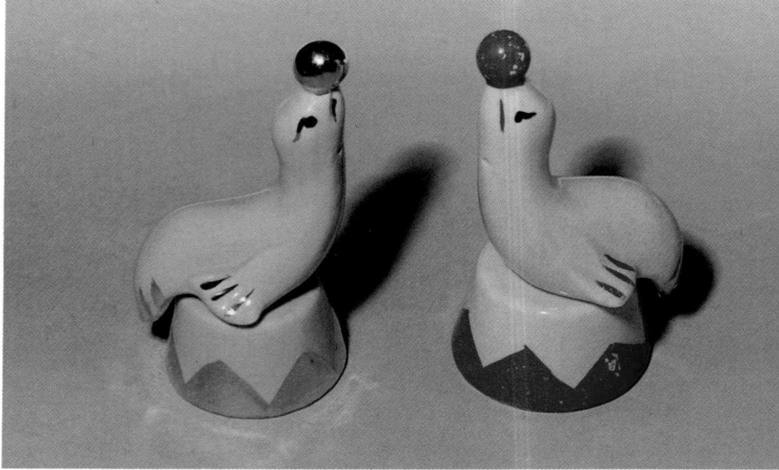

These seals represent two pair of shakers, not one. Each seal is a shaker, so is each stand. Height is 4-½ inches as shown. Only mark is an Elbee Art paper label. *Thornburg Collection.*

The rear set is 4-¾ inches high, has a "Japan" inkstamp. The two seals in front appear to be a pair based upon their clay and glaze, but they are made differently. At left, as you can see, the ball has a small brad in it that fits into a hole in the seal. At right, the ball has a hole that enables it to be slipped over the nose of the seal. The mark of this seal is shown below. *Thornburg Collection.*

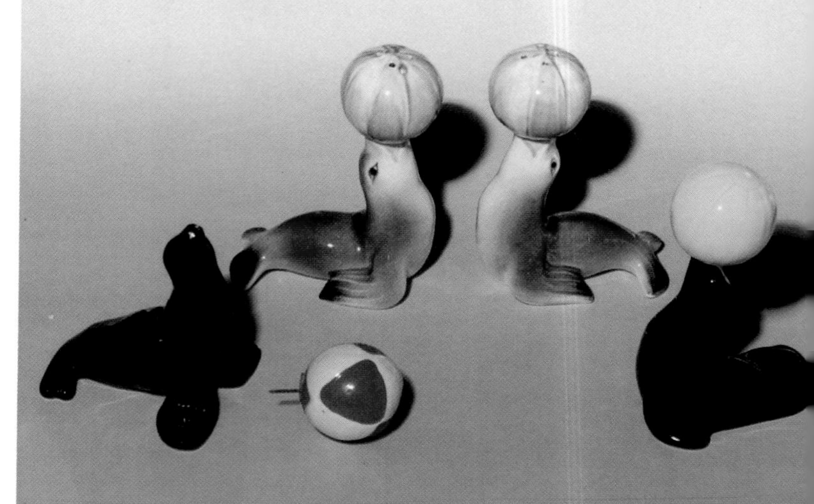

Harper Pottery inkstamp of black seals.

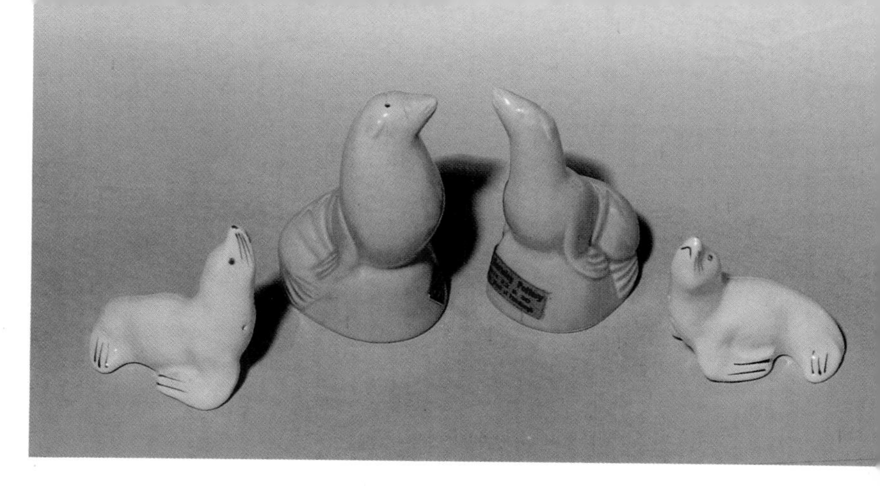

The yellow pair is 3-½ inches high. Paper label reads, "Gibson Calif. Arts." The white pair appears to be the same as at left, has a "Japan" inkstamp. *Thornburg Collection.*

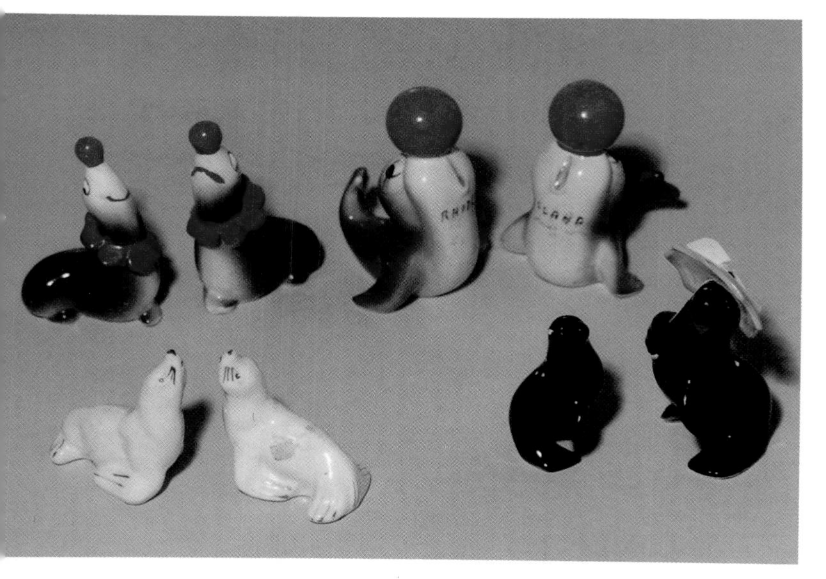

The left pair in the back row is 3-⅜ inches high, has a paper label that reads, "Regis. Pat. Off. / Norcrest / Japan" on three lines. The pair on the right in the back row is 3-¾ inches high. Both shakers have "Japan" paper labels. At left front the height is 2-⅜ inches, while the black pair is 2-½ and 4 inches high. Both pair in the front row have "Japan" inkstamps. *Carson Collection.*

These sheep are called modern lambs. The one on the left is 2-⅝ inches high. Both have Ceramic Arts Studio marks. While photographing the book, I found it surprising that I didn't run into more sheep. *Thornburg Collection.*

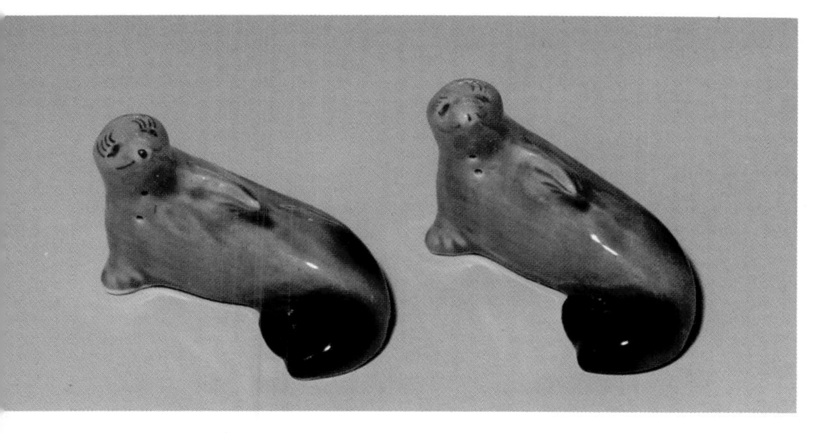

Size here is 3-¼ x 3-½ inches. While part of the inkstamp mark has been obliterated, what is left reads "New——land Ceramics Corp." *Carson Collection.*

These skunks are Goebel. They have an inkstamped "Germany" and a bee mark. The one on the right is 3 inches high. *Thornburg Collection.*

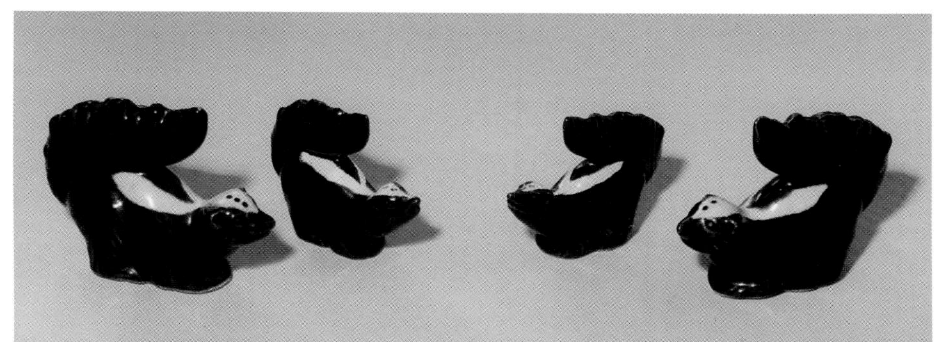

Unmarked Rosemeade. The outside pair is 3 inches high, the inside pair 2-½ inches. *Carson Collection.*

Skunk on the left is 2-3/8 inches long, skunk on the right is 2-1/2 inches high. Their paper label, shown below, indicates Japanese origin. *Carson Collection.*

The squirrel on the left is 2 inches long, the one on the right 2-1/8 inches high. They have an Elbee Art paper label. *Courtesy of Cloud's Antique Mall, Castalia, Ohio.*

Goebel squirrels. The pair on the left are 3 inches high, have "P119A" and "P119B" impressed. The squirrel on the right is 3-1/4 inches high, the tray is 4-5/8 inches long. Marks are shown below. *Thornburg Collection.*

Thrift Ceramics paper label of above skunks.

The squirrels are approximately 2-1/2 and 3 inches high. They have a Relco paper label. The out of place beavers are 4 inches high. Their Thrifco paper label is shown below. *Thornburg Collection.*

Skunks on stumps, each 1-1/2 inches high. The shakers are unmarked. *Private Collection.*

Thrifco Ceramics paper label of above beavers. The palette shape is the same as the Thrift Ceramics paper label on the skunks above, probably indicating same company with a slight name change.

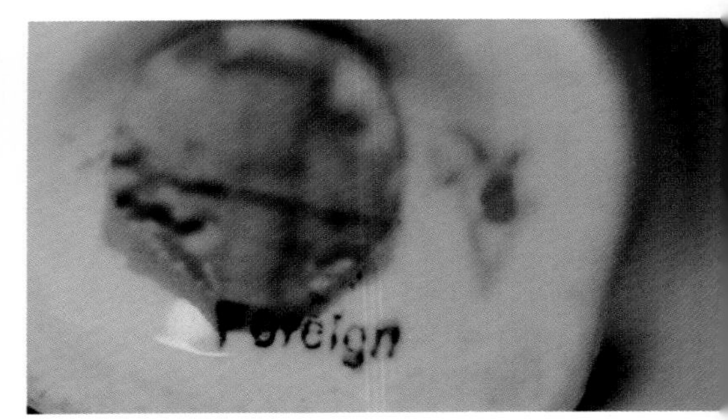

The full bee mark, somewhat blurry, of the squirrel on the left above. According to *M.I. Hummel—The Golden Anniversary Album,* edited by Robert L. Miller and Eric W. Ehrmann, and approved and authorized by Goebel, the full bee mark—a bee above a V—was used from 1950 through 1955. The mark was made to honor Sister Maria Innocentia Hummel, creator of the company's popular Hummel figurines. Hummel is the German word for bumblebee, according to the book, and the V stands for Verkaufsgesellschaft which means distribution company. The word "Foreign" designates a piece made for export to other than the United States.

Ceramic Arts Studio mother and baby skunk. Her tail is flattened so the baby may sit on it. Height of the mother is 3-1/2 inches, while the baby is 2-1/8 inches. Both have Ceramic Arts Studio marks. *Courtesy of Cindy Schneider.*

The squirrels are 2-3/4 inches high, the stand is 5 inches high. All three pieces are marked "Japan." Note the gold bees, one on the left near the roots, the other at right near the limb. *Carson Collection.*

This is the mark of the squirrel on the tray in the above photo. It is either the Baby Bee (1958), or V Bee (1959). They are so similar it is nearly impossible to tell them apart.

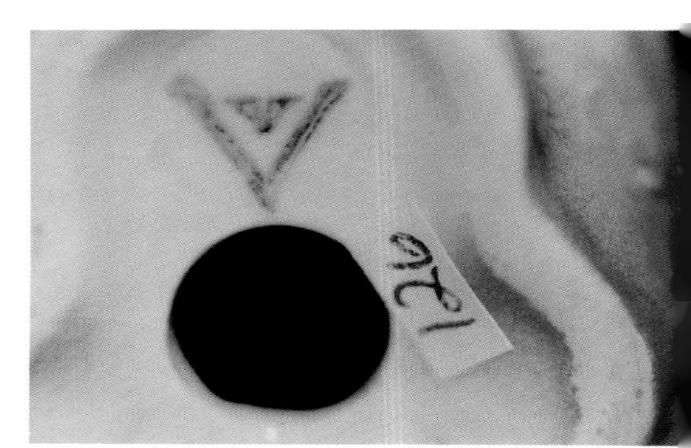

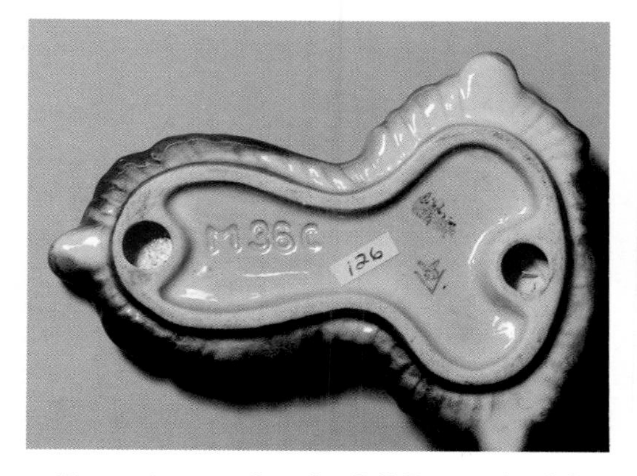

Clearer but smaller, the Full Bee mark of the tray, along with the impressed "M36C." Now the question is, how do we explain the tray having a Full Bee mark (1950 through 1955), while the squirrel sitting on it has a Baby V (1958) or V Bee (1959)? I don't have an answer. I'm just glad I stopped at the squirrel and didn't photograph the mark of the pinecone. That might have created more of a mystery.

Pilgrim squirrels, a limited edition of 25 by Sandy Srp. The squirrel on the right is 1-⅜ inches high. *Carson Collection.*

Another pair of Sandy Srp miniatures, this time shown in a truer perspective. The one on the left is only ⅞ inches high. *Carson Collection.*

Heights of both the squirrel and the stump are approximately 2-¾ inches. The set has a "Japan" inkstamp. *Lilly Collection.*

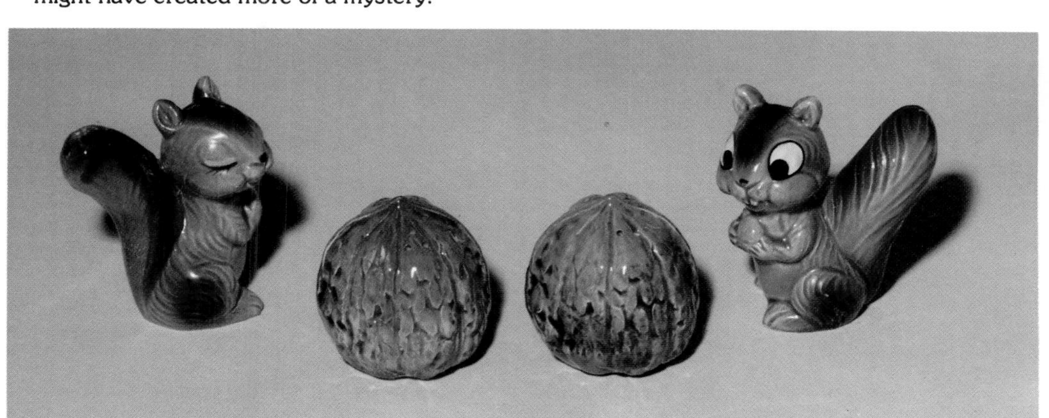

These squirrels are unmarked. They are 3-¾ inches high. The nuts have an Inarco paper label. They are 2-¾ inches high. *Courtesy of Millie Gisler and Joyce Beulah.*

Are these squirrels, or skunks with reversed colors? Whichever, they are 2-½ inches high and unmarked. *Oravitz Collection.*

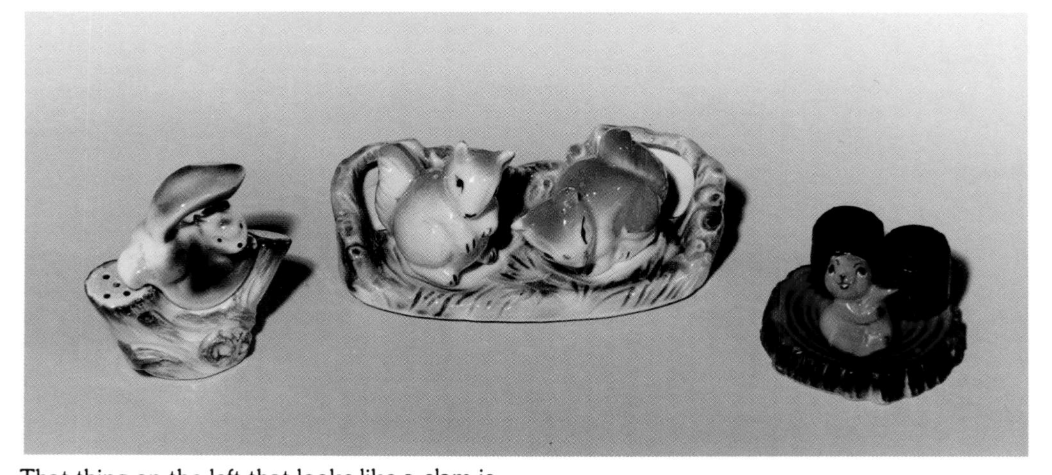

That thing on the left that looks like a clam is actually a squirrel on a stump. The gray object on top is his tail, which he is holding above his head. As shown, the set is 3-½ inches high. It has a "Japan" inkstamp. The set on the right has a "Japan" paper label, while the three-piece set in the middle is unmarked save for a "Handpainted" inkstamp. *Thornburg Collection.*

Not very likely looking squirrels, but what else could they be? Height is 2-⅜ inches. Marks are a "Made in Japan" paper label, and a "3004" inkstamp. *Thornburg Collection.*

The unmarked pair on the left is a true pair. The squirrel is 2-¼ inches high. The pair on the right is suspect. While both pieces have a "Japan" inkstamp, they are written in completely different styles. The squirrel is 2-¾ inches high. *Carson Collection.*

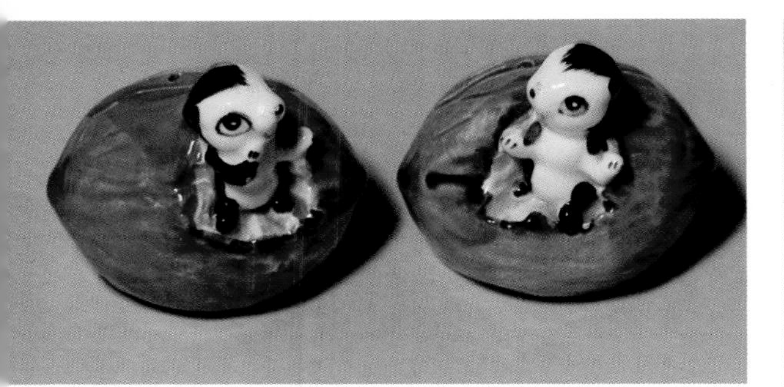

This plastic pushbutton set is 7 inches high to the top of the smile face, which is usually not seen. Mark is "Whirley Industries / Warren Pennsylvania USA / Des. Pat. 169526" on three lines. That design patent number dates to 1953. *Courtesy of Jazz'e Junque Shop, Chicago, Illinois.*

The height of these tigers is 6-¼ inches. Only mark is "© 1964." Whether this is an advertising set is unknown. *Carson Collection.*

The tigers in the drums are 3-⅛ inches high. They have "Made in Japan" inkstamps. The tallest giraffe stands 3-¾ inches. It has a "Japan" inkstamp. *Carson Collection.*

The pair of walruses on the left have "Made in Taiwan" paper labels. The pair on the right, 2-⅞ inches high, carry Seymour Mann paper labels, one of which is shown below. *Thornburg Collection.*

Seymour Mann paper label of above walruses.

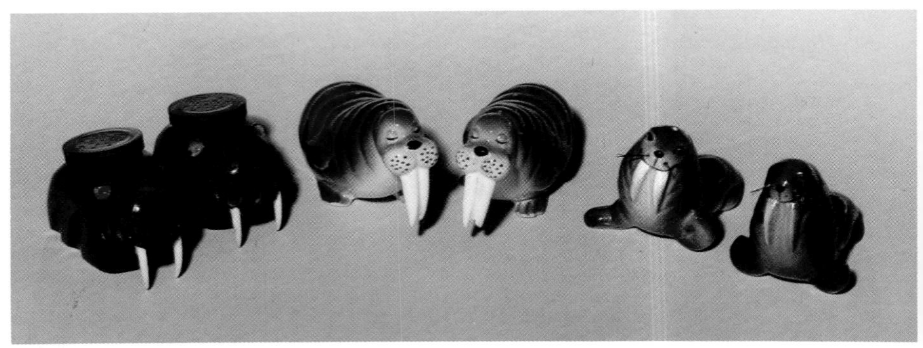

Wood is the material of the walrus shakers on the left. They have plastic caps, and are 2 inches high. They are not marked. The middle pair is 3-½ inches long, has "Japan" inkstamps. The pair on the right is unmarked, but interesting because it has real whiskers. *Thornburg Collection.*

As you might suspect from the glaze, the whales at left rear are made of redware. They are 4 inches long. Mark is "EFCCO / Imports / San Francisco / Made in Japan" on four lines. The pair next to them is 2-½ inches high, has a "Made in Japan" paper label. In the front row, left to right, lengths are 3 inches and 3-¼ inches. Neither pair is marked. *Carson Collection.*

The whales, on the left, are 2-½ inches high, and unmarked. That's a pair of fish that swam in on the right. They are 2-⅛ inches high, have a "Japan" inkstamp. *Lilly Collection.*

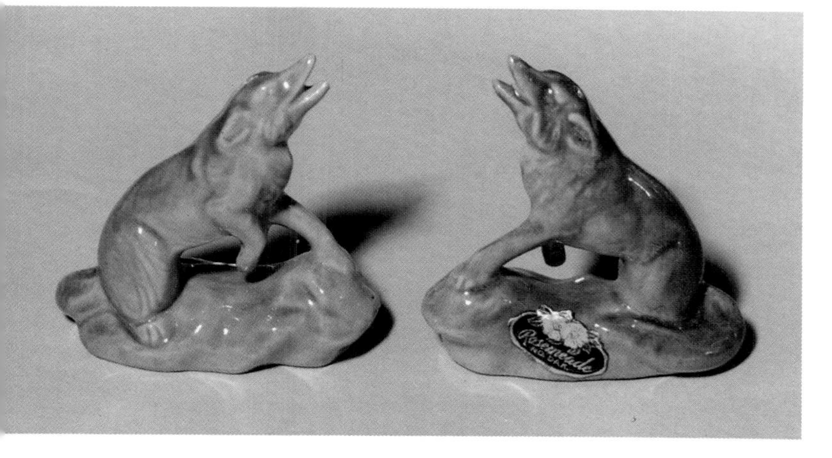

Rosemeade wolves, 3-⅛ inches high. *Thornburg Collection.*

The way he is howling this is probably a coyote. Perhaps he won't mind if we slip him in with the wolves. He is approximately 3-½ inches high. The cactus of the pair on the left is 4-½ inches high. Both sets have Vandor paper labels that indicate they were made in The Phillipines. *Thornburg Collection.*

While I had no reservations about making the coyote above hang out with the wolves, I wasn't about to put this woodchuck in with the gophers. If he took offense, ol' Punxsutawney Phil probably would have seen that I had six more weeks of winter for the rest of my life. The shakers are 4-½ inches high. The mark is shown below. *Thornburg Collection.*

Mar Mar mark of Punxsutawney Phil shakers.

My wife Cindy, who collects Ceramic Arts Studio figurines, about went crazy when she saw this fine pair of zebras; thought I had lost her for the rest of the day. But after a cup of strong coffee and a cigarette or two she regained her composure and we were able to continue the photo session. The shakers are 5-½ inches high. They have Ceramic Arts Studio marks. *Thornburg Collection.*

Rosemeade zebras with paper labels, but not marked in any other way. Height is 3-⅞ inches. *Thornburg Collection.*

Craftsman one-piece, 10-¼ inches long. Mark is shown below. *Thornburg Collection.*

Mark of Craftsman one-piece zebra above. Same mark, with a different number, appears on the one-piece rhinocerous above.

Height of these zebra shakers is 3-½ inches. They have a Relco paper label. *Carson Collection.*

The rearing zebras are 5-¾ inches high, have a Sarsparilla paper label. Heights of the others are 4 inches ("Japan" paper label), and 2 inches ("Japan" inkstamp). *Thornburg Collection.*

Height of the unmarked rocking zebras is 3-⅞ inches. While they make a nice pair, they are not a true pair as they should have a boy and girl on them. The set in front is 2-⅜ inches high, has a "Japan" inkstamp. *Thornburg Collection.*

Musical zebras. I wonder if there are any other members of this band floating around out there. Height on the left is 3-⅜ inches, height on the right is 3-¾ inches. Both have "Japan" inkstamps. *Carson Collection.*

Birds

The surprising thing about bird shakers is how many high quality ones were made by other than acknowledged industry leaders such as Goebel, Rosemeade and Ceramic Arts Studio. Many collectors could easily fill their allotted space with nothing but Pacific-Asian bird shakers, and not have a mediocre pair in the bunch.

POULTRY

Ceramic Arts Studio chickens, the one on the left being 4 inches high. They have typical Ceramic Arts Studio marks. *Carson Collection.*

The chicken on the right is the same as the one above on the right. But note the one on the left—different shape, different color. It is 3-½ inches long, has a Ceramic Arts Studio mark. Thornburg Collection.

Range size Shawnee chantecleer roosters, 5-⅜ inches high. These were also made in a 3-½ inch size. While they are not marked, the size of Shawnee's fill hole, shown below, can often be used for identification. *Smith Collection.*

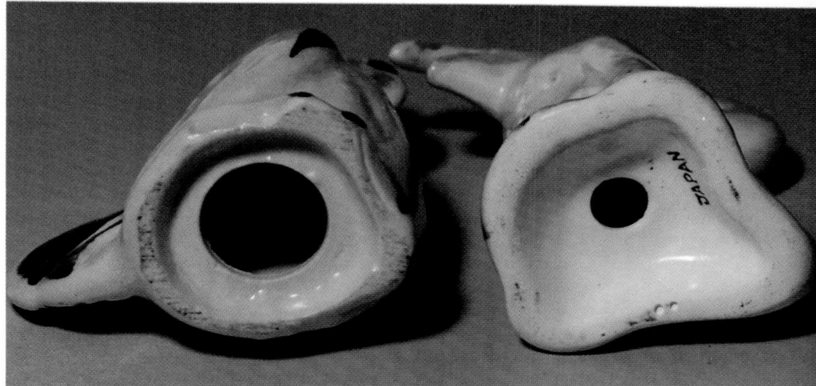

On the left is the large fill hole of the Shawnee chantecleer rooster shown above. On the right is the bottom of a shaker with what we might call an "average" size fill hole. All Shawnee shakers that I know of have the large size fill hole shown on the left. Once you've seen one it is instantly recognizable. But it is not a completely foolproof means of identification because I have seen a couple Japanese shakers with holes just as large. Still, when you consider that we closely inspected more than 4000 pairs of shakers while writing this book, and aside from Shawnee found only two nonconformists, it seems as close to foolproof as you can get.

Blue Ridge rooster and hen by Southern Potteries. The shaker on the left is 4-¾ inches high, the one on the right 3-⅞ inches. Like Shawnee, Blue Ridge shakers are not marked. Also like Shawnee, they become recognizable very quickly once you have encountered a few. *Carson Collection.*

Same Blue Ridge blanks as above, but not a true pair because of the different colors. *Carson Collection.*

Goebel Plymouth Rock chickens. The rooster is 3-¾ inches high.Impressed marks are "P100A" and "P100B." These were also made in a smaller size 2-⅞ inches. The smaller ones are P101A and P101B. The McHughes show the shorter pair with a tan glaze. *Thornburg Collection.*

Three sets of nodding chickens, all Japanese. Heights and marks were not recorded, but you can see the bases are the smaller size, and you can bet they have "Patent TT" or "Pat. TT" impressed. *Carson Collection.*

Height here is 3-¼ inches. One of these shakers has a "Japan" inkstamp, the other a "Japan" paper label. *Lilly Collection.*

Long and tall, 7-¼ inches to be exact. The shakers have a "Japan" paper label. *Courtesy of Allen and Michelle Naylor.*

Looks like a tea set, but is a stacked condiment set. The base is the mustard. The set is made of redware, usually a sign of Japanese origin, but marks were not recorded. *Private Collection.*

This is a one-piece, 10 inches long. It has a "Japan" inkstamp, and the original price tag which shows it sold for 69 cents. Sure would be nice to get it for that today. *Thornburg Collection.*

A Rick Wisecarver cookie jar and shakers, sold as a set by the artist in 1991 and 1992. Eventually, you will probably see the shakers popping up without the jar, and vice-versa. The shaker on the left is 4 inches high, the one on the right 4 inches long. The jar is 12-½ inches high. The jar is marked with typical Wisecarver marks, the mark of the shakers is shown below. The name of the set is "Sunday Dinner," the reason for which is also shown below. The address of Rick Wisecarver appears in Appendix I Sources. *Oravitz Collection.*

Back of Wisecarver "Sunday Dinner" cookie jar.

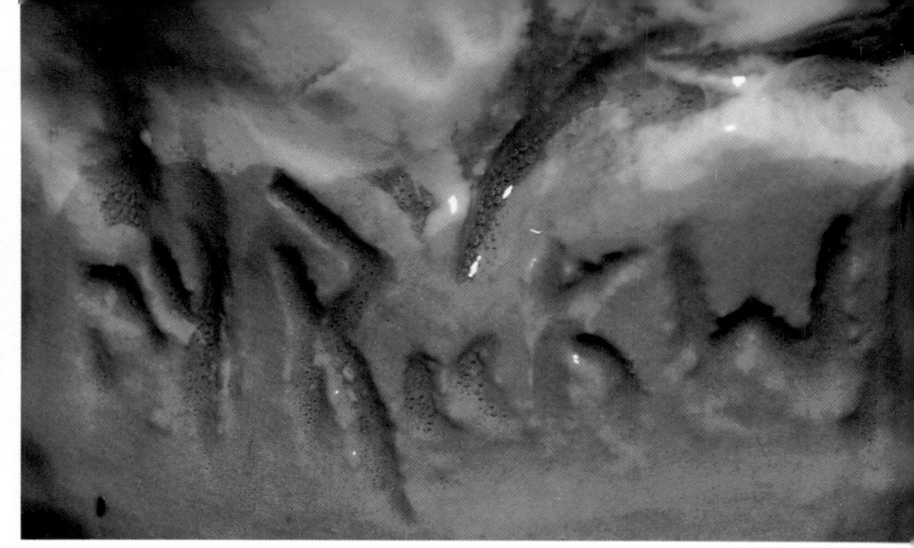

Mark of Wisecarver "Sunday Dinner" chicken salt and pepper shakers.

The rooster is 2-¼ inches high. Inkstamp reads "Made in Japan." *Lilly Collection.*

The large rooster is 4-⅞ inches high, the hen that goes with it 3-⅜ inches. They have "Made in Japan" paper labels. The smaller pair in front is 2 inches high. They are not marked. *Carson Collection.*

Rear left 3-⅛ inches high, "Japan" inkstamp. Rear center 5-¼ and 4-¼ inches high, "Japan" paper label. Rear right 4-⅜ inches high, "Japan" paper label. The sitter at front left is 3-¾ inches high as shown, has a "Taiwan" paper label, while the tallest member of the pair at front right is 3-⅝ inches high with an inkstamp that says "Germany." *Courtesy of Joyce Hollen and Millie Gisler.*

This set is unmarked. The tray is 6-¼ inches long. *Carson Collection.*

The unmarked pair on the left is 4-¾ inches high. The pair on the right, marked "Japan," is 3-¾ inches high. *Carson Collection.*

The pair on the right was either very popular in its day, or is very unpopular now. I see many of them at antique shows. Each shaker is 4-½ inches high, is marked "© 1960 / Holt Howard." The taller shaker of the pair on the left is 4-¾ inches high, has a "Japan" inkstamp. *Carson Collection.*

None of these are marked, but all have Rosemeade paper labels. The rooster in the center is 3-¼ inches high. Another set like this was photographed that had a black glaze. *Thornburg Collection.*

The hen is 3-¾ inches high. The set has a Vandor paper label that says "Pelzman Design." *Private Collection.*

Left, 2-¾ inches high, Enesco paper label. Middle, redware, no mark. Right, "Japan" inkstamp. Front, unmarked. *Carson Collection.*

The pair in the middle is 4-½ inches high, marks were not recorded. The pair on the left has a "Japan" paper label, the pair on the right a "Japan" inkstamp. *Carson Collection.*

Two go-withs, two go-togethers, two sitters—any name you prefer to label these two separate sets. Both pair are unmarked, both are 3-¼ inches high. *Carson Collection.*

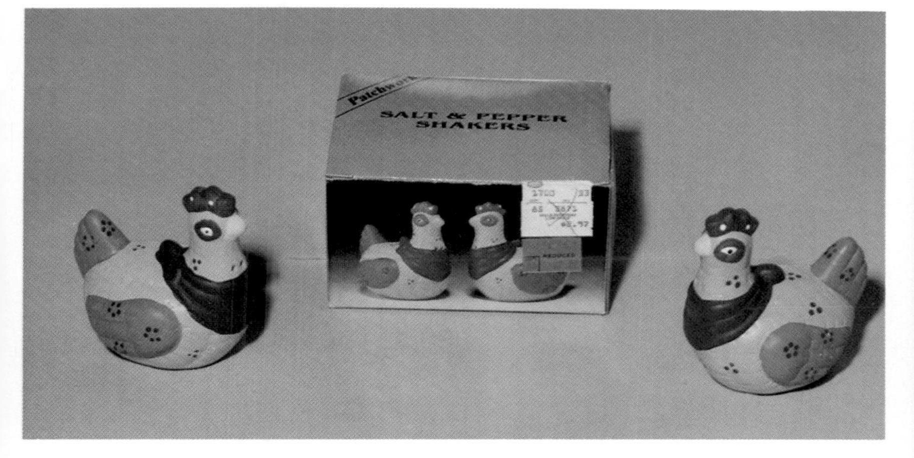

The shakers are 3-½ inches high. The soft sculpture technique used on these ceramic chickens gained popularity during the early 1980s, faded some years later in the decade. While the shakers are not marked, the box says "Dist. by C.R.H. Int'l. Inc., 33 North High Street, Columbus, Ohio, 43215 345-4500." *Lilly Collection.*

The chick and egg are 4-½ inches high. They have a "Japan" inkstamp with a three-leaf clover. The basket of the middle set is 3-⅜ inches wide, carries a "Taiwan" paper label. The pair on the right, 3 inches high, has a "Japan" paper label. *Courtesy of Joyce Hollen.*

Photographed to look like a one-piece, but far from it. The shaker on the left is 3-¼ inches high, the one on the right 3-⅛ inches high. The set has a "Japan" inkstamp. *Lilly Collection.*

The chicks on the left are 4-¼ inches high, have a "Japan" inkstamp. The chicks on the right stand 4-⅜ inches high, have an inkstamp that reads "7247." *Carson Collection.*

The taller shaker is 2-⅞ inches high. It has a Napco paper label. *Carson Collection.*

Each shaker is 3 inches high. Marks indicate Japanese origin. *Carson Collection.*

Hatching chicks, 4-½ inches high with a "Japan" inkstamp. *Carson Collection.*

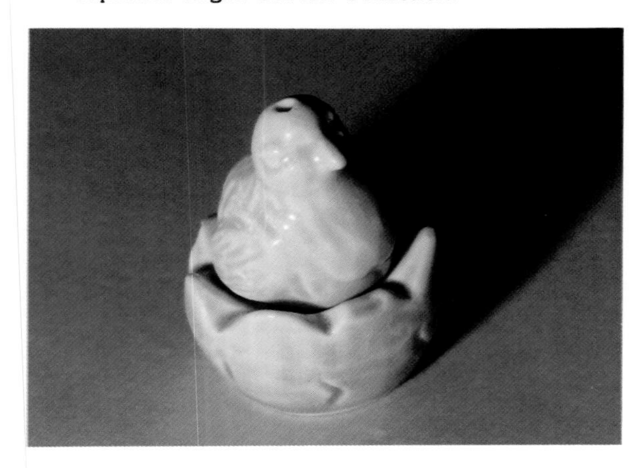

This is a small set, 1-½ inches high as shown. It is not marked. *Lilly Collection.*

Egg cups with shaker tops. Height is 4-¼ inches when together. Holt Howard paper label is shown below. *Courtesy of Joyce Hollen.*

WATERFOWL AND SHOREBIRDS

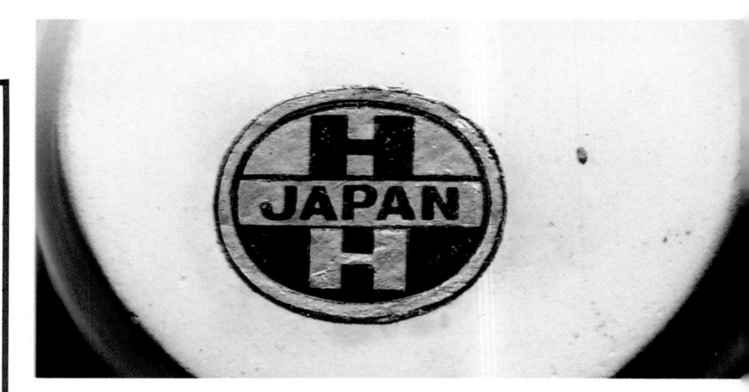

Holt Howard paper label of above egg cup/shakers. Note that it shows Japanese origin.

Two sets of Rosemeade ducks, their paper labels barely visible. The mallard drake is 3-⅝ inches high, while the blue shakers stand 2-⅝ inches. *Thornburg Collection.*

Goebel mallards, the drake being 3-¼ inches high. Each shaker has a Goebel mark. According to the McHughes, these were first marketed in 1949. *Thornburg Collection.*

The taller of the pair on the left is 4-½ inches high; information on marks was not recorded. The pair on the right have a paper label reading, "Handpainted / Royal / Japan" on three lines. *Carson Collection.*

Sandy Srp miniatures, the taller of the two just 1 inch high. *Carson Collection.*

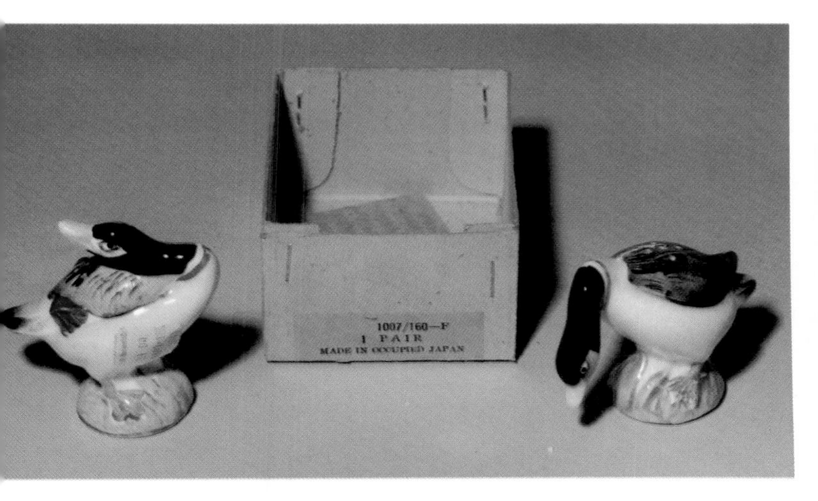

This pair illustrates why people should not get too carried away with Occupied Japan items. Contrary to popular belief, everything that came out of Japan during the American occupation did *not* have to be marked "Occupied Japan." In this case, the original box is marked "Occupied Japan," but the shakers' inkstamps say simply "Japan," negating any premium that might have been paid for them. The shaker on the left is 2-¾ inches high. Original price sticker, incidentally, showed 39 cents per pair. *Lilly Collection.*

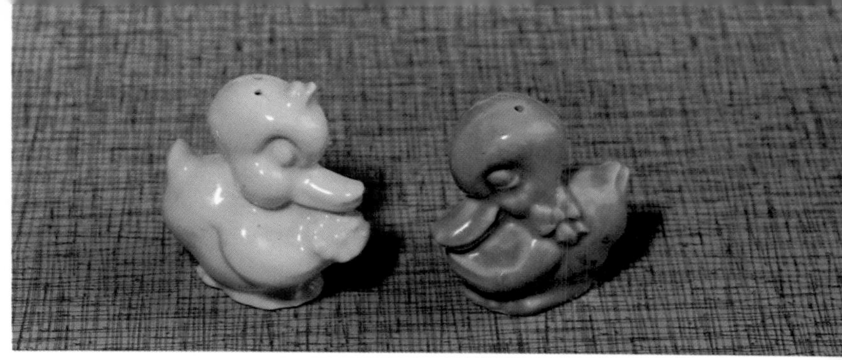

Height of these ducks is 2-¾ inches. They are not marked. *Oravitz Collection.*

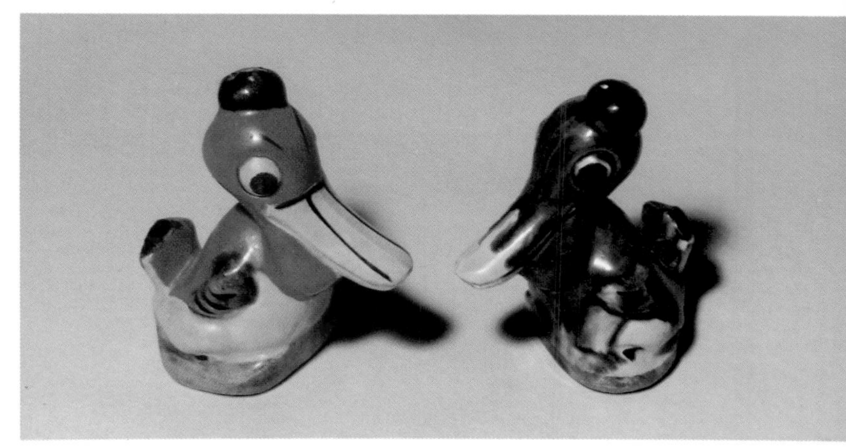

These shakers are 3-¼ inches high, have a "Japan" inkstamp. *Lilly Collection.*

Two sets of nodders. The set at left is 4 inches high including the base. It has a "Japan" inkstamp. The set on the right has a Victoria paper label with "Pat." impressed. *Thornburg Collection.*

Peppy and Salty, each 3-½ inches high. Neither shaker is marked. *Carson Collection.*

Donald Duck as sold by the Leeds China Company in the 1940s. The shaker at left is 3-¼ inches high. All have "W. Disney ©" impressed in their tails. *Thornburg Collection.*

The pair on the outside in the rear is 4 inches high, marked "© Donald Duck Walt Disney." Daisy is 2-⅞ inches high, marked "The Good Company," while her bag of groceries carries a "© Disney" mark. The mark of the sitters at rear center is shown below. *Thornburg Collection.*

Donald Duck and Ludwig Von Drake. Donald is 5-¼ inches high. This set has a "Japan" paper label along with a "© Walt Disney Prod." inkstamp. However, I recently saw the same shakers at an antique show with an "Inarco 1961" inkstamp. *Thornburg Collection.*

Mark of Donald Duck sitter set above.

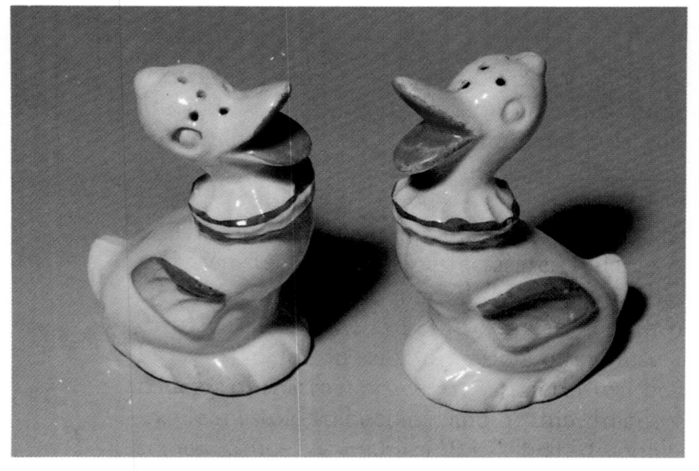

Height here is 3-¼ inches. The set has a "Japan" inkstamp. *Lilly Collection.*

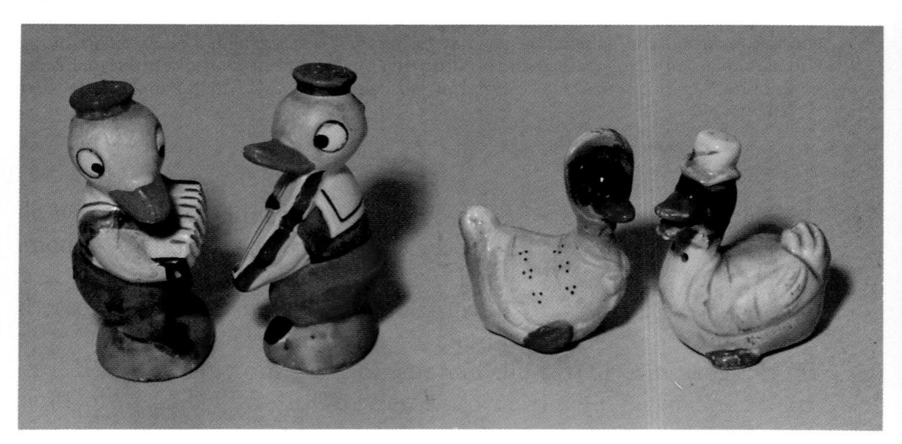

The pair on the left is 3-½ inches high; the shakers carry "Japan" inkstamps. The pair on the right is 2-¾ inches high. It is not marked. *Carson Collection.*

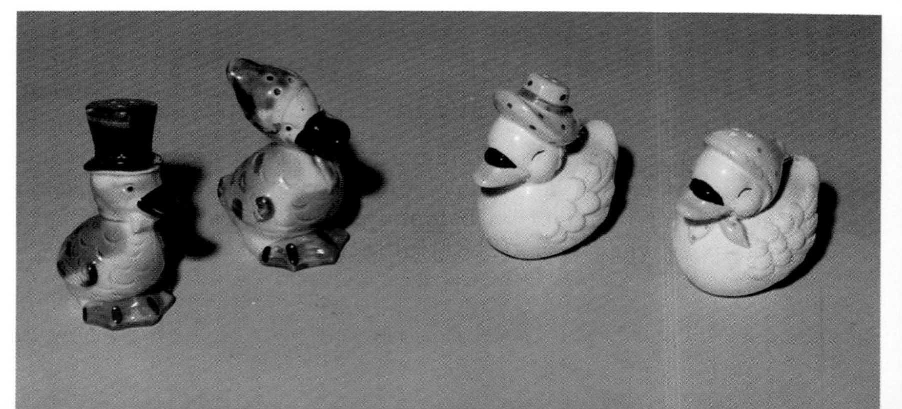

The drake of the couple at left stands 4-⅛ inches high, has a "Japan" inkstamp. The hen of the couple on the right is 3-⅜ inches high, is unmarked. *Lilly Collection.*

Ducks and geese. The ducks are 3-⅞ inches high, the geese 2-¾ inches. Both sets have "Japan" inkstamps. *Oravitz Collection.*

Geese, 3-½ inches long and 3-⅝ inches high. They have a "Japan" inkstamp. *Carson Collection.*

Regal China huggie ducks, the one on the left in each pair standing 4 inches high. Impressed mark on all four pieces is "© Van Tellingen." *Thornburg Collection.*

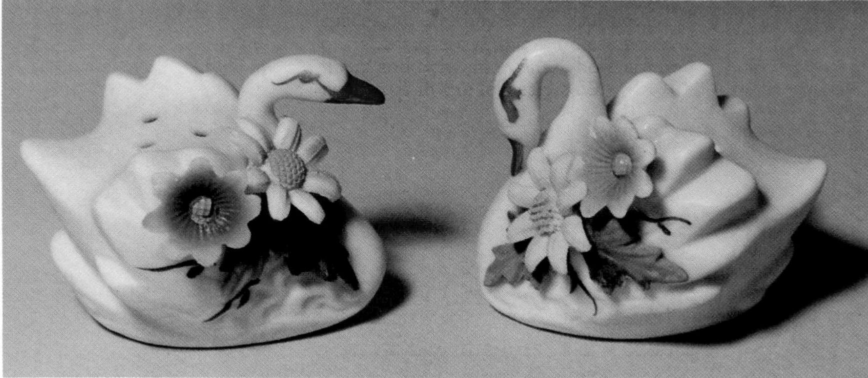

Each swan is 2 inches long. Each has a "Japan" inkstamp. *Carson Collection.*

Similar to the above Regal China Van Tellingen huggie ducks, but clearly marked "Occupied Japan." Height of the one on the left is 4-¼ inches. *Oravitz Collection.*

The swan on the right is 2-⅛ inches high. It has a "Japan" inkstamp. Its mate is unmarked. *Carson Collection.*

White swans and black swans, both sets unmarked Rosemeade. Height is 2-½ inches. *Thornburg Collection.*

Big swans and little swans, both sets unmarked and unknown. The smaller of the pair in the rear is 3-⅞ inches high, while the larger of the pair in front is 1-⅞ inches high. *Courtesy of Joyce Hollen.*

More Rosemeade. The pelicans are 3 inches high, the flamingoes 3-⅞ inches. None of the pieces are marked except for paper labels. *Thornburg Collection.*

The pelicans are 3-⅜ inches high, have a "Made in Japan" paper label. *Lilly Collection.*

Rosemeade flamingoes in a darker color. *Carson Collection.*

The pair at left stands 3 inches high, the pair at right 3-⅞ inches. The pair on the left has a "Japan" inkstamp. Paper label of the pair on the right is shown below. *Carson Collection.*

The flamingoes from above, showing the bottom of one and the sandy color pottery that Rosemeade collectors depend on to identify the company's products.

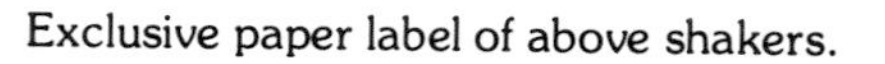

Exclusive paper label of above shakers.

Pelicans, 3-¾ inches high, "Japan" impressed. *Lilly Collection.*

Sandy Srp shakers, 1-⅜ inches high on the right. *Carson Collection.*

These shakers are 3-½ inches high, have a "Japan" inkstamp. *Carson Collection.*

The shakers on the outside are 4-½ inches high. They have "Japan" paper labels. The tray of the condiment set is 6 inches long. It is marked "Maruri" with a "M R" symbol below, and "Made in Japan" below that. *Carson Collection.*

UPLAND GAME BIRDS

These pheasants all have Lego paper labels that indicate Japanese origin. Each member of the pair in front is 5 inches high, 6 inches long. *Thornburg Collection.*

The bird on the right in the back row is 4 inches high, has a "Made in Japan" paper label. The pair in the center is from the same mold but is decorated differently. It has a "Japan" inkstamp. The shaker at front right is 6 inches long, also carries a "Japan" inkstamp. *Thornburg Collection.*

Ever hear of a golden pheasant? Here's a pair. They measure 3-¼ x 5-¼ inches. They are unmarked. However, if you compare them to the pair of dogs on page 97, you might come to the conclusion there is a pretty good chance they were made by Grindley. *Oravitz Collection.*

Nodder pheasants, the marks and measurements of which were not recorded. One thing that should be said of nodders with nonfigural bases is that you seldom know for sure if the base and the shakers were originally together. Most collectors routinely replace broken bases, or purchase pairs of shakers separately at antique shows and flea markets to put in bases they have available. It usually makes no difference, as with the sets in this picture. On the other hand, you might think twice before sticking a pair of pheasants into a base that shows a beach scene, or using a base with a woods scene for a pair of dolphins. *Carson Collection.*

Three pair of Rosemeade pheasants, like most Rosemeade all unmarked with the exception of paper labels. The shaker at far left is 2-¾ inches high, the one at far right is 5-½ inches long. The pair on the left must have been the pottery's most popular, as it is the most prevalent on today's secondary market. *Thornburg Collection.*

Red Wing quail to match the company's popular quail pattern dinnerware. The bird on the left is 4 inches high, the bird on the right 4 inches long. Marks, or lack of them, were not recorded. *Courtesy of Jazz'e Junque Shop, Chicago, Illinois.*

Rosemeade again, this time a pair of shakers and an hors d'oeuvres holder. The shakers are 4 inches high, the hors d'oeuvres 4-¾ inches. None of the pieces are marked. The common name for this set is strutting pheasant. Rosemeade also made strutting chickens and strutting turkeys. *Thornburg Collection.*

Rosemeade quail. The bird on the left is 2-½ inches high, the one on the right 1-½ inches. No marks except for the paper label. *Thornburg Collection.*

Believed to be partridges, at least that is what we are going to call them. The shaker on the left stands 2-¾ inches high, has a "Japan" paper label in addition to a Holt Howard paper label. *Carson Collection.*

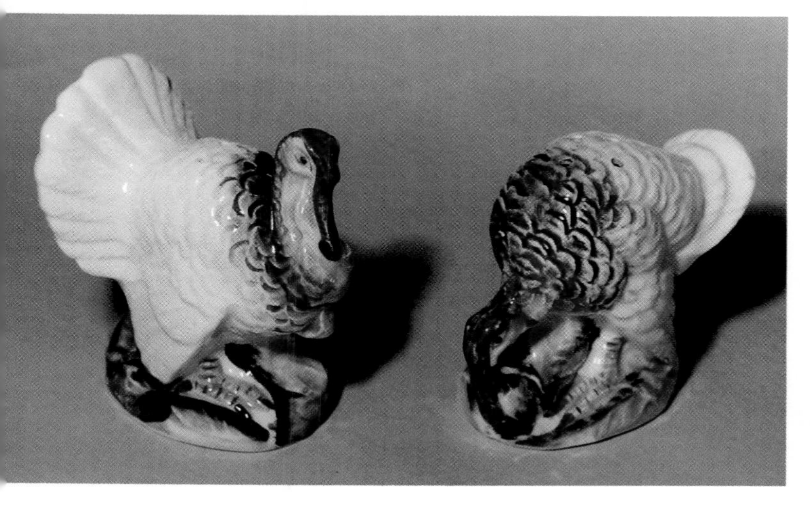

The gobbler of this pair is 3-½ inches high, has a "Japan" paper label. *Thornburg Collection.*

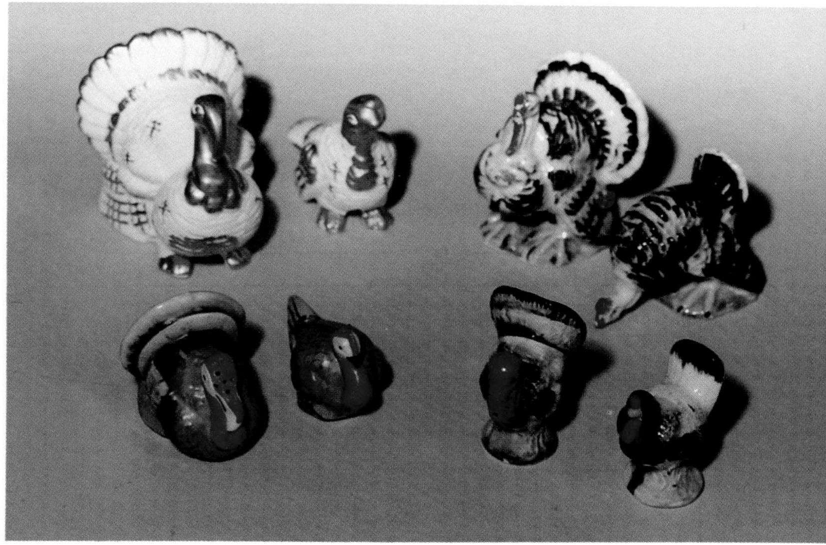

The gobbler at left rear is 4 inches high, has a "Made in Japan" paper label, while his counterpart in the pair at right front is 2-¼ inches high, has a "Japan" inkstamp. So does the pair at front left. The pair on the right in the rear row has a Chase paper label, which is shown below. *Thornburg Collection.*

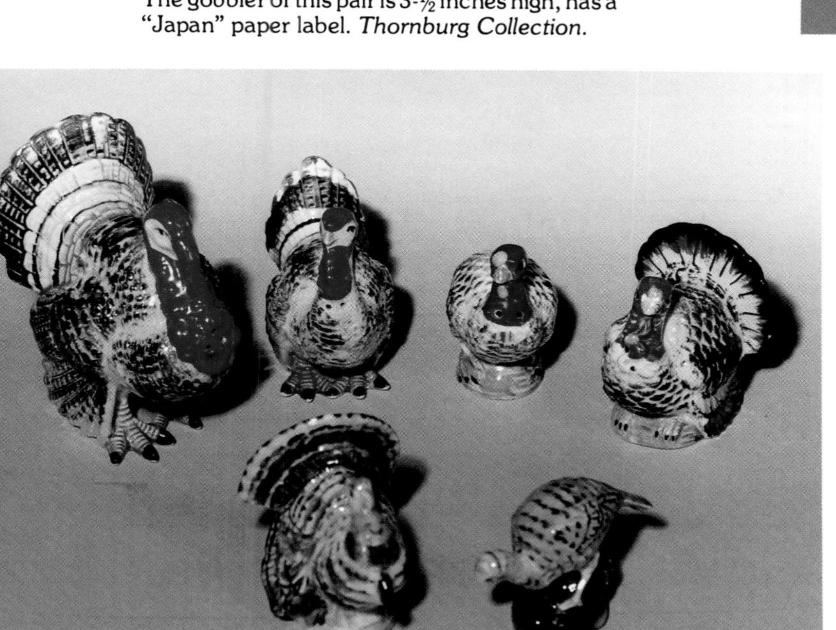

The big gobbler at left rear is 5-½ inches high, has a "Japan" inkstamp. The blue hen in front stands 2-½ inches high, is unmarked. The pair at right rear carry "Japan" paper labels. *Thornburg Collection.*

Chase paper label of turkeys at right rear above.

The shakers in front are two inches high. Like the pair at left, they are unmarked, and I have reservations about calling either set a true pair as they are composed of toms without hens. The tom of the green pair is 3-½ inches high, has a "Japan" inkstamp. *Thornburg Collection.*

The mallards were shown under waterfowl, but since this is the only picture I shot of Rosemeade turkeys, we'll look at them again. The male turkey is 3-⅝ inches high, has a Rosemeade paper label. *Carson Collection.*

Height on the right is 2-¾ inches. The birds have the same Chase paper label as shown above. *Carson Collection.*

The pair on the left is 3 inches high, has a "H-549" inkstamp. The pair on the right is 3-¾ inches high, has a "Japan" paper label. The set in the middle carries an Enesco paper label. *Thornburg Collection.*

A nicely painted set for which no one took credit except to place a "Japan" paper label on it. The bird on the left is 3-⅞ inches high, the one on the right is 4-¾ inches long. *Courtesy of Cloud's Antique Mall, Castalia, Ohio.*

The gobbler here is 3-½ inches high, has a "Made in Japan" inkstamp. *Oravitz Collection.*

Here is what seems like an appropriate set to close out the section on turkeys. The platter is 3-⅝ inches long, neither piece is marked. *Oravitz Collection.*

None of these are marked. The bird on the left is 3-¾ inches high, the bird on the right 2-¼ inches high. *Courtesy of Joyce Hollen.*

The two pair in the middle of the photo are 3-¼ inches high. Neither is marked. The set at left rear has a "Made in Taiwan" paper label, the one at right rear a Knobler paper label that indicates Japanese origin. *Carson Collection.*

Two pair of Shawnee owls, along with a pair of Shawnee ducks. The owls are 3 inches high. Those on the left, with more intricate decoration and gold trim, are considerably more expensive than those on the right. None of the pieces are marked, but all have Shawnee's unusually large fill hole. *Smith Collection.*

Plated metal owls, 2-½ inches high. They are not marked. *Carson Collection.*

Height is 2-½ inches. There are no marks. *Courtesy of Joyce Hollen.*

These unmarked owls are 2-¼ inches high. *Lilly Collection.*

Goebel owls, 2-¾ inches high. Mark is an unidentifying "West Germany" inkstamp. *Thornburg Collection.*

Height is 3-⅛ inches, paper label reads "Brinn's / Bone China / Taiwan" on three lines. *Carson Collection.*

Size of this pair was not recorded, but they have a mark that reads "© AGIFTCORP" along with a paper label that says "Genuine / Bone China / Taiwan" on three lines. *Carson Collection.*

The owl on the left is 2-¾ inches high. Neither piece is marked. *Lilly Collection.*

Woodsy Owl, 4 inches high. Paper label reads "Made in Japan." *Thornburg Collection.*

SCAVENGERS

The prickly pear cactus is 4-½ inches high, the buzzard approximately 3-½ inches high. The set has a Vandor paper label. *Thornburg Collection.*

Peacocks or pheasants, whichever you wish to call them. They are metal, 8 inches long, and unmarked. *Courtesy of Cloud's Antique Mall, Castalia, Ohio.*

SONGBIRDS AND EXOTIC BIRDS

Ceramic Arts Studio parakeets, 4 inches high. Laumbach shows them in blue and yellow. They have Ceramic Arts marks that include the names Twirp and Chirp. *Thornburg Collection.*

The birds on the left are 4-½ inches high. They have bellows inside to make them tweet while being used. They also have a "Japan" paper label. Height on the right is 4-½ inches, mark is a "Made in Japan" inkstamp. The small pair of geese in front is 2-¼ inches high, has a "Japan" inkstamp. *Carson Collection.*

As you can see from the paper label, these are Rosemeade. They are 2-¾ inches high. *Thornburg Collection.*

Birds on a log. The shakers are 2-⅛ x 4-⅛ inches, the log 5-½ inches long. Paper label reads "A Quality / Product / Japan" on three lines. *Lilly Collection.*

Wooden birds, 3-¼ inches high including the base. The wooden stoppers, upon which the birds sit, are threaded. They have a Napco paper label. *Carson Collection.*

Parrots, 3 inches high, unmarked. Bisque is a sandy color a shade or two lighter than Rosemeade's. *Private Collection.*

These birds are 2-⅜ inches high. The impressed mark is "(unreadable) / of Calif." on two lines, with the first line in cursive and the second line printed. *Oravitz Collection.*

Hanging bluebirds, the stand being 3-⅛ inches high. The birds are marked with a "Japan" inkstamp. *Carson Collection.*

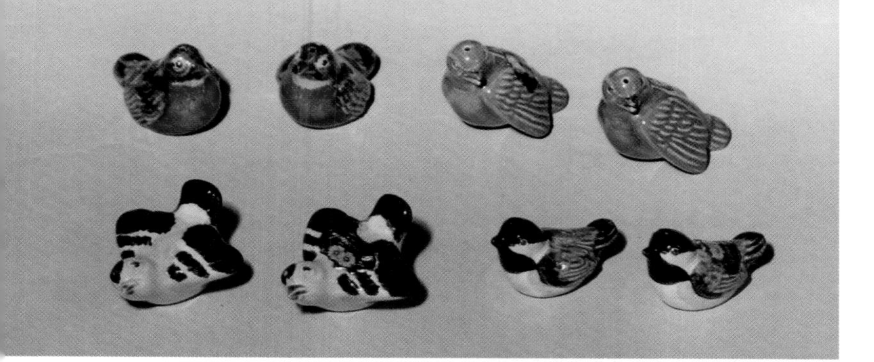

Rosemeade songbirds—bluebirds, robins, goldfinches and chickadees. The bluebirds are 1-¾ inches high, the chickadees 1-¼ inches high. Unmarked other than paper labels. *Thornburg Collection.*

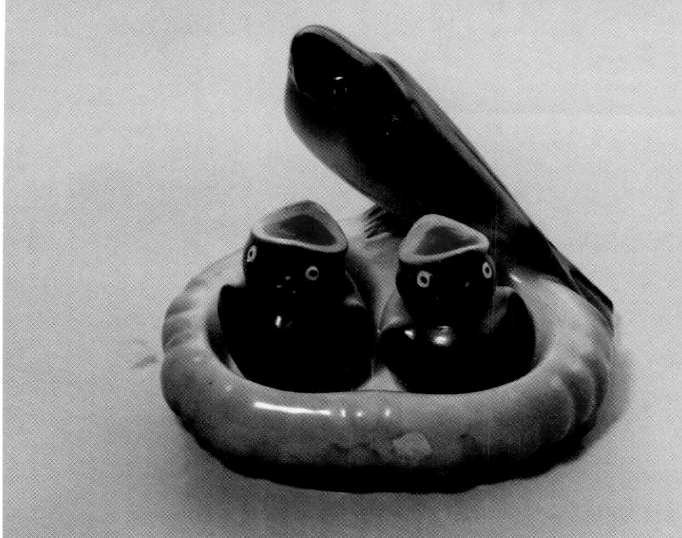

Robins at feeding time. The base is 4 inches high, has a "China" inkstamp. *Courtesy of Allen and Michelle Naylor.*

Bluebirds, unmarked, and 2-⅜ inches high. *Carson Collection.*

This condiment set appears somewhat questionable. The green of the pepper and the tray are no doubt a match, but the green of the mustard doesn't seem to have as much blue in it. Also, the baby birds look very much unlike pheasants, and the salt, which is the lid of the mustard, lacks the printing on the pepper. But, I measured the mustard and tray as one item (4-½ by 5-¼ inches), so perhaps they were glazed together. Inkstamp on the tray says "Japan." *Carson Collection.*

Plastic bluebirds, also 2-⅜ inches high, also unmarked. *Carson Collection.*

As the boxes say, these two sets are handpainted porcelain. The birds on the left are 1-⅝ inches high, their tray is 3 inches long. The birds are unmarked but the tray has a "China" paper label. The birds on the right are 1-⅜ inches high, have a "China" paper label. *Lilly Collection.*

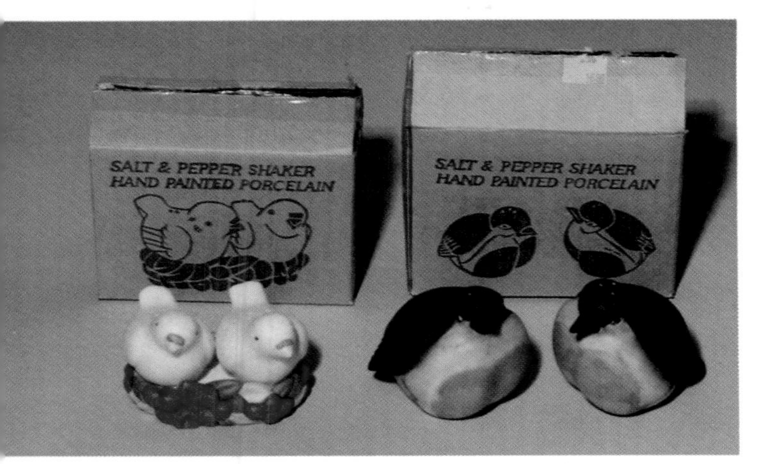

The birds are 2-⅛ inches high, the tray 5 inches long. The mark of the tray is shown below. *Lilly Collection.*

Mark of the tray of the above birds.

Every piece in this set is plastic. The parakeets are 3-¾ inches high. There are no marks. *Carson Collection.*

Ceramic on the left, plastic on the right. The base of the cardinals is 3-¾ inches long. The base on the right is 5-¼ inches long. Neither set is marked. *Courtesy of Millie Gisler.*

Nodder parakeets in a base that matches the subject. The shakers are 3-¼ inches long. The base is marked "Patent TT" in raised letters, and also has a Victoria Ceramics paper label. *Thornburg Collection.*

Unmarked plastic nodders. Height to the top of the bird is 5 inches. *Thornburg Collection.*

The birdhouse is 3-¼ inches high, both pieces are unmarked. The gnome sitter stands 4-½ inches high as shown. It has a mark that indicates Japanese origin. *Carson Collection.*

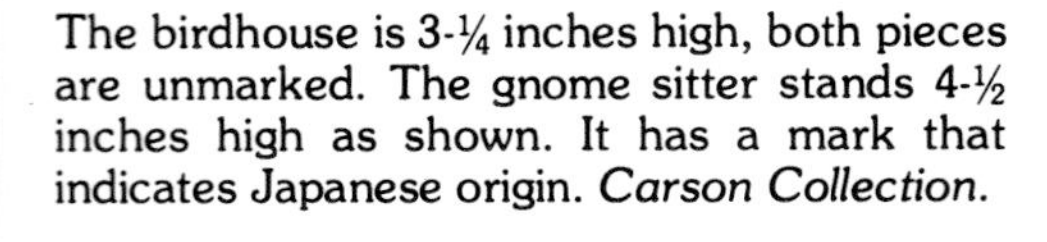

Nodder condiments, the sizes and marks of which were not recorded. Note that the base of the birds has squared ends, while the base of the horses and riders has rounded ends. *Carson Collection.*

This is a good size set, the height to the top of the holder being 8-¼ inches. There is a "Japan" paper label. *Carson Collection.*

These penguins are 4 inches high. They are marked Ceramic Arts Studio. *Thornburg Collection.*

Here is a plastic set, 4-¾ inches high. Marks, or lack of them, were not recorded. *Carson Collection.*

The pair on the outside is 3-¾ inches high. "© WLP" is marked on the back. Woody of the inside pair stands 5-¾ inches. A paper label reads "W. Lantz / © 1990" on two lines. *Thornburg Collection.*

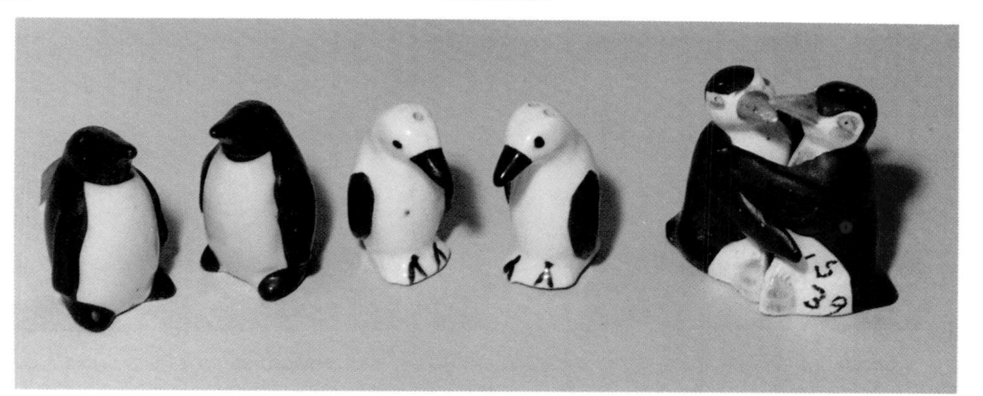

Left, 2-⅜ inches high, "Japan" inkstamp. Middle, 2-¼ inches high, no mark. Right, 3 inches high, "Japan" inkstamp. *Carson Collection.*

Woody is 4 inches high, has an inkstamp mark that reads "IC3635 / © 1958 / Walter Lantz Prod." on three lines Winnie carries a Napco paper label. Thornburg Collection.

This pair is metal, 2-¾ inches high, with machine screw stoppers. The mark indicates it was a product of Occupied Japan. *Carson Collection.*

Keeping warm inside the Antarctic Circle. The shaker on the left is 2-⅝ inches high. Both are marked with "Japan" inkstamps. *Lilly Collection.*

PENGUINS

The penguins on the left, 4-½ and 4 inches high, are not marked. Height in the middle is 3 inches, mark is a "Japan" inkstamp. On the right, heights are 3-⅝ inches. Like the pair on the left, this pair is unmarked. *Carson Collection.*

Amphibians and Reptiles

Reptiles and amphibians were not as popular with shaker manufacturers as some other types of animals. However, many different sets of frogs have been made, and collectors also have a decent variety of turtles from which to choose. Although some snakes are very beautiful, they have been used sparingly as shaker subjects.

Each of these frogs is 3-½ inches long. Each has "Japan" impressed. *Carson Collection.*

The set on the left is plastic. The frogs measure 2-¾ inches high, the stand 4-¼ inches. "Hong Kong" appears in raised letters. The pair on the right is ceramic, 2-¾ inches high, with a "Japan" inkstamp. *Lilly Collection.*

The frog is 1-⅞ inches high, the mushroom 2-⅜ inches high. Marks are Ceramic Arts Studio. *Thornburg Collection.*

This frog is 4-½ inches high. The pair has a "Japan" inkstamp. *Courtesy of Jazz'e Junque Shop, Chicago, Illinois.*

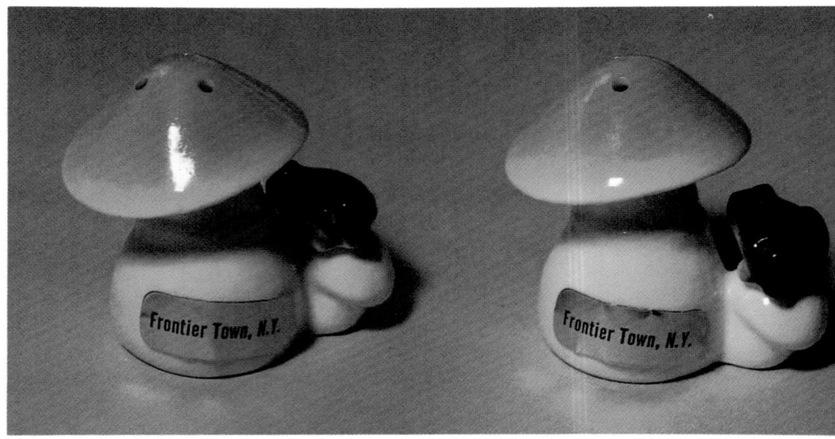

Each shaker is 2 inches high, each has a paper label stating "Bone China / Japan" on two lines. *Carson Collection.*

The mushrooms, left to right, are 2-¼ and 2-¾ inches high. The set on the left is unmarked, the set on the right has a Lefton paper label. *Carson Collection.*

The frog at the extreme left is 3-½ inches high, has a "4-923" inkstamp, and a "Japan" paper label. The frog at the extreme right stands 4 inches, has a "Japan" inkstamp. *Carson Collection.*

The sitters on the left measure 3-¼ inches high as shown. Inkstamp on the lily pad reads "© Sears Roebuck & Co. 1976 / Japan" on two lines. The pair of frogs on the right are marked "Japan" in raised letters. *Thornburg Collection.*

The set on the outside measures 3-¼ inches in height. It is unmarked. The shakers of the tray set are 2-¼ inches high. All three pieces carry "Japan" inkstamps. *Thornburg Collection.*

The frogs are 1-⅞ inches high, the tray 5-⅛ inches long. Mark says "Japan." *Carson Collection.*

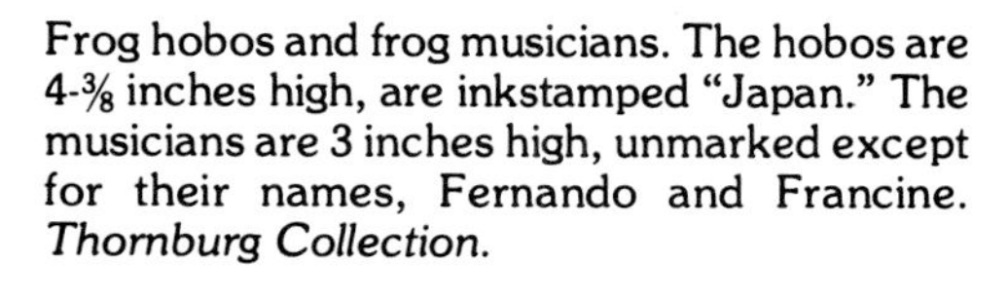

Frog hobos and frog musicians. The hobos are 4-⅜ inches high, are inkstamped "Japan." The musicians are 3 inches high, unmarked except for their names, Fernando and Francine. *Thornburg Collection.*

And another country is heard from—Thailand. The frogs stand 1-⅜ inches high. The tray measures 4-⅛ inches long. The mark is shown below. *Lilly Collection.*

Frog condiment set, 3-½ inches high. It has two paper labels. One says "North Bay, Canada," the other "Made in Japan." *Carson Collection.*

These shakers are 2-½ inches high with a mark that indicates they were made in Occupied Japan. An identical pair was photographed that had a "Japan" inkstamp. As with the glaze on this pair, the glaze on the second pair exhibited slightly different colors. *Carson Collection.*

Cellophane decal on frog tray above.

Two inches high, marked "Japan." *Carson Collection.*

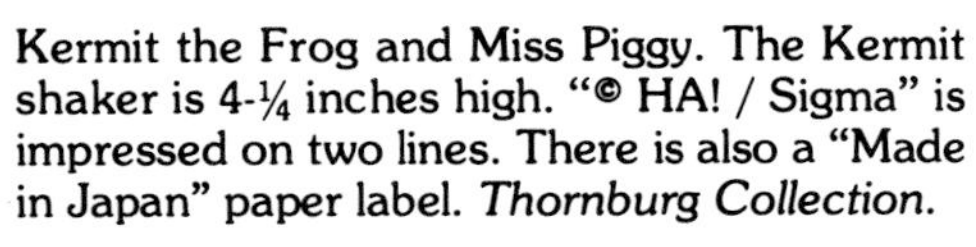

Kermit the Frog and Miss Piggy. The Kermit shaker is 4-¼ inches high. "© HA! / Sigma" is impressed on two lines. There is also a "Made in Japan" paper label. *Thornburg Collection.*

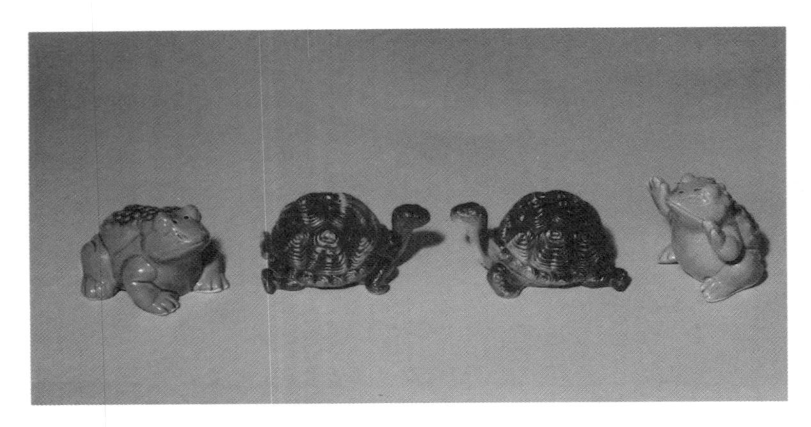

With this picture we say goodby to frogs, hello to turtles. The frog on the left is 3-½ inches long, the frog on the right 2-½ inches high. They have "Made in Japan" paper labels. The unmarked turtles measure 1-¾ x 3-¼ inches. *Courtesy of Millie Gisler.*

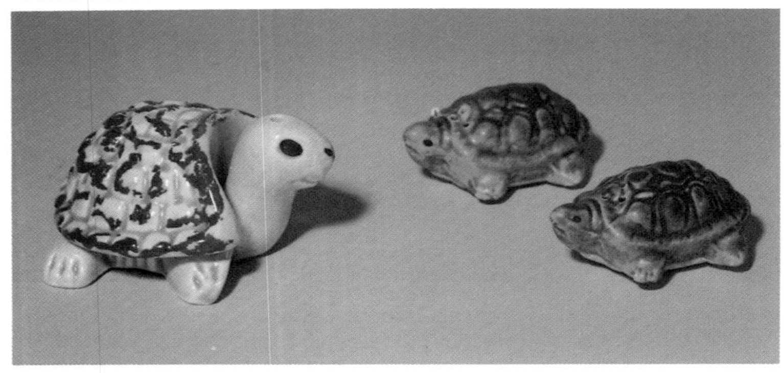

The turtle on the left is two pieces. It measures 2 inches high as shown. It is not marked. The turtles on the right are 2-⅜ inches long. They, also, are not marked. *Carson Collection.*

These guys are 2-½ inches high. They are unmarked. *Oravitz Collection.*

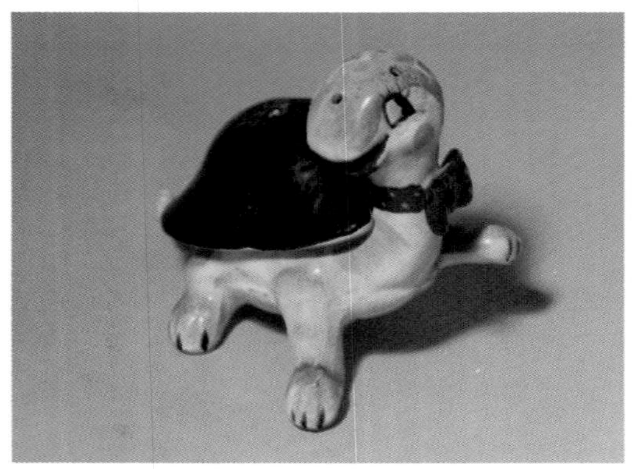

The body is the salt, the shell the pepper. Height is 3-½ inches. There is a "Japan" paper label. *Carson Collection.*

The shakers on the left are 2-⅞ and 3-¼ inches high. They are marked "Japan." The ones on the right average 2-13/16 inches high, have a "Japan" paper label. *Carson Collection.*

A rather lengthy one-piece, 8 inches long. It is not marked. *Thornburg Collection.*

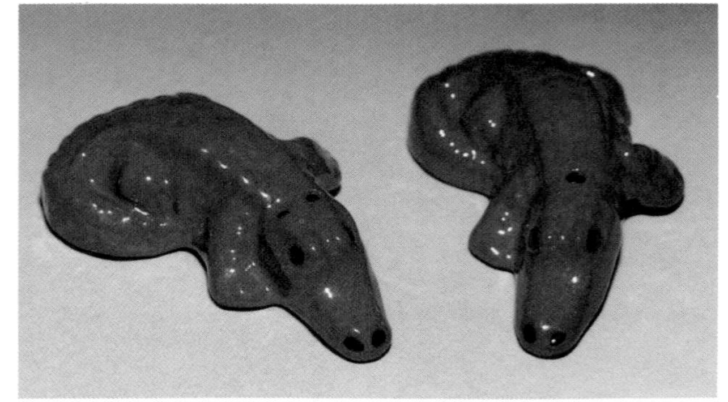

Miniature alligators by Sandy Srp. Length is 1-⅜ inches. *Carson Collection.*

Same set, different decoration. Heads up measure 2-⅝ x 4-⅝, heads down 1-¼ x 3-⅝. All are marked "Japan." *Carson Collection.*

The pair in front is unmarked, 4-½ inches long. The pair in back, also unmarked, is 2-⅜ inches high. *Carson Collection.*

The larger dragon is 3-⅛ inches high. Neither piece is marked. *Carson Collection.*

Height is 3-⅛ inches. Mark is a "Japan" inkstamp. *Carson Collection.*

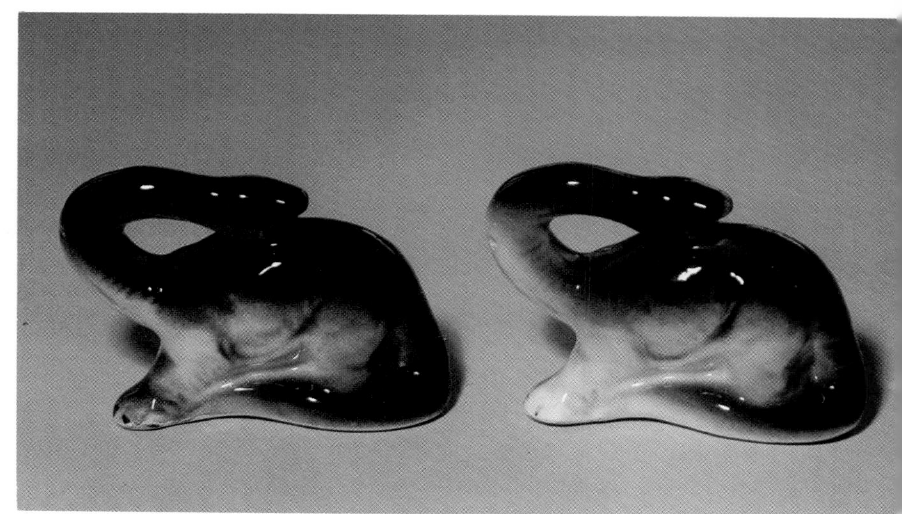

Dinosaurs, 2-¼ inches high. They have "Japan" inkstamps. *Carson Collection.*

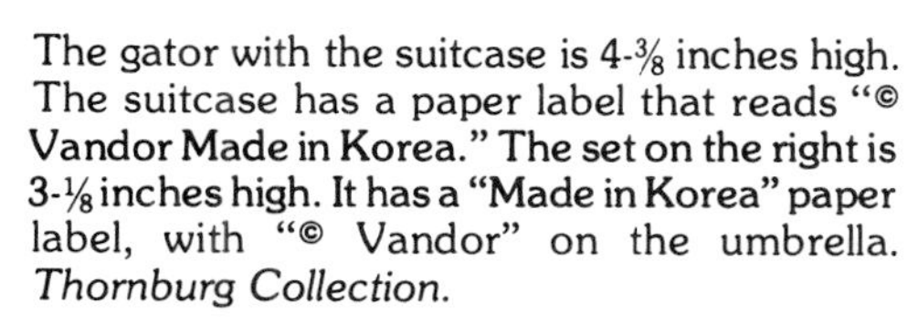

The gator with the suitcase is 4-⅜ inches high. The suitcase has a paper label that reads "© Vandor Made in Korea." The set on the right is 3-⅛ inches high. It has a "Made in Korea" paper label, with "© Vandor" on the umbrella. *Thornburg Collection.*

Insects and Other Little Creatures

Bugs seem to be the favorite in this category. Like amphibians and reptiles above, it is an area that is wide open to manufacturers in the future. Spiders would be great for Halloween, walking sticks or preying mantises might go well with summer cookouts, and fishing worms would seem appropriate as lakeside cottage accessories.

The tray for this bee and beehive set is 4 inches long. It has a paper label that reads "Brinn's / Pittsburgh PA / Made in Japan," in addition to a "T-1489" inkstamp. *Carson Collection.*

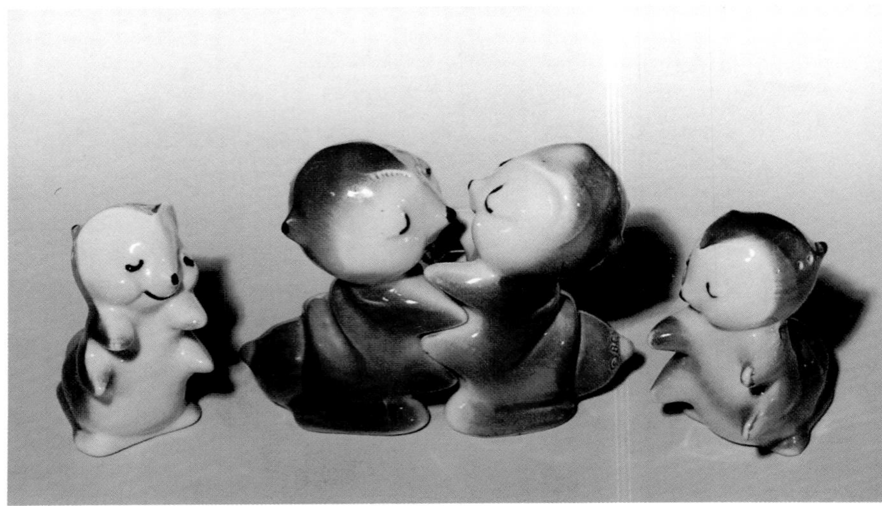

Regal China Van Tellingen lovebugs, 4-⅜ and 3-⅝ inches high. They are marked with an impressed "©Van Tellingen." *Thornburg Collection.*

The unmarked bees measure 3-½ inches in length. The ladybugs are 2-½ inches high, have a "Japan" inkstamp. *Carson Collection.*

These ladybugs measure 1-½ x 3-½ inches. The pair on the left has a VG paper label shown below. The pair on the right has a "Made in Japan" paper label. I have also seen these shakers in pink, with the VG paper label. *Thornburg Collection.*

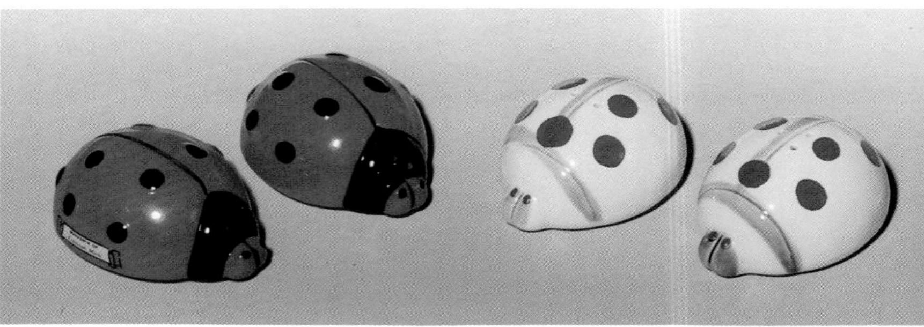

Lovebugs, 3-⅜ inches high with a "Made in Japan" paper label. *Carson Collection.*

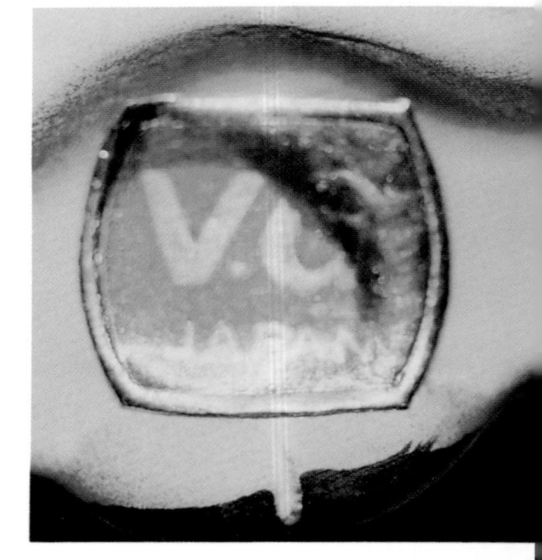

VG paper label of above ladybugs.

The pair on the left is 1-⅞ inches high, has a Vallona Starr paper label, shown below. The pair in the center is 2-½ inches high. It appears to be the work of a hobby ceramist, a theory that is reinforced by the impressed mold company mark, "© Duncan Ceramic, Inc. 1975." The pair on the right is unmarked. *Thornburg Collection.*

Paper label of above shaker.

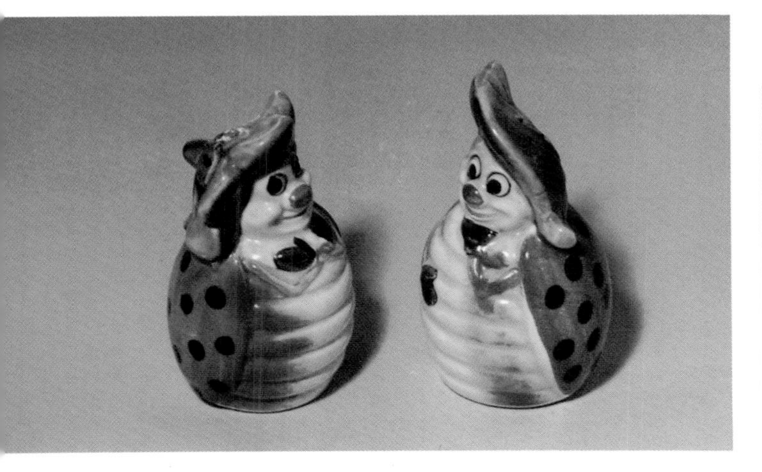

The shaker on the right is approximately 3-¼ inches high. It has a "Japan" inkstamp. *Carson Collection.*

A bug band straight from Japan. Four pairs of shakers are shown. The bug in front is 3 inches high. All pieces have "Japan" inkstamps. *Thornburg Collection.*

As above, each of these shakers has a "Japan" inkstamp. The one at left stands 2-⅝ inches high. *Thornburg Collection.*

The personified set in the middle is 3-½ inches high. It has a "Japan" paper label, plus a "50/103" inkstamp. (The slant is included in the mark.) Marks of the set on the outside were not recorded. *Thornburg Collection.*

The set on the left is 3-½ inches high as shown. The pair on the right was not measured. All pieces in this picture have "Japan" inkstamps. *Thornburg Collection.*

All between approximately 2-½ and 3-⅛ inches high, all marked with "Japan" inkstamps. *Thornburg Collection.*

Probably one of the best places to find bugs is in the pigpen, or so this picture would indicate. The bugs are 3-¼ inches high, mark information was not recorded. The out-of-place pigs stand 2-⅜ inches high, have a "Japan" paper label. *Carson Collection.*

Another pair of music makers, 2-⅞ inches high. One has a "Japan" inkstamp, the other is unmarked. *Lilly Collection.*

Grasshoppers and watermelon, what a combination. The shakers are 3 inches high, have a "Japan" inkstamp. What appears to be a paper label on the shaker at left is an inventory sticker. *Private Collection.*

We are calling these garden lizards, don't know what else to do. The one on the left is 3-⅜ inches high, has a "Japan" inkstamp. *Carson Collection.*

The snails at the rear measure 3-¼ inches to the top of the head. The antennae are plastic; the set has an Enesco paper label that indicates Japanese origin. The pair at front left has a "Japan" inkstamp, while the pair at front right carries a Meridan paper label. *Thornburg Collection.*

The pair on the left stand 2-⅛ inches high. They are not marked. The pair on the right have a "Made in Japan" paper label, while the two in front have an old Fitz and Floyd paper label which is shown below. *Thornburg Collection.*

Fitz and Floyd paper label of above snail shakers. Color variation is due to improper exposure.

The snails are 3-½ high. Their Enesco paper label reads in part, "Gourmand Stoneware Escargot." The lobsters are not marked. *Carson Collection.*

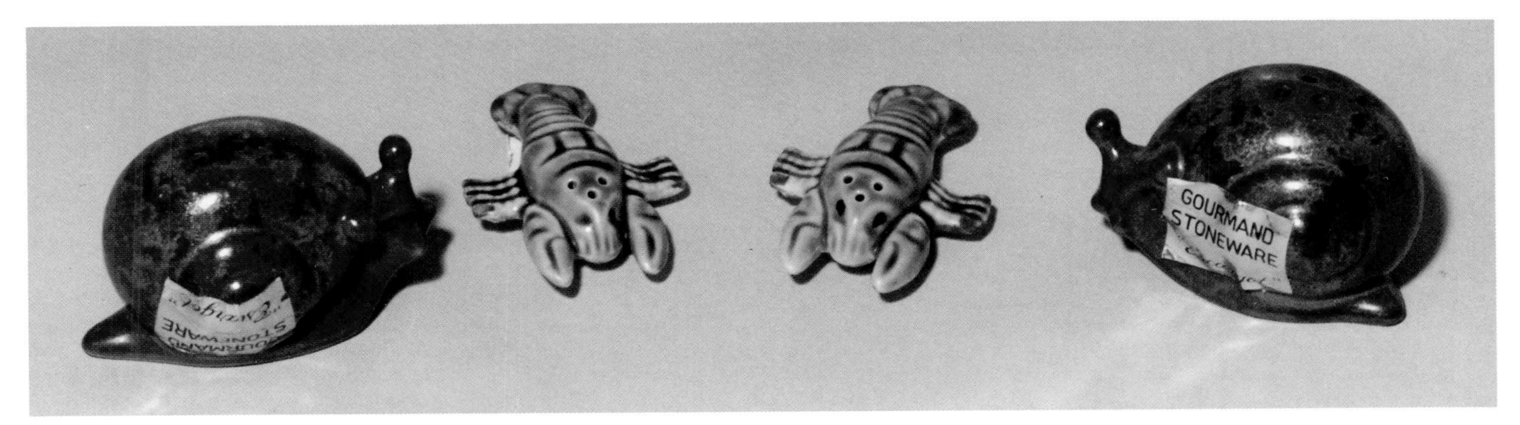

Creatures of the Deep.

This section covers fish, shellfish, shells and seahorses. Also included are dolphins and porpoises which, although mammals, seem to fit here much better.

As far as individual companies, Rosemeade clearly led the fleet in attempting to net the fish buyers. By and large, most collectors I interviewed agreed that you have to go a long way to beat the quality of Rosemeade's fish shakers, both naturalistic and stylized versions. But, as with all areas of shaker collecting, some of the imports come pretty close.

The red, yellow and blue fish are 3-¾ inches long, are not marked. The others are 2-½ inches high, marked "Bernard Studios, Fullerton, Calif." *Carson Collection.*

The size of these fish is 3 x 4-⅝ inches. Their mark reads "Sun / Fish / Japan" on three lines. *Carson Collection.*

Rear 5-½ inches long, middle 4-¼ inches long, front 3-⅛ inches long. Marks were not recorded. *Carson Collection.*

All the fish shakers in this photo are the same length, 4-¼ inches, yet each pair came from a different mold. The ones at left rear are unmarked, the ones at right rear have a Napco paper label, and the only mark on the pair in front is a "Rock Bass" inkstamp. *Carson Collection.*

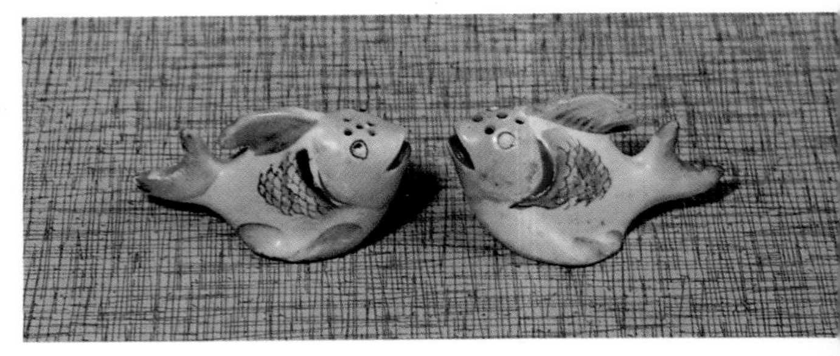

This pair of fish has "Japan" inkstamps. Size is 1-½ x 2-¾ inches. *Oravitz Collection.*

Trout shakers, 3-¼ x 4-¾ inches. Gold paper label reads "A Quality / Product / Japan" on three lines. *Carson Collection.*

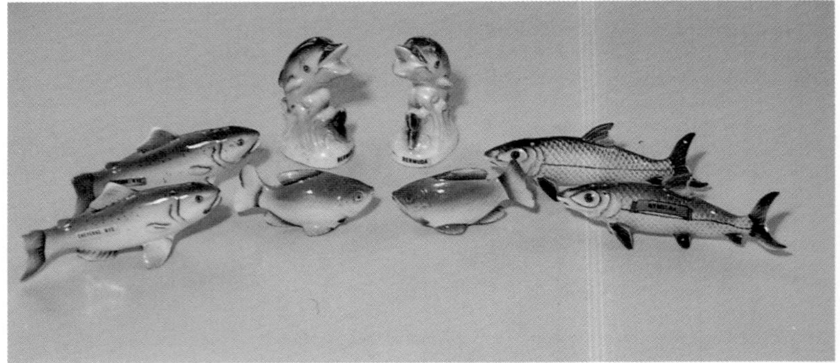

The pair on the left has a Victoria paper label. The porpoises, at rear center, are 3-½ inches high, have a "Made in Japan" paper label. The bonefish, on the right, measure 5-¼ inches in length. They have a "Made in Japan" paper label. The pink set is not marked. *Carson Collection.*

Lengths left to right are 4-¾ and 5 inches. None of the shakers are marked with the exception of a "Speckled Trout" inkstamp on the pair on the left. *Carson Collection.*

Less realistic Rosemeade fish, 2-⅝ inches high, marked by paper labels. *Carson Collection.*

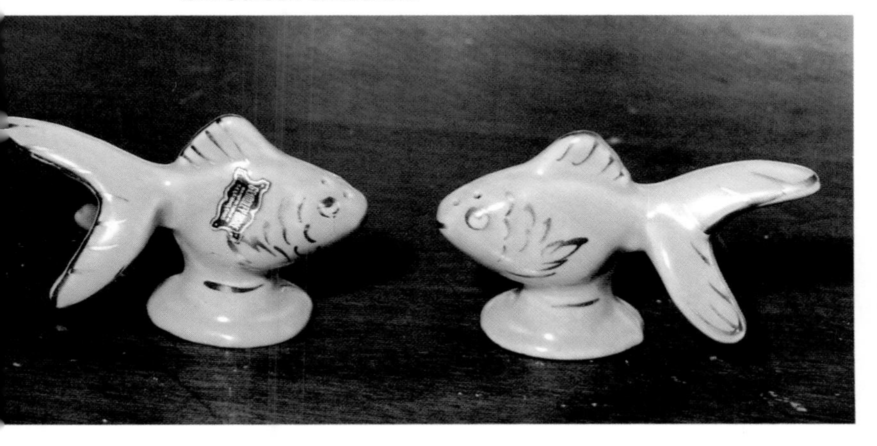

A pair of Grindley fish shakers, the paper label of which is shown below. The fish measure 2-¼ x 3-½ inches. This picture was incorrectly exposed. True color is closer to that in the picture below. *Oravitz Collection.*

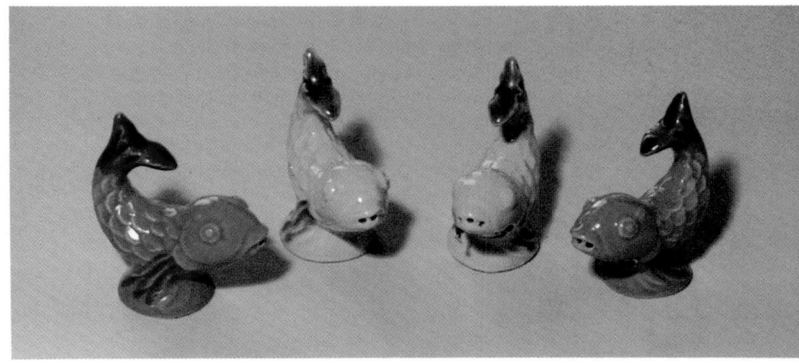

Rosemeade fish above in other colors. *Carson Collection.*

Paper label of Grindley fish shakers above.

All Rosemeade, walleye at top right, trout below them, either perch or smallmouth bass at left. The trout are 3-⅞ inches long. *Thornburg Collection.*

Rosemeade muskies and northern pike, unmarked except for the paper label on the one pike. Length of the muskies is 5-¼ inches. *Thornburg Collection.*

Rosemeade panfish. The bluegills, on the right, measure 3-⅞ inches in length. *Thornburg Collection.*

The fish on the left is 4 inches high while the pair in the middle stands 3-⅞ inches. All have Ceramic Arts Studio marks. *Thornburg Collection.*

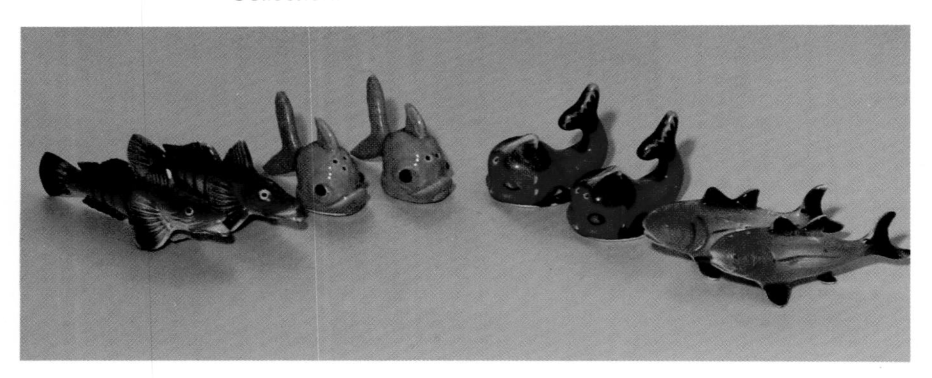

Quite a variety in this picture. Lengths left to right are 5, 3, 3 and 4-⅛ inches. The two pair on the ends have "Japan" paper labels. The two pair in the middle are unmarked. *Carson Collection.*

Nice examples of kitchen art made in Japan. Size is 2-½ x 4-¾ inches, paper label indicates they were marketed by Napco. *Carson Collection.*

Orange fish, 2 inches high, with a "Japan" inkstamp. *Lilly Collection.*

Goebel fish, 2-¾ inches high. Their mark is shown below. *Thornburg Collection.*

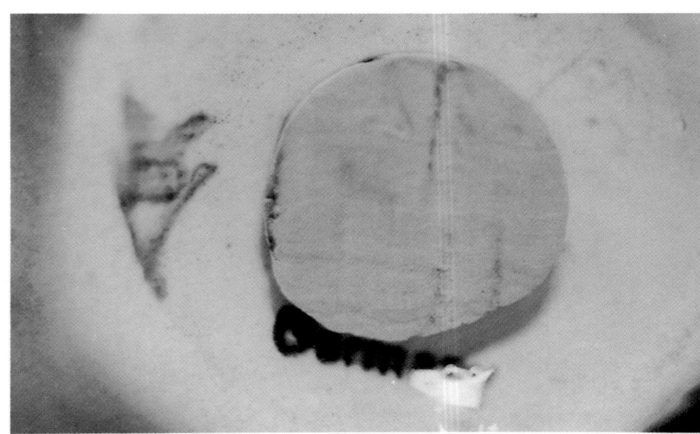

Mark of above Goebel fish.

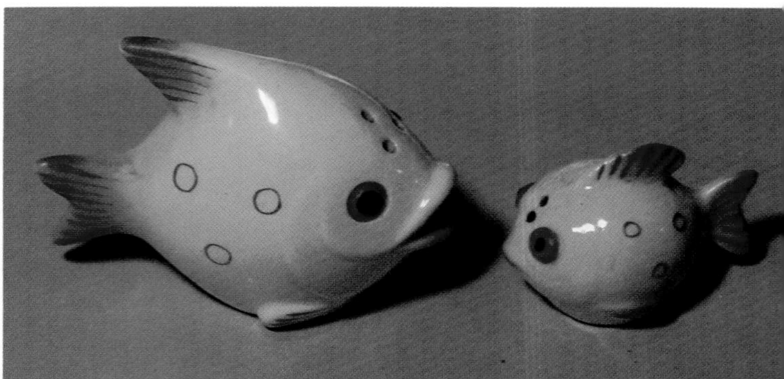

Another pair of Goebel fish. The shaker on the left is 2-¼ inches high, has an impressed mark, "P6981A." *Thornburg Collection.*

The pair in front is an advertising set, "Chicken of the Sea" being impressed in their bottoms. Each is three inches long. A matching mustard, about 6 to 7 inches long was made to accompany the shakers. I have also seen this set in maroon, and have often wondered if they might have been made by Homer Laughlin as the blue and yellow glazes are identical to Fiesta and the maroon matches a Rhythm color. The pair at rear left is 1-5/8 inches high, unmarked. The pair at rear right is also unmarked, and stands 3-1/8 inches high. *Carson Collection.*

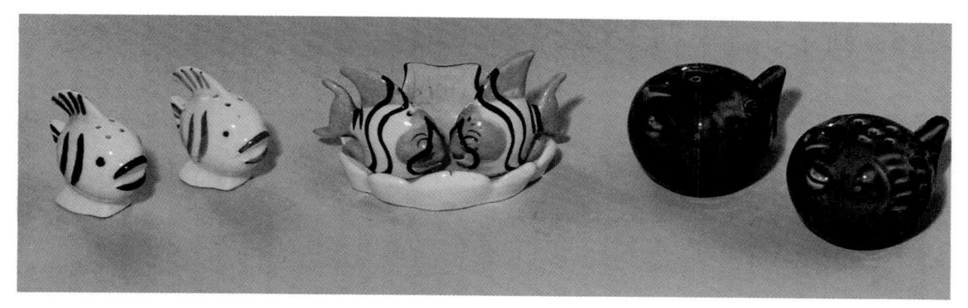

The unmarked pair on the left is 2-1/4 inches high, while the unmarked pair on the right is 2-1/8 inches. The tray of the set in the middle is 4 inches long, the shakers are 2 inches high. The tray has a paper label, "Poinsettia Studios California." *Carson Collection.*

The pair on the outside is 2-7/8 inches high. Mark information was not recorded. The inside pair measures 3-3/4 inches in height, mark is "Made in Japan." Although made of ceramic, their fill holes are threaded. *Carson Collection.*

Height on the left is 2-1/4 inches. Mark is "Japan." Same mark on the right where the height is 1-5/8 inches. *Carson Collection.*

The white fish with black stripes stand 2-1/8 inches high, have no identifying marks. The blue shakers are 2 inches high, say "Occupied Japan." *Carson Collection.*

None of these shakers are marked. Heights left to right in the rear are 1-7/8 and 2-1/8 inches, while the pair in front stands 1-5/8 inches. *Carson Collection.*

These angel fish are 2-⅝ inches high, are marked "Japan." *Carson Collection.*

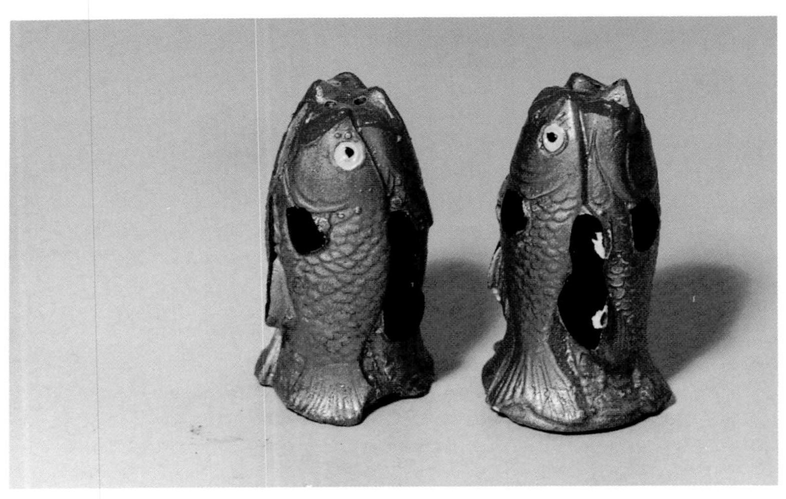

These are chalkware, each shaker composed of three fish. Height is 2-½ inches. *Carson Collection.*

Metal, 2-⅝ inches high, unmarked, and not in the best of shape. *Carson Collection.*

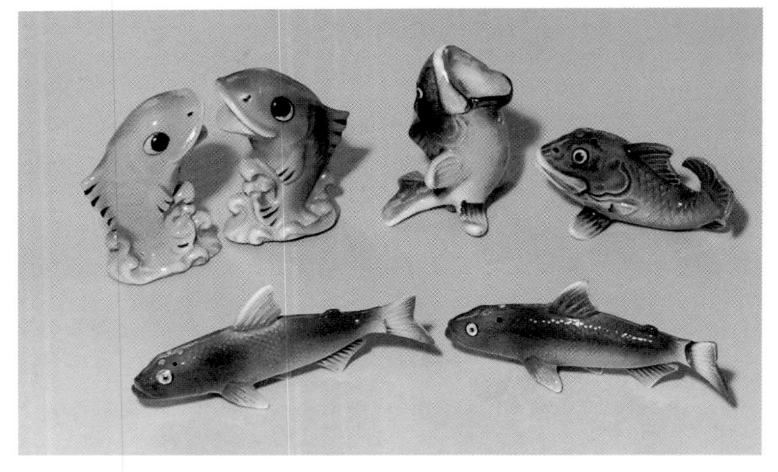

The pair at left rear is unmarked, 3-¼ inches high. The pair at right rear, the right member of which measures 3-⅝ inches in length, is marked "Japan." The pair in front, 4-⅝ inches long, is also marked "Japan." *Carson Collection.*

In the back, 2-¾ inches high with an unreadable mark. In the front, 2-⅜ inches high with no mark. *Carson Collection.*

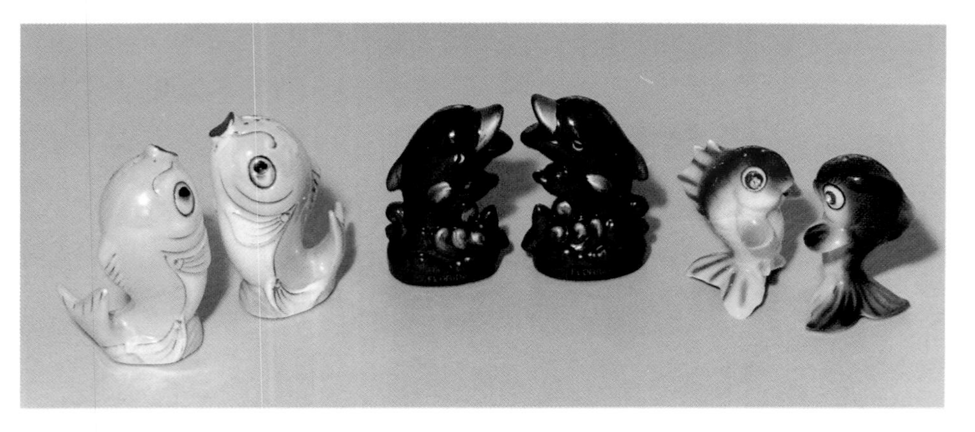

The pair on the left is 4-¼ inches high, is marked "Made in China." The middle set is made of redware, stands 3-½ inches, and has a "Japan" inkstamp. The pair on the right is 3-⅛ inches high, has a "Made in Japan" paper label along with a "D43" inkstamp. This pair has rhinestone eyes. *Carson Collection.*

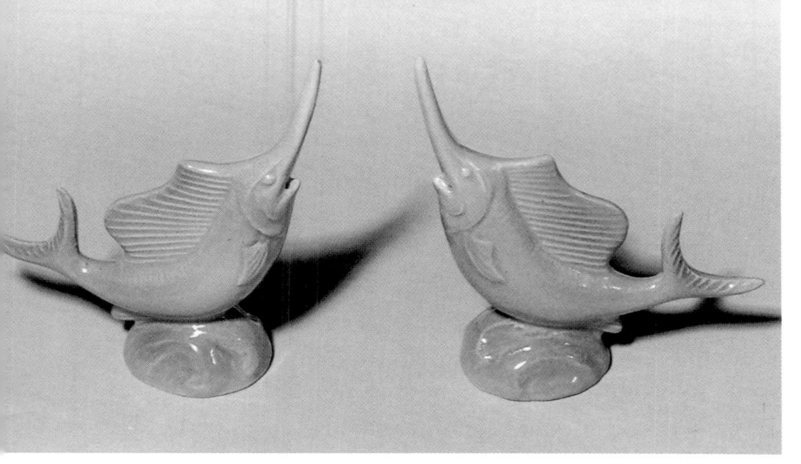

Here are some Rosemeade ocean fish, 4-⅛ inches high. They are not marked. *Thornburg Collection.*

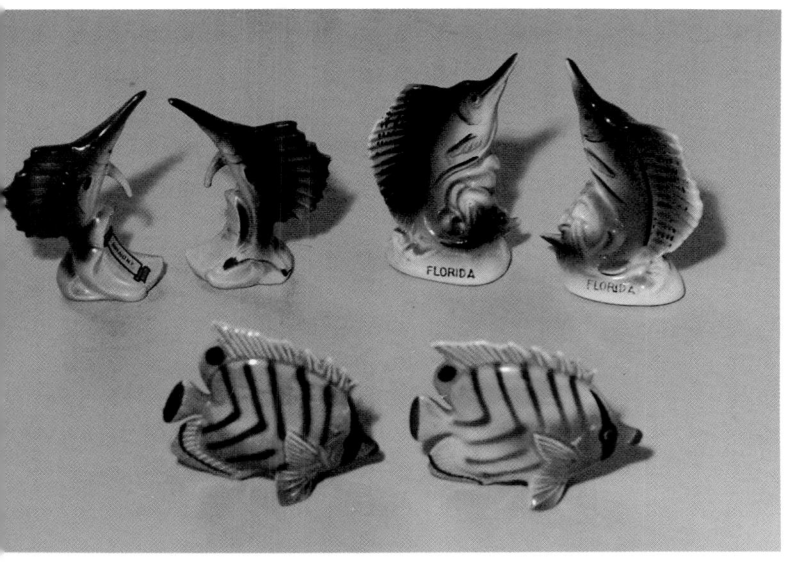

The fancy pair in front are 3-¼ inches high, have "Japan" inkstamps. The pair at left rear are 3-¾ inches high, also have "Japan" inkstamps, while the pair at right rear stands 4-½ inches and has a "Japan" paper label. *Carson Collection.*

The whale on the left is 2-⅞ inches high, the one on the right 3-¼ inches. Paper label reads, "Kelvins / Treasures / Japan" on three lines. The dolphins are 3-½ inches high and are made of chalkware. A paper label indicates they were sold as a souvenir at Sea World. *Carson Collection.*

Height here is 4-¼ inches. Inkstamp reads "Japan." *Carson Collection.*

People fish. Or fish people, if you prefer. The shaker at far left measures 4-⅜ inches high, has a "Made in Japan" paper label. The shaker at far right is 2-⅞ inches high, has a "Japan" inkstamp. *Thornburg Collection.*

Fish sitters, the base of which is 3-¾ inches long. The inkstamp mark says "Japan," with a stylized three-leaf clover above. *Carson Collection.*

Goebel condiment set with a full bee mark and another inkstamp that says "Germany." The mustard is 3 inches high. *Thornburg Collection.*

Treasure Craft fish, a souvenir of Hawaii. The base is 5 inches long. *Courtesy of Joyce Hollen.*

Lobster musicians, probably not a true pair judging by the color but possibly made by the same company on two different runs. The one on the left is 3-¼ inches high, has a "Made in Japan" inkstamp. The inkstamp of the one on the right reads "Made in Occupied Japan." *Oravitz Collection.*

These nodders weren't measured, but the bases are the smallest of the three common sizes. Each is marked with "Pat TT" in raised letters. *Thornburg Collection.*

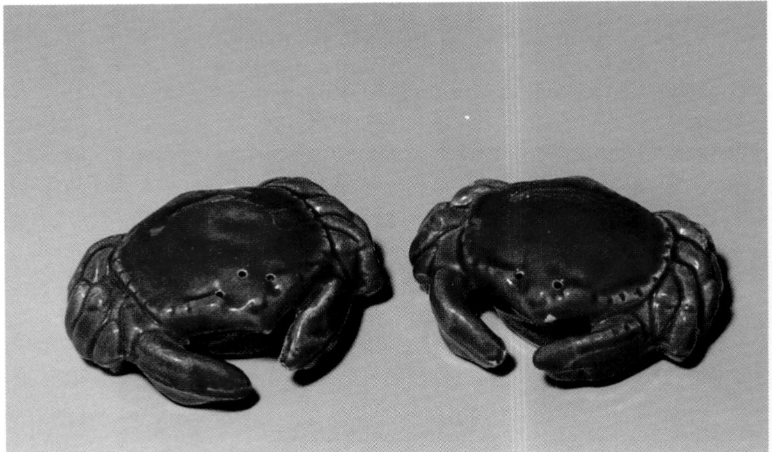

These crabs are 3-⅜ inches wide, are not marked. *Carson Collection.*

The fish is 2-¼ inches long, the creel 1-¾ inches high. Neither is marked. *Carson Collection.*

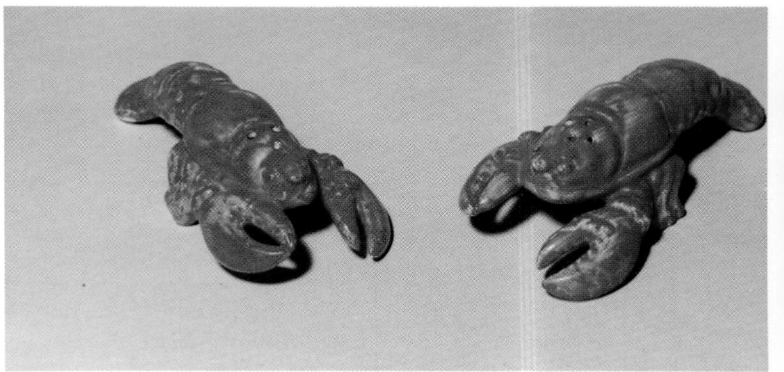

Lobsters, 3-⅞ inches long. Inkstamp reads "Japan." *Lilly Collection.*

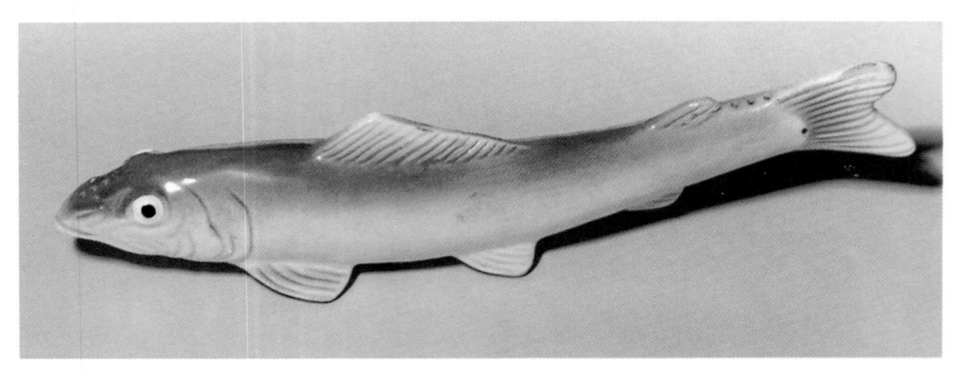

A very long one-piece fish shaker measuring 9-½ inches in length. It has a "Japan" paper label. *Thornburg Collection.*

The set on the left is 4-⅞ inches long, has a "Made in Japan" paper label. The set in the middle is 4-⅝ inches long, carries a "Japan" inkstamp. The set on the right measures 4-¾ inches in length, has a "Japan" paper label. *Carson Collection.*

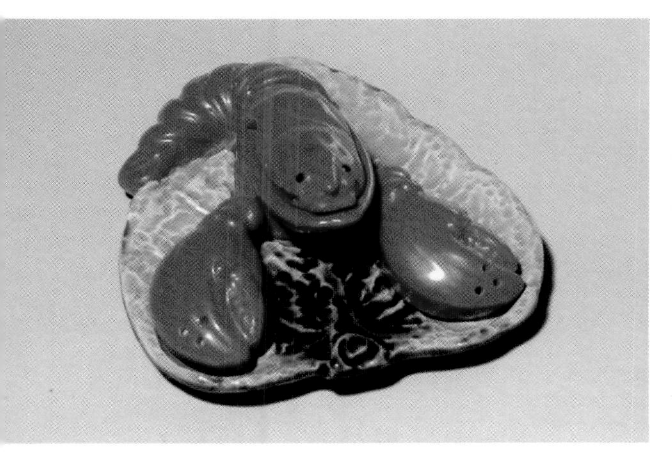

The claws of this condiment set are the shakers, the body is the mustard. The base is 5-½ inches long. All pieces have "Made in Japan" inkstamps. *Lilly Collection.*

Length on the left is 3-¾ inches. The pair is unmarked. Length on the right is 2-¼ inches, inkstamp under the glaze is "PY" in an ellipse. The middle pair is 2-⅝ inches high, has a "Japan" inkstamp. *Carson Collection.*

Doesn't the pair on the left look like real shells! They are 4 inches long, have a "Kenmar / Japan" paper label. The pair on the right is 4-¾ inches long, is unmarked. The pair in the middle has a "Japan" inkstamp. *Carson Collection.*

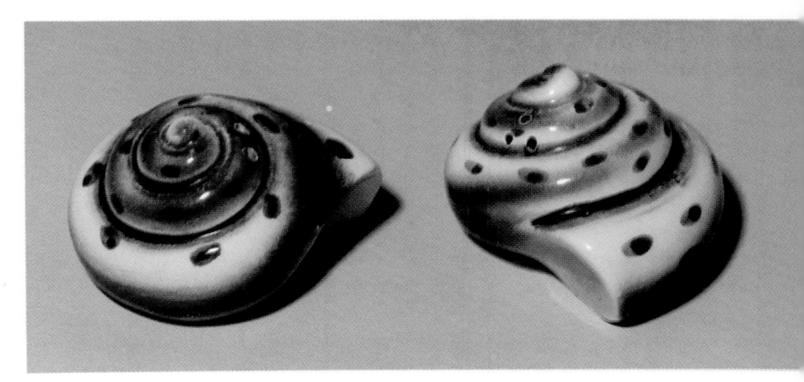

Snail shells, 2-⅝ inches across, unmarked. *Carson Collection.*

A hodgepodge of real shells and glitter glued to plastic shakers. Height of the shakers is 2-¼ inches. There are no marks save for a raised "1" (one) on the bottom of one, a raised "6" on the bottom of the other. *Carson Collection.*

These seahorses were made by Goebel, have a V and bee inkstamp in addition to impressed marks, "P141A" on one, "P141B" on the other. Height of the larger shaker is 3-½ inches. *Thornburg Collection.*

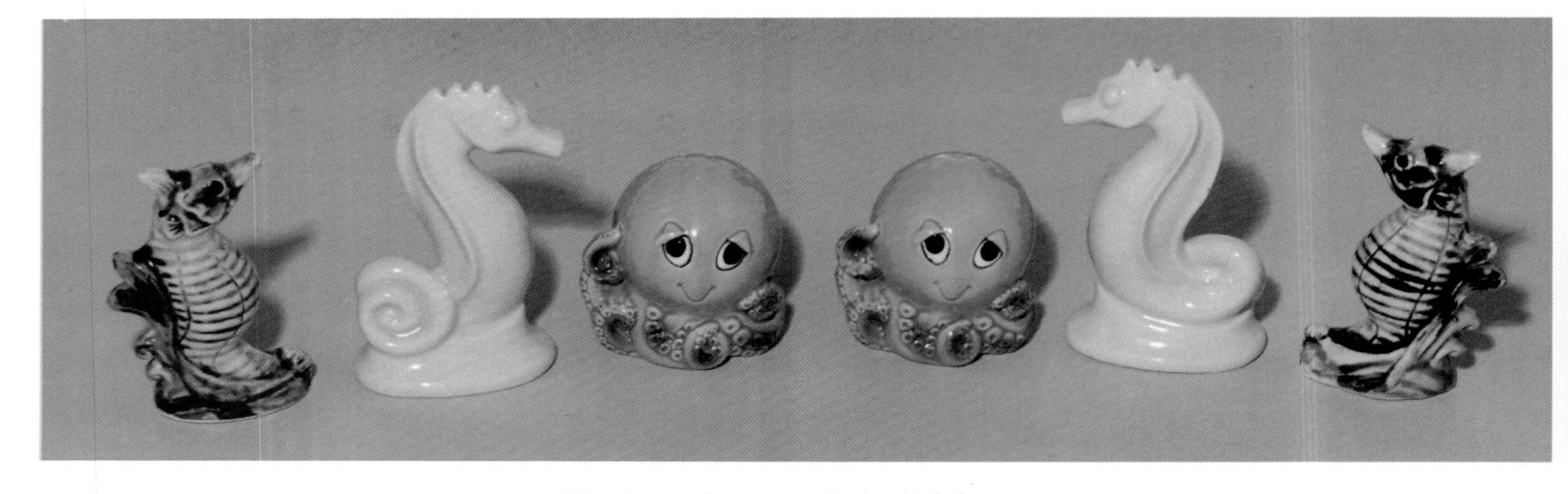

The white seahorses are 4 inches high, have an inkstamp that reads "Fine / China" on two lines. The darker seahorses are 3-⅜ inches high with a "Japan" inkstamp. The octopi are 2-½ inches high, have a "Japan" paper label and an "H986" inkstamp. *Carson Collection.*

These are both Treasure Craft. Mark on the squirrel and nut is "Treasure Craft / © Made in U.S.A." on two lines. Mark on the seahorses is "Treasure Craft / © U.S.A." on two lines. *Carson Collection.*

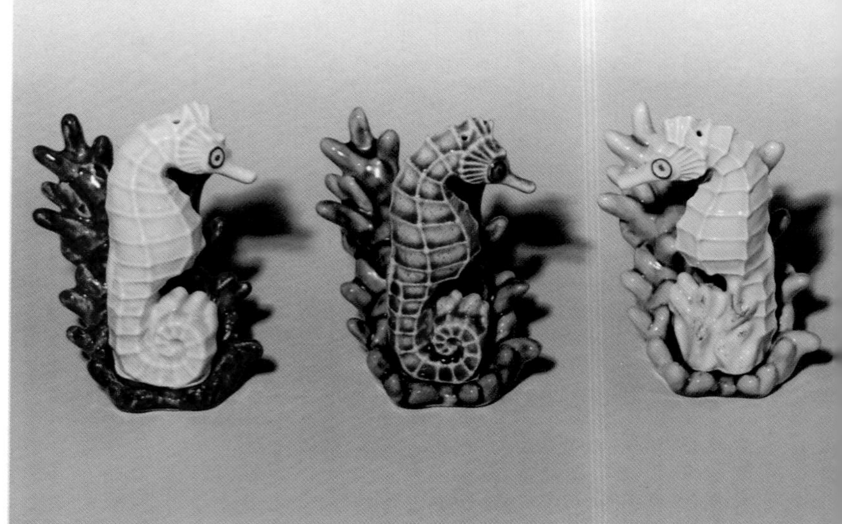

Each of the seahorses is 3-¾ inches high. Each base has a Ceramic Arts Studio mark. *Thornburg Collection.*

Chapter 7
Plant Life

Chapter 7 is divided into four parts—plants, fruits, fruits and vegetables, and vegetables. The plants section includes flower shakers.

The fruits and vegetables section is my own creation, a result of poor planning. Originally, I had intended to show all fruits and vegetables in one section, the section having a mixture of both. Then about one-third of the way through the photography phase of the book I decided each deserved it own section. Thus, the fruits and vegetables section is made up of those early pictures that included both. If you cannot find a pair of fruit shakers you are attempting to identify in the fruits section, try the fruits and vegetables section. Likewise for vegetable shakers.

Many of the shakers in this chapter are brightly colored. Considering the subject matter, they are just the thing to liven up the decor of any kitchen.

Plants

Trees, cacti and various flowers nearly come to life in the world of figural salt and pepper shakers. Like the fruit and vegetable shakers below, many flower shakers have charming happy little faces painted on them. I find them especially appealing at the breakfast table, where the first face you see in the morning wears a smile and helps set the tone for the rest of the day.

The pinecones on the outside are 1-⅛ inches high, unmarked. The pinecones sitting on the tray are 2 inches high, while the tray is 4-¼ inches long. All three pieces have "Japan" inkstamps. *Lilly Collection.*

The palm tree is 3-⅛ inches high, the coconut 1-½ inches. Neither piece is marked. *Lilly Collection.*

More pinecones, this time 2-⅜ inches high with "Made in Japan" paper labels. *Carson Collection.*

The palm trees on the outside are 3-½ inches high, made of redware, and unmarked. In the center is a Holt Howard set. The light post, 4-½ inches high, is marked with a "© HH" inkstamp. The trees are also seen in the same color as the light post, and in white. *Thornburg Collection.*

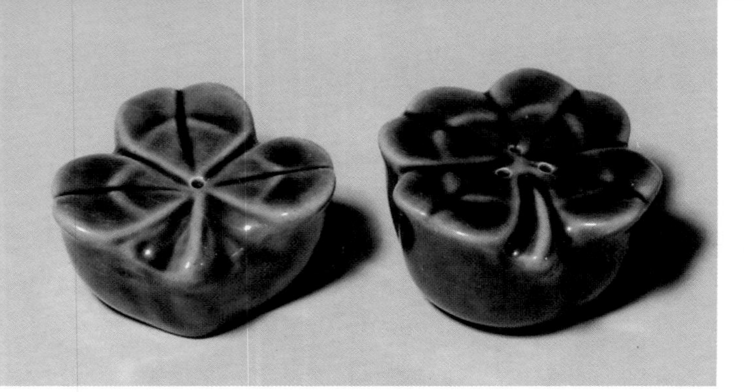

These cloverleafs are 1-⅛ inches high. They are unmarked. *Thornburg Collection.*

Cacti made of metal, 2-½ inches high and unmarked. *Carson Collection.*

The cactus on the left is 2-¾ inches high. Both shakers have Goebel marks. *Thornburg Collection.*

The shaker at far left stands 3-¾ inches high. It has "101" impressed. The pair in the middle have "Japan" inkstamps. The pair on the right has a "Japan" paper label. *Thornburg Collection.*

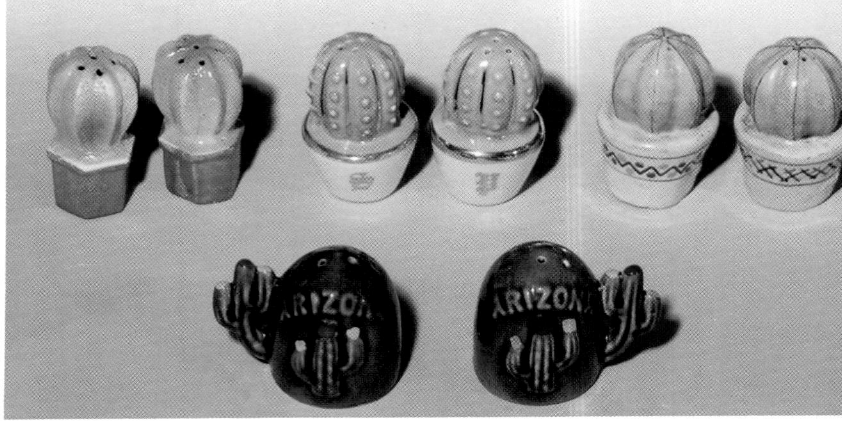

Height of the shakers with the red flower pots is 2-½ inches. They have a "Japan" inkstamp. All others in the picture are unmarked. *Thornburg Collection.*

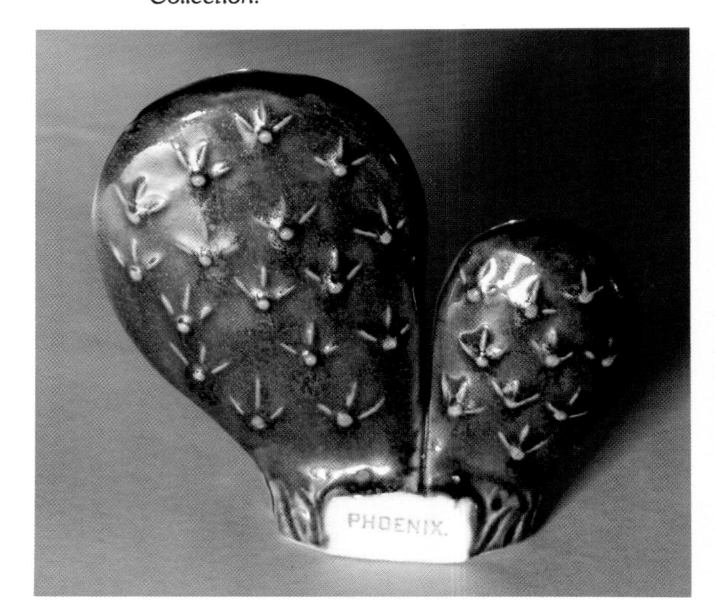

This one-piece shaker is 3-¼ inches high. It is not marked. *Thornburg Collection.*

Height here is 3-⅞ inches. Inkstamp reads "Japan." *Carson Collection.*

The flower pot is the salt, the cactus the pepper. Height together is 3-½ inches. There is no mark. *Thornburg Collection.*

A condiment set, the tray of which is 7 inches long. An inkstamp mark indicates the set originated in Japan. *Carson Collection.*

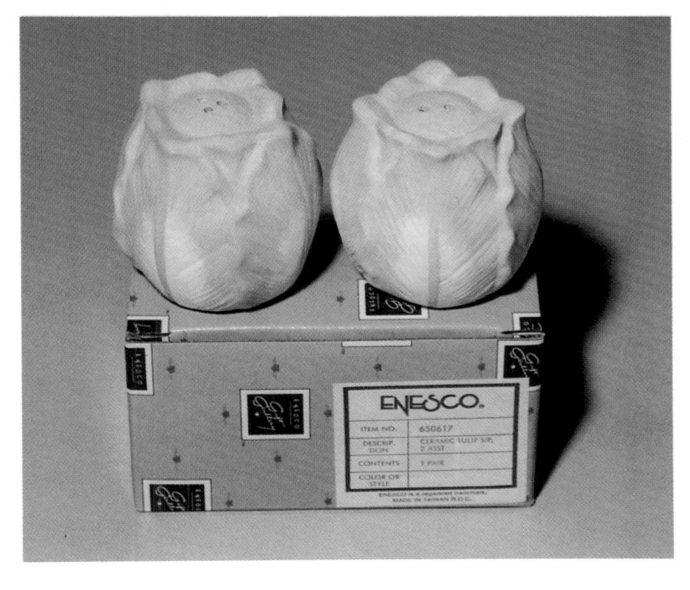

Enesco tulips with a matte glaze, 2-¾ inches high, and a paper label that reads "Made in Taiwan." *Lilly Collection.*

Tulips with bees, 2-¼ inches high. Both pieces have "Made in Japan" inkstamps. *Lilly Collection.*

The basket of the set on the left is 4-¼ inches high. The mark is "Relco / Creations / Japan" on three lines. The flowers of the set on the right are each 2-¼ inches high. They are unmarked but the base is marked "Occupied Japan." *Courtesy of Joyce Hollen.*

The flowers are each 1-⅞ inches high. They lift out of the leaves. The base is 3-½ inches high, has a "Japan" inkstamp. *Lilly Collection.*

Plastic flowers, each 2 inches across, on a plastic base. Mark on the base is "Davis Products / Brooklyn NY / Made in USA" on three lines. *Courtesy of Joyce Hollen.*

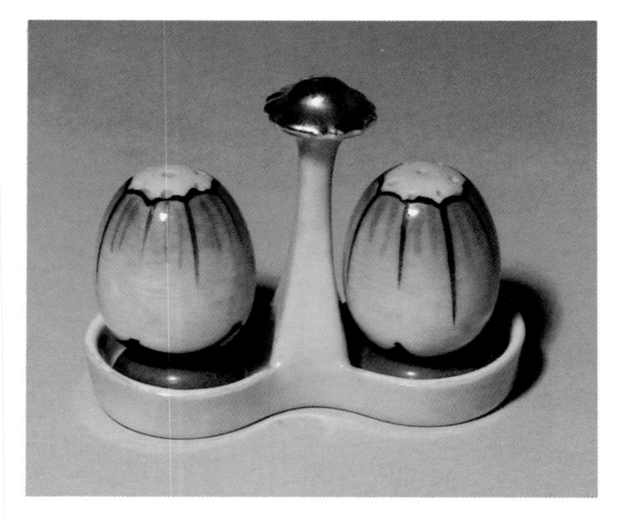

The shakers are 1-⅞ inches high, the tray 3-¼ inches long. Partially legible inkstamp reads "Chik........Made in Japan." This is probably the mark of Chikaramachi. *Lilly Collection.*

The roses measure 1-½ x 3-½ inches. They have an Elbee Art paper label. *Oravitz Collection.*

The flower face on the left is 2-½ inches high, has a "Japan" inkstamp. *Oravitz Collection.*

Roses with faces. Height is 3-⅜ inches, mark is "PY / Japan" on two lines; the PY is encircled by an ellipse. *Carson Collection.*

Shawnee flower pots, 3-¾ inches high. They are unmarked. *Smith Collection.*

The flower faces are 3 inches high and unmarked. The flowers with the bees are 2-⅞ inches high, have a "Japan" inkstamp. *Carson Collection.*

Shawnee stove set, 5-⅛ inches high, unmarked. Like other Shawnee shakers, both this set and the set directly above can be identified by their large fill holes. *Smith Collection.*

A Wade condiment set in the company's Bramble pattern. The shakers stand 2-¼ inches high. Neither the shakers nor the mustard are marked. Mark of the tray is shown below. *Thornburg Collection.*

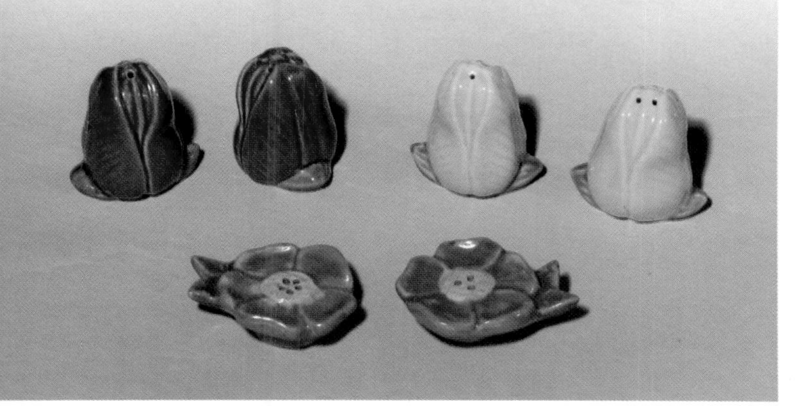

Three pair of Rosemeade shakers, 2 inches high in the rear, 2-⅜ inches across in the front. None of them are marked. *Thornburg Collection.*

Mark of the above Wade Bramble tray.

Blue Ridge shakers, made by Southern Potteries. Height is 5-⅛ inches. They are not marked. *Carson Collection.*

Fruits

I didn't look up any records, but just by living, talking and shopping for 40-odd years I would guess that strawberries are our most popular garden fruit, apples our most popular orchard fruit, and that bananas are the most popular overall. But in the exciting subculture of salt and pepper shaker collecting, one seems to be no more popular than another. At every antique show you will see grapes, lemons, limes, pears, peaches and others right alongside the top three.

Holt Howard Bananas with faces, 2-⅜ inches high. Inkstamp mark reads "© HH," paper label reads "Made in Japan." *Carson Collection.*

Both these sets are chalkware. The bananas are 3-⅞ inches long. The others, perhaps strawberries, are 1-⅝ inches high. As with most chalkware, neither set is marked.

Here the bananas are 2-½ inches high, and are unmarked. The shaker at far left is 4-¼ inches high, has a "© Napco" inkstamp plus a Napco paper label. *Oravitz Collection.*

More pineapple people, 3 inches high and unmarked. Another one that was photographed, however, had an Enesco paper label which is shown below. The plain pineapples on the left are 2-¾ inches high with a "Japan" paper label. *Carson Collection.*

The pair on the left stands 3-¼ inches high, the pair on the right 2-¾ inches. Both have "Japan" inkstamps. *Carson Collection.*

Enesco paper label on pineapple person exactly like those above.

The pineapple heads are 3-⅛ inches high, are not marked. The orange people are 3-¾ inches high, have a "Japan" inkstamp. *Carson Collection.*

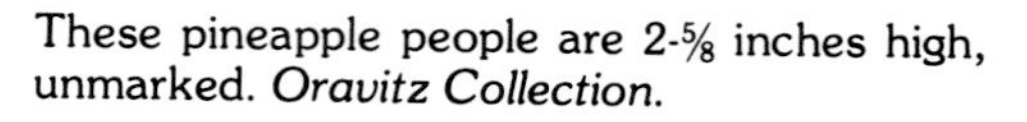

These pineapple people are 2-⅝ inches high, unmarked. *Oravitz Collection.*

Unmarked plastic oranges, 2-⅜ inches high. *Carson Collection.*

A plastic orange tree, 5-½ inches high. The baskets are the shakers, and the tree itself acts as a napkin holder, as shown below. Only mark is a raised "Z." *Courtesy of Allen and Michelle Naylor.*

Above orange tree from the side, showing how it is also designed to hold napkins.

Hanging oranges. The base measures 5-⅞ inches high, has a "Made in Taiwan" paper label. Carson Collection.

The center set is plastic, the oranges measuring 2-¼ inches high. The mark is "Hong Kong." The pair on the outside is made of chalkware. Height is 1-½ inches. *Courtesy of Joyce Hollen.*

Oranges with faces, 2-⅝ inches high. They have "Japan" inkstamps. *Oravitz Collection.*

A salt and pepper tree bearing two oranges, two lemons and two strawberries. The tree is 9 inches high, has a "Japan" paper label. *Carson Collection.*

Let's start with the oranges in the middle, obviously Artmark, and 2-⅜ inches high. Mark information for the shakers themselves was not recorded. The oranges at right, and the lemons, are both Empress, a picture of the paper label on them appearing below. The Empress lemons are 3-⅛ inches high, the Empress oranges 2-⅞ inches. *Lilly Collection.*

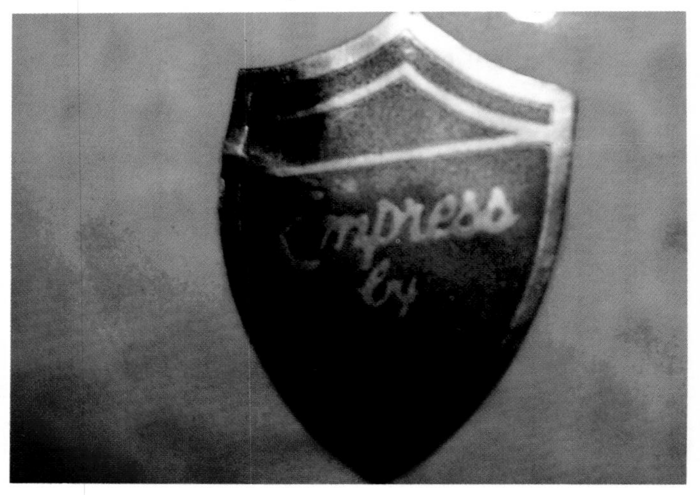

Paper label of the above oranges on the right. The same label is on the lemons.

All chalkware. The lemon at far left is 2-½ inches high, the lemon at far right 1-⅜ inches. *Courtesy of Joyce Hollen.*

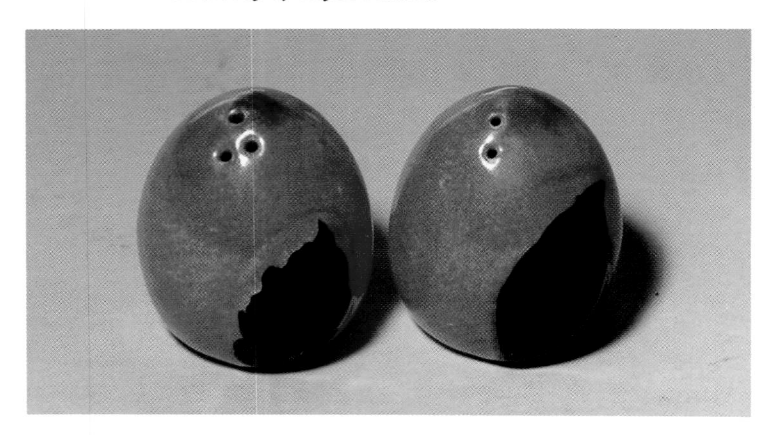

Look like kumquats to me. They are 2 inches high, and unmarked. *Carson Collection.*

Wooden apples, 1-½ inches high not including the stems. They are unmarked except for the black printing around the bottoms which says "Yellowstone Nat'l Park." *Courtesy of Stan and Karen Zera.*

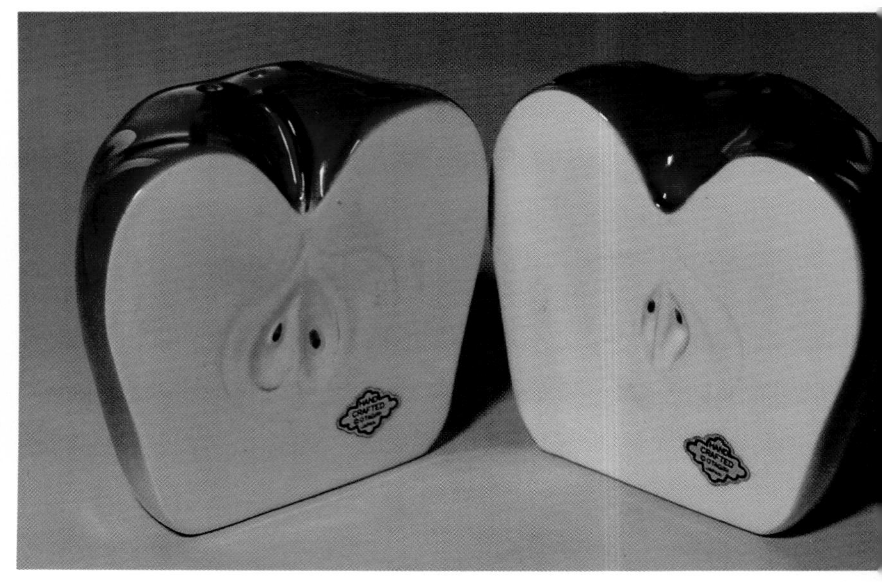

These shakers are 3-¼ inches high. They can be displayed like this, or with the halves together to form a whole apple. The paper label reads "Hand / Crafted / © Otagiri / Japan" on four lines. Additionally, there is an impressed mark, "© Otagiri / 1984 / Japan" on three lines. *Carson Collection.*

Nectarines on the left, apples on the right. The nectarines are 3 inches high, have a "429" inkstamp. The apples are 2-⅛ inches high, have a "665" inkstamp in addition to a "Japan" inkstamp. *Courtesy of Jazz'e Junque Shop, Chicago, Illinois.*

Height of these unmarked apples is 2 inches. They are accessories for Gladding, McBean & Company's Franciscan apple pattern china. *Carson Collection.*

The hanging apple is 6-½ inches high, the shakers are 2-⅜ inches. There is no mark on any of the three pieces. *Courtesy of Stan and Karen Zera.*

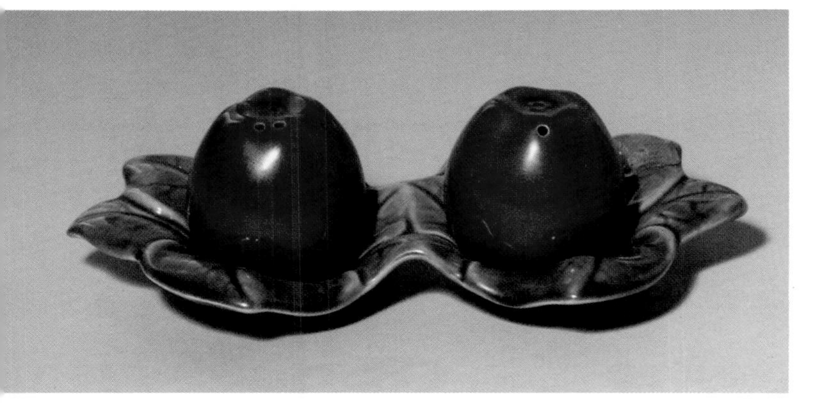

The tray measures 6-¾ inches in length. It is marked "Arcadia / Ceramics / Arcadia / California." The shakers were not measured. *Carson Collection.*

School books and an apple for the teacher. The height of the books in this Sandy Srp miniature set is ⅝ inches. *Carson Collection.*

These unmarked apples are 3-¾ inches high. A matching cookie jar was also made. *Courtesy of Mark Taylor.*

Peaches with faces, all pieces bearing "Japan" ink stamps. The pair on the left is 2-⅝ inches high, the pair on the right 2 inches. *Oravitz Collection.*

Strawberries and apples, apparently by the same company. Both baskets are 4-½ inches long, both are marked "Occupied Japan." All four shakers are marked "Japan."

The pears in the rear are 2-¾ inches high, the ones in the front 2-½ inches high. Although different sizes, they are obviously the same set, the discrepancy occurring because the stems of the front set were cut shorter. The apples are 2-½ inches high. The set in the rear has a "Japan" paper label, the others have "Japan" ink stamps. *Carson Collection.*

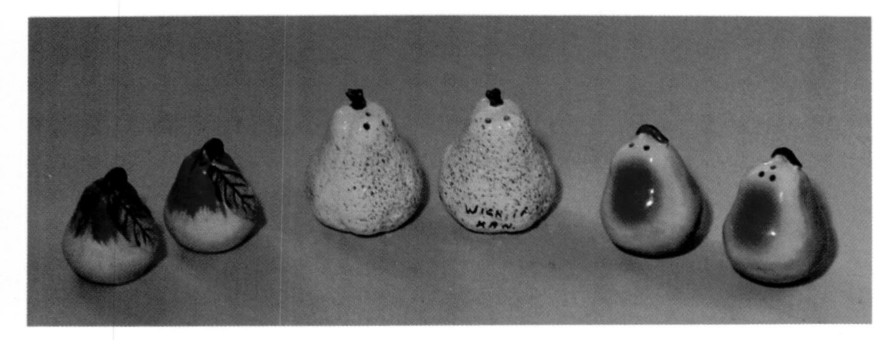

Three pair of pears. Heights left to right are 2-¼, 3-⅛ and 2-½ inches. Mark information in the same order is unmarked, "Japan" ink-stamp and "Japan" paper label. *Carson Collection.*

The pear is 3-¼ inches high, the worm 2-½ inches long. Neither piece is marked. You also see the same worm crawling out from under an apple. *Lilly Collection.*

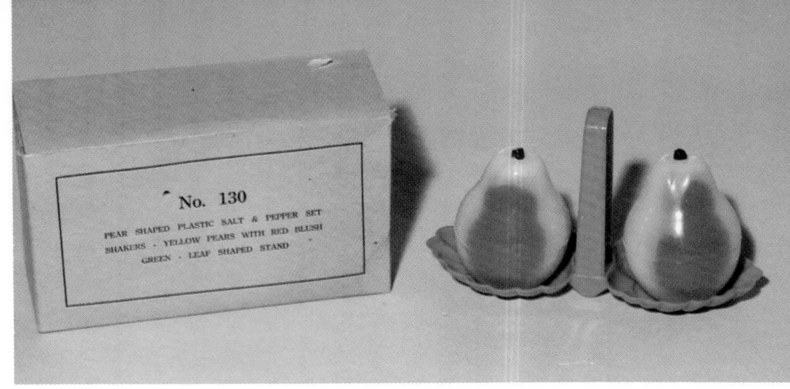

Unmarked plastic set with original box. The pears are 2-¾ inches high, the tray 6-¼ inches long. Printing on the box is shown below. *Lilly Collection.*

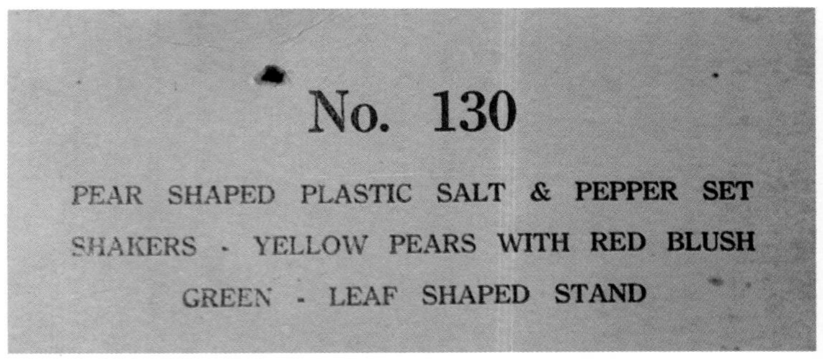

No. 130
PEAR SHAPED PLASTIC SALT & PEPPER SET
SHAKERS - YELLOW PEARS WITH RED BLUSH
GREEN - LEAF SHAPED STAND

Printing on the box that held the above pears. It almost makes you think they were government issue.

The tray for the apples is 5-½ inches long. It has a "Japan" paper label. The peaches and pears have Enesco paper labels. *Carson Collection.*

An unmarked plastic advertising set issued by Stokley's. The strawberries are each 1-¾ inches high. *Carson Collection.*

Pretty much miniatures, only 1-½ inches high. Mark is "Arcadia / Ceramics / Arcadia / California" on four lines. *Carson Collection.*

A strawberry set that, like the apple above, may be displayed open or closed. Height is 2-¼ inches. Neither piece is marked, but assumed to be Otagiri. *Carson Collection.*

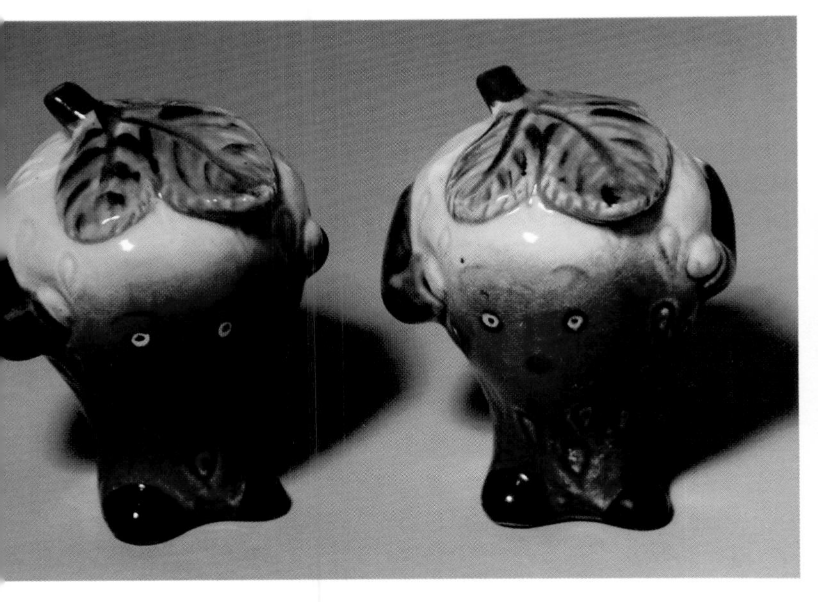

Strawberries with faces. They are 3-¼ inches high, have a "Japan" paper label. *Carson Collection.*

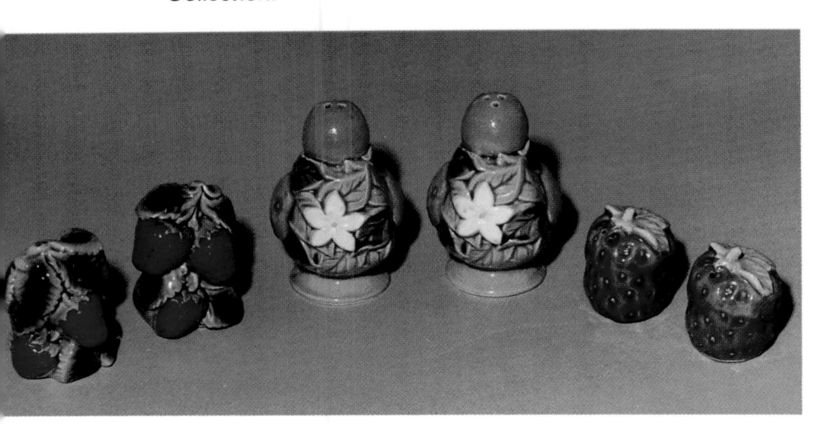

The strawberries on the left in the middle row are the same as those in the above picture. Unlike those above, however, they sport a "Made in Japan" ink-stamp. The set next to them, plus the two sets in the front row, have "Japan" ink stamps. The bowl of strawberries in the rear is 2 inches high with a "Japan" paper label, while the creamer is 3-⅝ inches high and unmarked. *Carson Collection.*

The pears on the left are 2-¼ inches high, unmarked, and made of chalkware. The pears on the right are the same size, made of ceramic, and carry an "Occupied Japan" inkstamp. The strawberries are 3 inches long, have an "E3183" inkstamp. *Carson Collection.*

Here the unmarked bananas are 4 inches long, while the strawberries are 3 inches high and marked with a "Japan" ink-stamp. *Carson Collection.*

The Holt Howard strawberries on the left stand 3 inches high, are marked "© HH." The strawberries on the right measure 2-⅜ inches in height, are unmarked. The oranges are 4-⅜ inches high, have a "Made in Japan" paper label. *Courtesy of Joyce Hollen.*

The slice of watermelon in the center is actually two pieces. The pair of shakers on the left can be put together the same way. Sizes left to right are 2-⅞ inches high, 3-¾ inches long, and 2-⅝ inches long. The middle pair is not marked. The others are marked "Japan." *Carson Collection.*

Height is 1-¾ inches. One has a "Japan" paper label. *Carson Collection.*

Watermelon people, 2-⅛ inches high, "Japan" ink-stamp. *Oravitz Collection.*

Plastic stand and tops, glass grapes. The stand is 4-¾ inches high, marked "Made in Hong Kong" in raised lettering underneath. *Courtesy of Cloud's Antique Mall, Castalia, Ohio.*

These bowls of fruit are 1-½ inches high. Paper label reads "Coventry / Fine / Porcelain" on three lines. *Carson Collection.*

The fruit stacks are 4-¾ inches high, have a paper label that reads "Quality Product of Japan." The grapes are 2-¾ inches high with a "Japan" inkstamp. The apples, which also have a "Japan" inkstamp, are 1-⅝ inches high. *Carson Collection.*

The watermelons and pineapple people are repeats of shakers shown above. The fruit bowl is made of plastic, the two top pieces lifting off as the shakers. The fruit forms the lid of the sugar bowl. Diameter of the bowl is 4-¾ inches. Mark inside the lid reads "Pat. Pend. ©1958 / M. Gaby" on two lines. Mark on the bottom of the bowl reads "Stark Design, Inc. / Bklyn NY / Pat. Pend. / Made in USA" on four lines. *Lilly Collection.*

Salt, pepper, and apparently whatever you want to put in the tiny cup. The tray is 6-¼ inches long, which makes the cup less than 2 inches across, certainly too small for a sugar. None of the pieces are marked. *Carson Collection.*

These cornucopias are 3-¾ inches high. They are not marked. *Lilly Collection.*

A one-piece, 2-⅝ inches high. The shaker is unmarked. *Oravitz Collection.*

Turned wooden nuts, souvenirs of the Block House in Pittsburgh, Pennsylvania. The shakers are 2-⅝ inches high. *Carson Collection.*

Fruits and Vegetables

Here is my mistake we talked about earlier. Actually, it did not create as big a problem as I thought it would since this can be considered an extension of the fruit section and a prelude to the vegetable section.

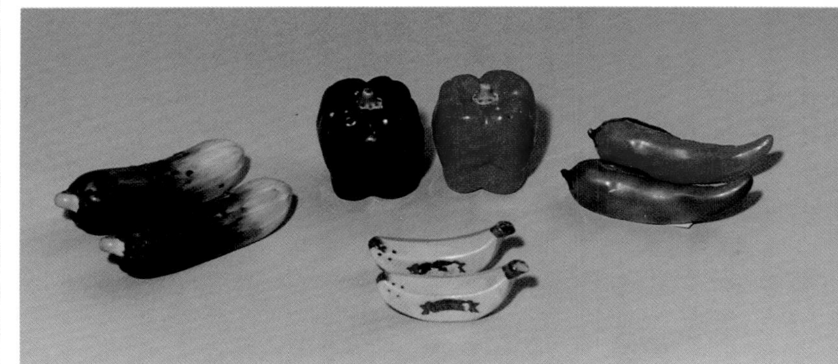

The pair of sweet peppers in the center are 2-¼ inches high. The other three examples are one-piece sets. The cucumbers measure 4-¼ inches in length, the hot peppers 4 inches, and the bananas 2-¾ inches. All of the shakers in this picture have "Japan" ink stamps. *Courtesy of Bev Urry.*

The mushrooms are 2-¾ inches high, unmarked. The pears are plastic, 3 inches high, and also unmarked. *Carson Collection.*

The acorns on the left are made of wood composition. They are 2-¼ inches high, unmarked. The fruit stacks are 3-¾ inches high. An inkstamp says "Japan / E21715" on two lines, while the paper label proclaims "Cool Claret / by Enesco" on two lines. The height of the tomato people is 2-⅛ inches. Each has a "Japan" inkstamp. *Carson Collection.*

The unmarked pineapples are 3-½ inches high. The banners across their fronts say "Virgin Islands." The taller gourd is the same height, also unmarked. The cherry peppers stand 2 inches high, have a "Japan" inkstamp. The grapes are 4 inches long. They are not marked. *Courtesy of Joyce Hollen and Bev Urry.*

The tomatoes are 2-⅝ inches high. The whole watermelon is 2-⅝ inches long, while the slice is 3-¼ inches long. Note that they are a true pair, the rinds and the eyes (barely visible on the whole melon) being of the same colors and style. All four shakers are marked "Japan." *Courtesy of Cloud's Antique Mall, Castalia, Ohio.*

With shakers like these you tend to wonder whether they are true pairs, or singles from a larger series that eventually found their way to each other. Height of the corn is 4-¼ inches. The mark is "Grantcrest / Japan" on two lines. *Carson Collection.*

Apples, peas, and what appears to be lettuce. The left member of the apple pair is 1-⅞ inches high, sports a "Japan" inkstamp. The peas are 4 inches high, have a "Japan" paper label. The lettuce shakers are 2-¼ inches high, have a "Japan" inkstamp. *Carson Collection.*

Four members of a larger series. The peach (or apple) is 4 inches high. All have Napco paper labels. Davern, in Book I, pictures a beet, orange, lemon, carrot, onion and pickle in the same series. *Carson Collection.*

Heights left to right in the rear row are 2-⅞ inches (unmarked), 3 inches ("Japan" inkstamp), and 3-⅛ inches ("Japan" inkstamp). Front row left to right measures 2-⅜ inches ("Made in Japan" paper label), 2-¾ inches ("Japan" inkstamp), and a 1-⅞ inches (unreadable paper label). Note that the pair on the right in the front row is bisque. *Carson Collection.*

Another obvious series. The apples on the left are 2-½ inches high. At least one member of each pair has a "Japan" paper label. *Carson Collection.*

A one-piece set, but what? Maybe a pear and an apple. In any event, it is 3-¼ inches long with a "Japan" inkstamp. *Carson Collection.*

Wouldn't this be great in the center of the patio table during your next cookout. The shakers in this set are 2-⅜ inches high, and are not marked. The mustards are 3-¼ inches high, have "Made in Japan" paper labels. *Thornburg Collection.*

Vegetables

Vegetable shakers are great not only for collecting but also for the table if you coordinate them with the vegetables you serve. When corn is on the menu, make corn shakers part of your table setting. When eggplant is being served, use eggplant shakers. If you look long enough and hard enough, you will find shakers to go along with just about every vegetable in your diet.

Five one-piece sets. The cabbage heads are 2-⅛ inches high, the carrots 4-¾ inches long. All have "Japan" ink stamps. *Thornburg Collection.*

Possibly a pair, perhaps not, but certainly worth showing. The onions are 4-⅛ inches high, carry a "Made in Japan" paper label. The unmarked artichoke is 2-½ inches high. *Carson Collection.*

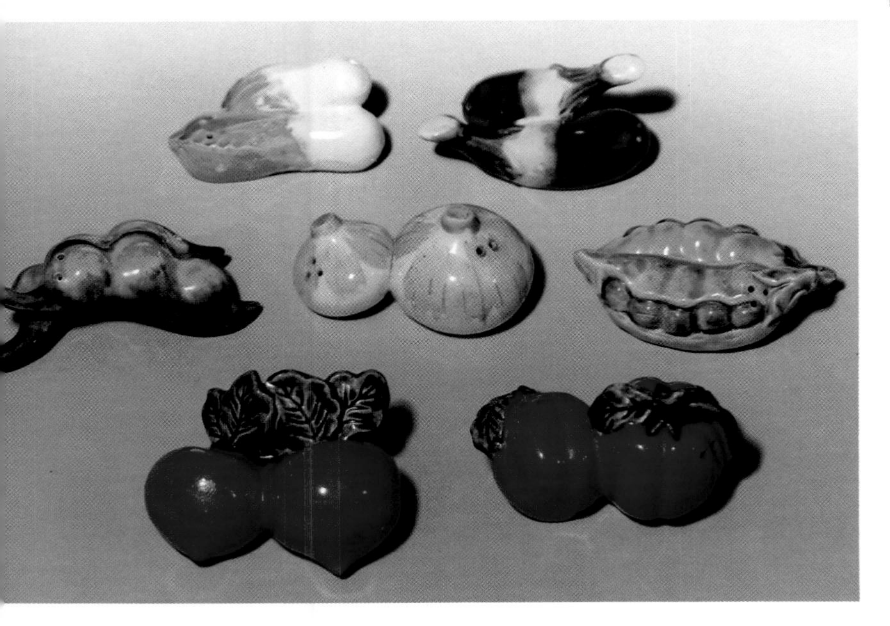

More one-piece vegetables. One set has its original $0.25 price sticker! The peas on the right measure 3-¾ inches in length, the beets at left front are 2 inches high. Mark information was not recorded. *Thornburg Collection.*

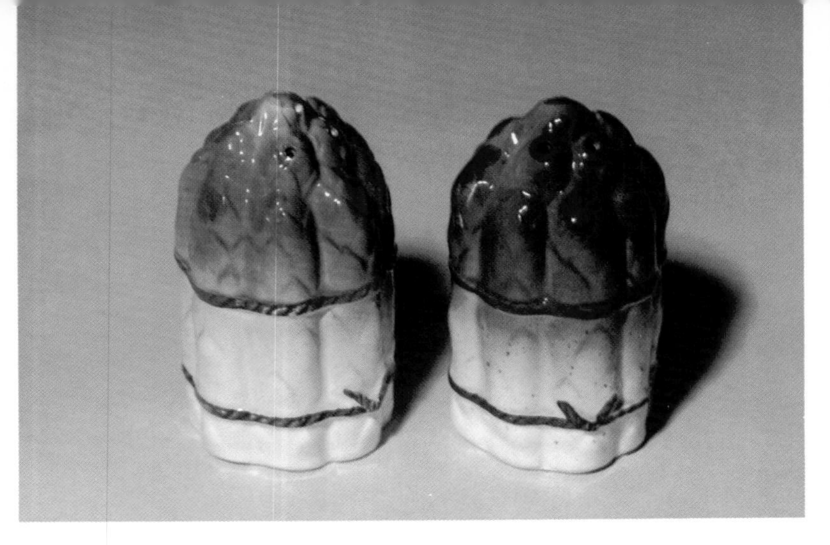

Asparagus spears, 3-¼ inches high, unmarked. *Lilly Collection.*

Marks of above McCoy shakers.

Beans anyone? The plastic shakers are 3 x 2-½ inches. The beans are real. Marks are "Made in Hong Kong," and "©Victor Goldman, Inc." *Courtesy of Allen and Michelle Naylor.*

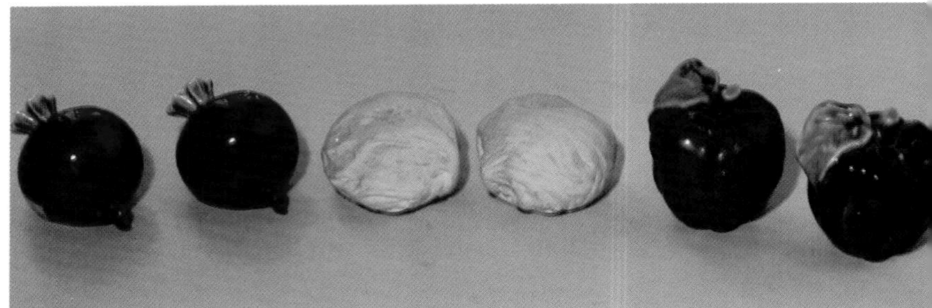

Beets, cabbage and peppers. The beets are 2-⅛ inches high, have a Holt Howard paper label. The cabbage, 2-¾ inches high, carries an Enesco mark. The peppers have a "Japan" ink stamp. They are 2-½ inches high. *Carson Collection.*

These shakers were made by McCoy. Their height is 4-½ inches. Their mark is shown below. *Lilly Collection.*

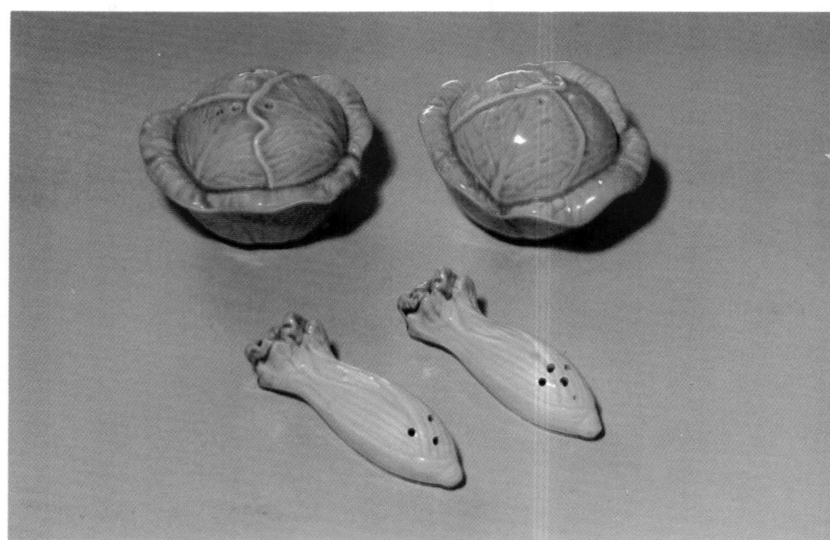

The cabbage is 2-¼ inches high, unmarked. The celery is 3-⅛ inches long, has a "Japan" ink stamp. *Carson Collection.*

The shaker in front measures 5-¼ inches long, the one in back 5-⅛ inches. Paper label reads "UCAGCO / Ceramics / Japan" on three lines. *Carson Collection.*

The rabbit-and-carrot shakers are 4-½ inches long, have "Made in Japan" paper labels. The other two sets are marked "Japan." *Carson Collection.*

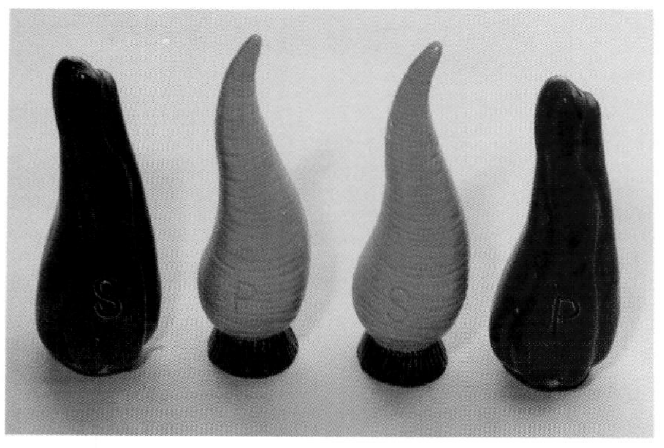

These shakers appear to be the work of a hobby ceramist, and are an excellent example of why not to exclude hobby ceramics from your collection. They are very neatly finished, and the glazes are outstanding. Height of the carrots is 7-¼ inches, marks are shown below. *Courtesy of Allen and Michelle Naylor.*

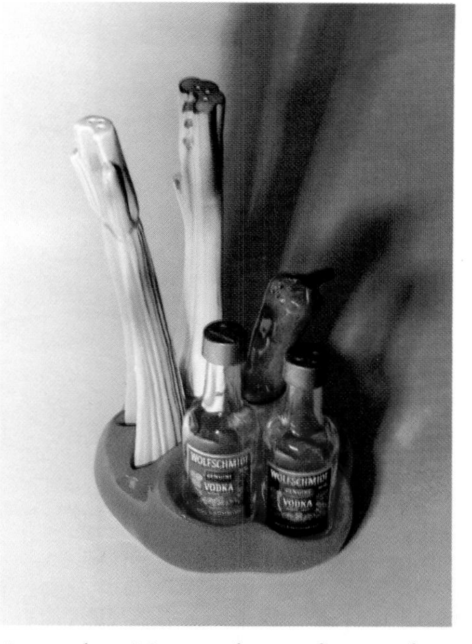

Here is an advertising set that ought to make any shaker collector envious. The celery stalks are 8-½ inches high. Impressed mark on the bottom of the base is "Wolfschmidt / Vodka" on two lines. It also has two paper labels. One reads "Product of U.S.A. Distilled from grain 80 and 100 proof Wolfschmidt Baltimore, MD." The other reads "Made in Japan." *Thornburg Collection.*

These carrots are 4-⅝ inches high. "Japan" is incised. *Carson Collection.*

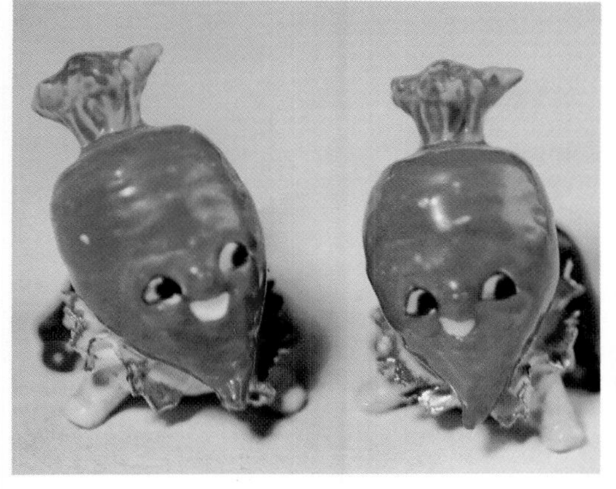

Height of these carrots is 2-½ inches. They have a Victoria paper label. *Oravitz Collection.*

Each of the above four shakers have this same "LC" scratched in the bottom.

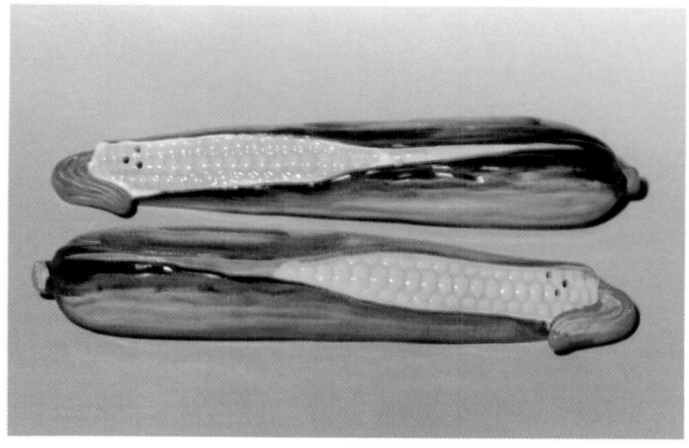

These ears of corn are long, 8-½ inches to be exact. They are not marked. *Thornburg Collection.*

This hanging set has an "Artmark / Japan" mark. Size was not recorded. *Carson Collection.*

Again, is it a pair or not? A better question, at least for people who want to use them at the table, might be who cares? They go well together. The carrot is 3-½ inches high. Neither piece is marked. *Carson Collection.*

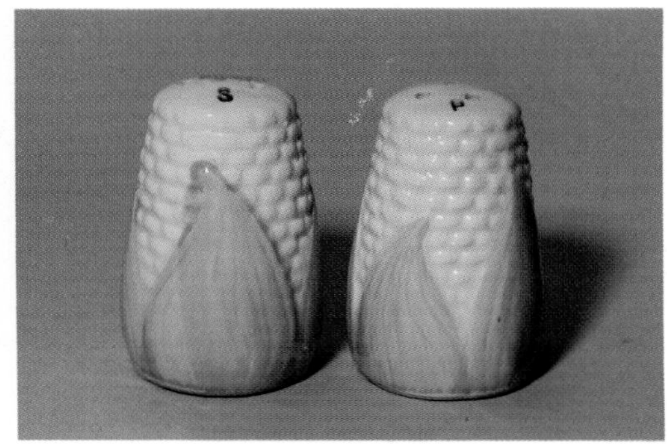

Unmarked corn, 3 inches high. *Courtesy of Cloud's Antique Mall, Castalia, Ohio.*

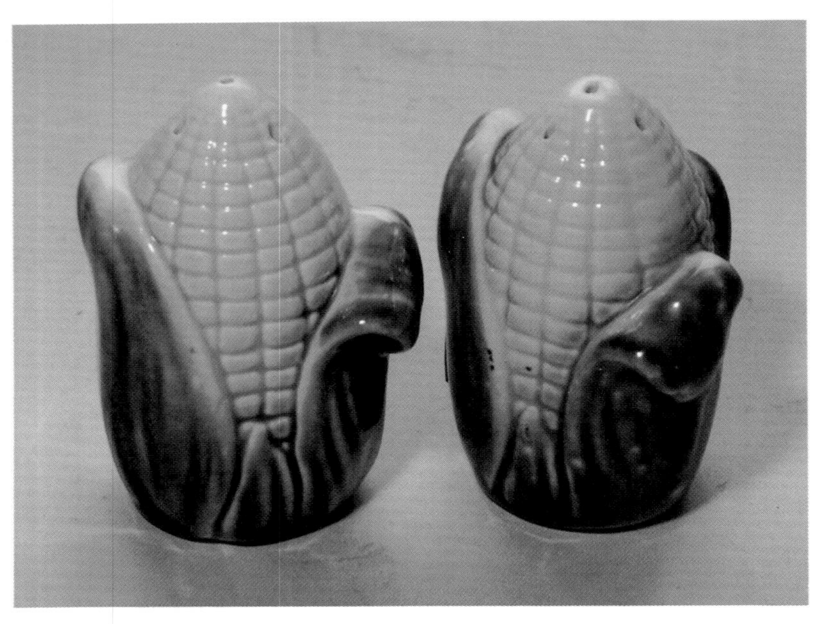

Stanfordware corn shakers, 4-¼ inches high, unmarked. The company also made a matching cookie jar, creamer and sugar, and other pieces. *Courtesy of Jazz'e Junque Shop, Chicago, Illinois.*

Goebel corn shakers with a full bee mark. They are 3-¼ inches high. *Carson Collection.*

More corn shakers, 6-¼ inches high, with a "Japan" inkstamp. *Carson Collection.*

The gold trim around the bottom of these shakers didn't show up very well in the photograph. They are 3 inches high, have a "Japan" inkstamp. *Oravitz Collection.*

Three pair of Artmark shakers, the boxes for which are shown below. The shakers were purchased at the Carolina Pottery a couple years ago for $1.98 per pair. The corn shakers are 3 inches high. *Lilly Collection.*

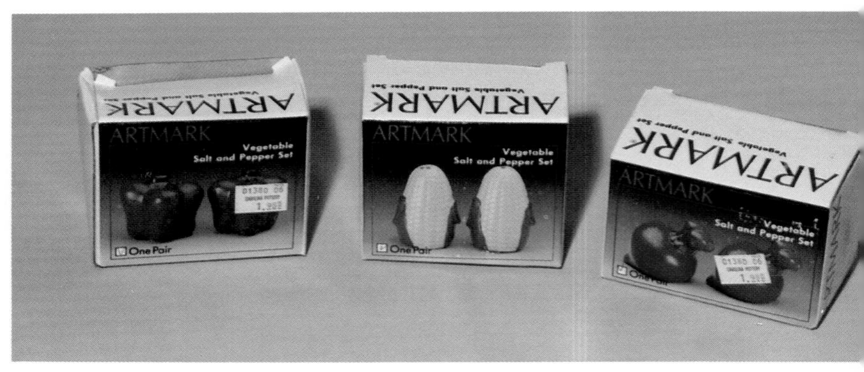

Boxes for the above Artmark shakers. The reason they are shown is that if you ever purchase Artmark shakers in this type of box, you have to be careful to make sure you get what you want. The company uses the same box for different shakers. On the back of the peppers box is a picture of carrots, on the back of the corn box a picture of eggplant, and on the back of the beets box a picture of tomatoes. Each package has a small box at lower left that will be checked on the side showing the shakers that are enclosed.

Four sets of Rosemeade vegetables. The corn is 2-⅛ inches high. None of the shakers are marked. *Thornburg Collection.*

The cucumbers with faces on the left are 2-¾ inches long. Mark information was not recorded. The peppers are 4 inches long, have a "Japan" ink stamp. The cucumbers on the right are a repeat. Length is 4-¼ inches, with a "Japan" ink stamp. *Carson Collection.*

The peppers are 3-½ inches high, have a "Japan" ink stamp. The corn set is marked "Duncan," the name of a mold company that caters to home ceramists but also sells molds to commercial potteries. The shakers in this set are 4-¼ inches high. *Courtesy of Joyce Hollen.*

Cucumbers, 3-⅛ inches long, marked "Made in / Occupied / Japan" on three lines. *Courtesy of Cloud's Antique Mall, Castalia, Ohio.*

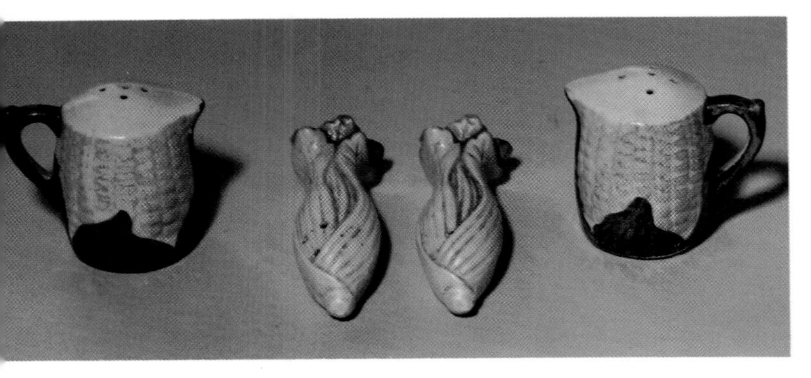

Here the celery is 3-⅛ inches long, has a "Japan" ink stamp. The corn pitchers stand 2-⅛ inches high, are unmarked. *Carson Collection.*

These shakers are very similar to the corn above, apparently by the same company. Height is 3 inches. They are unmarked. *Oravitz Collection.*

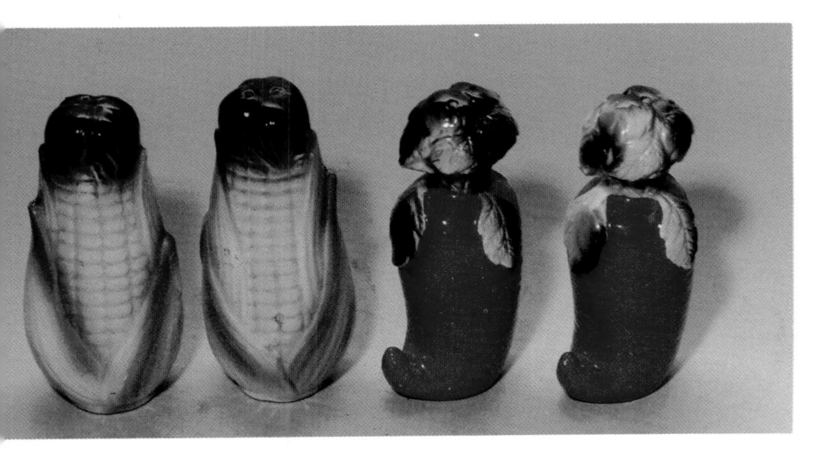

Corn, 5-⅝ inches high with a "Japan" ink stamp; carrots 5-½ inches high with "Japan" impressed. *Courtesy of Jazz'e Junque Shop, Chicago, Illinois.*

Another set of cucumbers/pickles with faces, this time by Arcadia Ceramics. Height is two inches. The pepper has only the eyes for pour holes while the salt has an additional hole in the top. The paper label is shown below. *Private Collection.*

Paper label on the bottom of cucumber/pickle shakers above.

Eggplant with faces and bodies, 2-⅝ inches high, unmarked. *Oravitz Collection.*

The beets are 3-⅛ inches high, have a "Japan" paper label. The squash stand 1-½ inches, are not marked. The eggplant and tomatoes are 5 inches long and 1-⅞ inches high, respectively, and are not marked. *Carson Collection.*

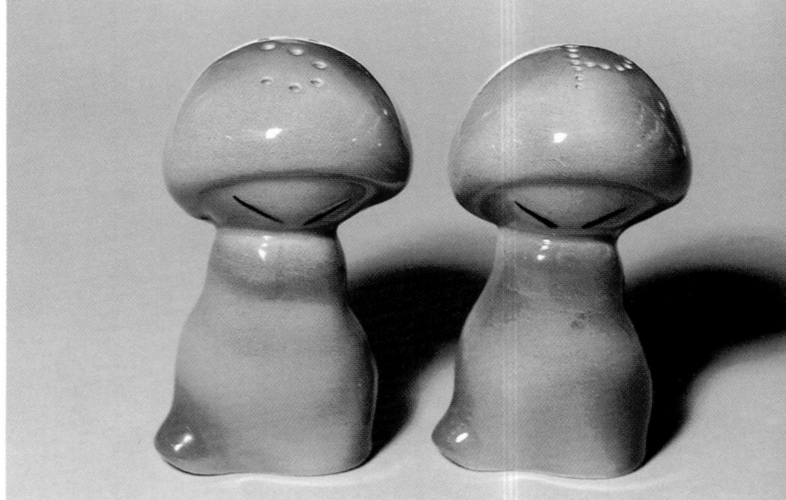

Here are Hop and Lo from the Walt Disney classic *Fantasia.* They are 3-¼ inches high. Their Vernon Kilns mark is shown below. *Private Collection.*

Inkstamp mark of Hop and Lo.

The mushrooms with faces are 2-½ inches high. They have Napco paper labels. Mark information of the other set was not recorded. *Carson Collection.*

Plastic onions in their own bag. Size wasn't recorded, but as I remember they are around 3 inches high. Mark is shown below. *Courtesy of Joyce Hollen.*

Pepper and peanut, apparently two singles. The peanut is 3-½ inches high. Both have "Japan" inkstamps. *Carson Collection.*

Mark of above onions, showing they were made in Hong Kong for GIFCO of Northbrook, Illinois, and copyright 1981.

A pair of peanuts, 3-½ inches high, with "Japan" paper labels and "TM-694" impressed. *Carson Collection.*

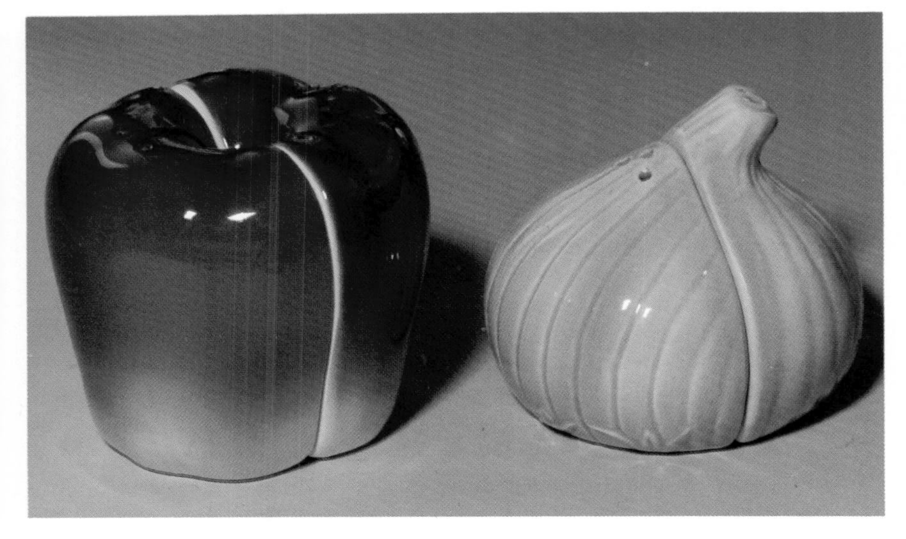

The 3-¼ inch high apple is a repeat of the Otagiri apple shown open above. The onion is 3-½ inches high, unmarked but assumed to have been imported by the same company. *Carson Collection.*

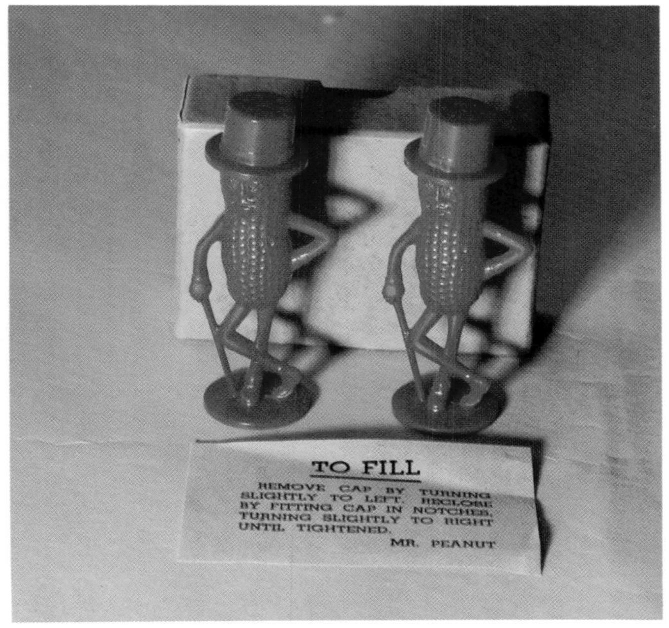

Planter's plastic Mr. Peanut, 3-¼ inches high, unmarked, with original box and instructions. *Private Collection.*

TO FILL

REMOVE CAP BY TURNING SLIGHTLY TO LEFT. RECLOSE BY FITTING CAP IN NOTCHES, TURNING SLIGHTLY TO RIGHT UNTIL TIGHTENED.

MR. PEANUT

Close up of Planter's peanut instructions.

Here the blue set is 4-⅛ inches, the gold 3-¼ inches. Both are plastic. *Lilly Collection.*

Plastic Mr. Peanut in different colors. *Carson Collection.*

Planter's, or someone else's version of Mr. Peanut? This set is made of ceramic, stands 4-½ inches high, has a rhinestone monocle, and a "©" inkstamp. *Thornburg Collection.*

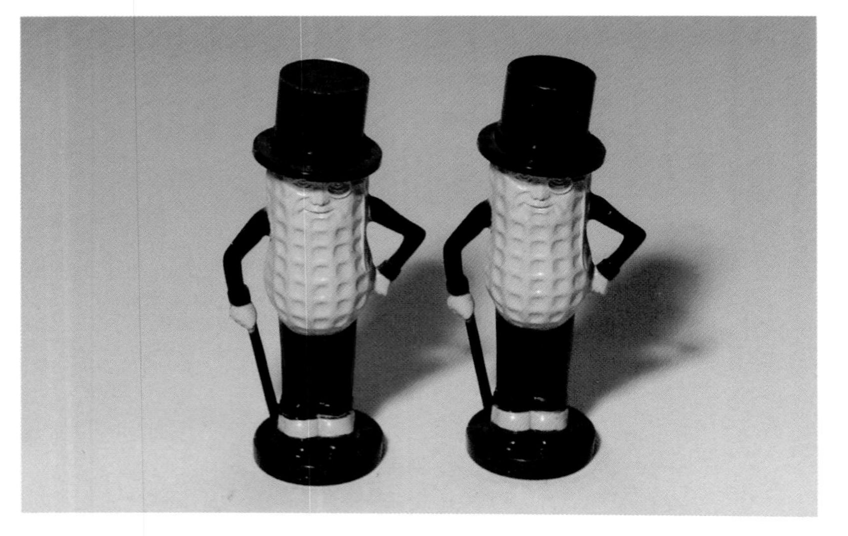

Larger plastic Mr. Peanut, 4-⅛ inches high. Although they don't show up on the black, "Planter's" is printed across the base, "Mr. Peanut" across the hatband. *Lilly Collection.*

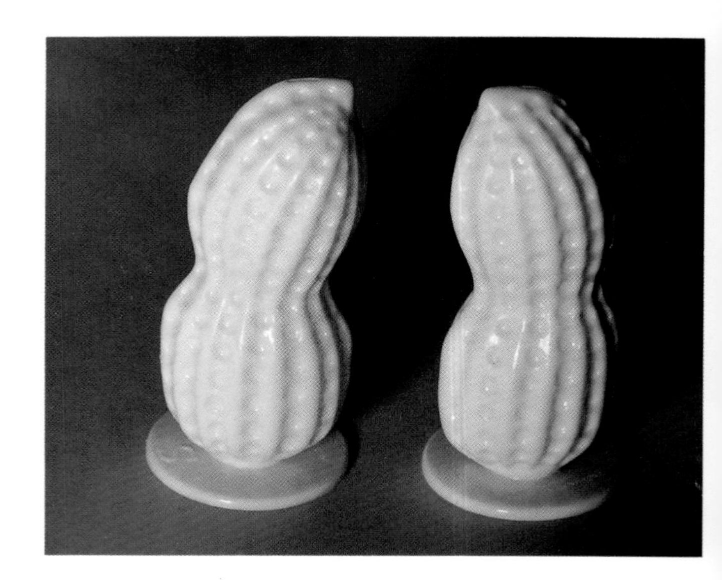

These peanuts are made of a very thick and heavy plastic. They are unmarked, 3-½ inches high. *Carson Collection.*

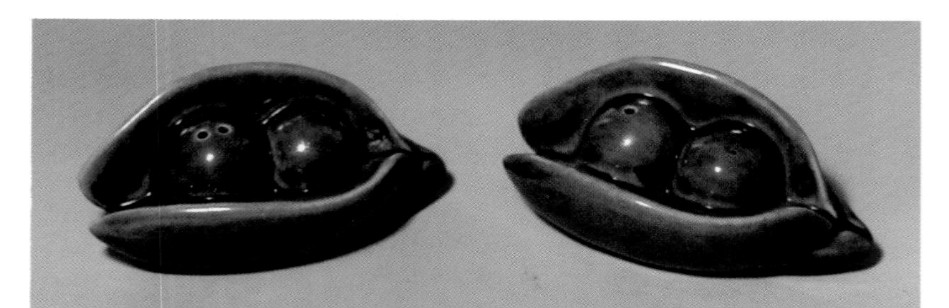

A curious set for two reasons. First, there are only two peas per pod, hardly a show specimen. Second, the are marked with a rather strange impression, "104 ©." Length of each is 3 inches. *Carson Collection.*

People peas, 2-⅞ inches high, with a "Japan" inkstamp. *Oravitz Collection.*

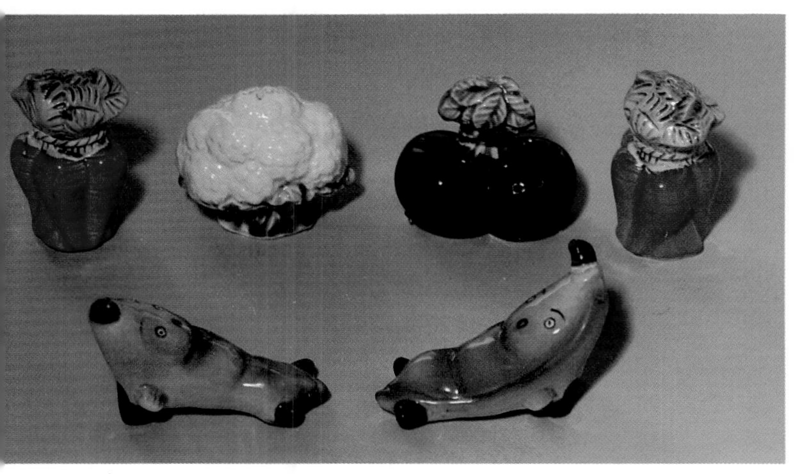

The carrots are 3-¼ inches high, their Giftcraft paper label is shown below. The cauliflower has a "Japan" paper label, while the beets and peas are unmarked. The beets and cauliflower are not a pair. *Carson Collection.*

Giftcraft paper label of above carrots.

Heights left to right in the rear are 2-⅜, 1-½ and 1-⅞ inches. The pickles in front are 2-¾ inches long. The only marked set is the pimentos on the left, which have a "Japan" inkstamp. *Carson Collection.*

The peppers are plastic, 1-⅝ inches high, and unmarked. The tray of the other set is 5-⅜ inches long, has an unreadable mark. The cabbage is 2 inches high. Like the turnip, it is unmarked. *Lilly Collection.*

Goebel peppers. Height on the left is 2-¾ inches. The shakers have "Germany" inkstamps with full bee marks. *Thornburg Collection.*

The potatoes are 2-⅝ inches high, unmarked. *Courtesy of Stan and Karen Zera.*

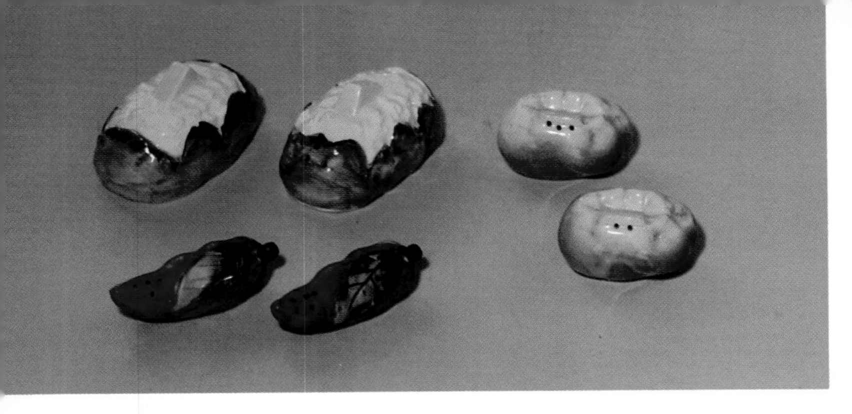

The baked potatoes at left rear are 3-½ inches long, have a "Made in Japan" paper label. The baked potatoes on the right are unmarked, while the sweet potatoes are marked "Japan." *Carson Collection.*

The appropriately filled sacks stand 3-⅝ inches high. Their paper label reads "Our Own Import." The potatoes have an Enesco paper label. *Carson Collection.*

The tray for the pumpkins is 5 inches wide. It has a "Made in Japan" inkstamp. The tray of the beets has "Germany / 6263" impressed on two lines. *Carson Collection.*

The tomatoes have a diameter of 3-½ inches. They are not marked. *Courtesy of Mark Taylor.*

The only piece measured here was the tray, which is 5 inches across. It has a "Made in Japan" paper label. *Carson Collection.*

This is three complete sets, two sitters and a one-piece. The squash sitter is 4-¼ inches high as shown. It has a "2" impressed, and a paper label which is shown below. The other two sets are not marked. *Oravitz Collection.*

Paper label of the squash sitter above, showing a rooster with the words "Handpainted," and "Japan."

The tray of this set is 6 inches wide, none of the pieces are marked. I'll leave it to you to name what the shakers represent. *Carson Collection.*

Chapter 8
Transportation and Fuel Vehicles

Vehicles

Transportation, in regard to salt and pepper shakers, is a perplexing subject. For decades automobile manufacturing was the backbone of our economy. There are probably few men who at one time or another have not been more in love with their car than with their girlfriend or wife. And, likewise, you will not find very many women who have not at least once in their life fully or partially based a decision to date or marry upon the type of automobile a would-be suitor drove. Yet this fascination with cars has somehow eluded the movers-and-shakers of the shaker manufacturing business. Sure, you will find a few cars in most every collection, but for the most part they are mediocre at best, and seldom based upon the real thing. Where are the '55 T-Birds, the '57 Chevys, the '32 Ford five-window coupes, the Avantis, the Jaguars and the Porsches? Even an Edsel would probably be a top seller. Listen up Vandor, Knobler and United China and Glass; the collecting public is waiting for you to act.

The locomotive is 4-½ inches long, the caboose 2-½ inches high. Neither is marked. *Lilly Collection.*

The yellow engine is 4-¼ inches long, the brown engine is 3-¼ inches long. The brown engine toward the center has "Salt" printed on the opposite side, just as its mate has "Pepper." None of these four pieces are marked. *Courtesy of Joyce Hollen.*

Same locomotive and caboose as above but in different colors. Neither the automobile nor the trailer is marked. The auto is 4 inches long. *Carson Collection.*

The tray of the set on the left measures 5 inches. The tender of the brown set is 2-½ inches high. Printing on top of the brown set reads "Georgia's Stone Mountain Scenic Railroad." Paper labels of both sets are shown below. *Carson Collection.*

The locomotives on the left are 2-⅛ inches high. They have "Made in Japan" paper labels. The set on the right has "Made in Japan" inkstamps. *Carson Collection.*

Knobler paper label of the black and white train set.

Charles Products paper label of the brown train set.

All metal. The locomotives in front are 2-⅝ inches high, have "1875" in raised numerals on their bottoms. The locomotive with the tender is marked "Made in Japan." The street car set is not marked. *Carson Collection.*

Look closely and you'll see this is *A Streetcar Named Desire*. Length is 3 inches. There are no marks. *Courtesy of Joyce Hollen.*

Greyhound buses, 3-¼ inches long, with "Japan" paper labels. *Thornburg Collection.*

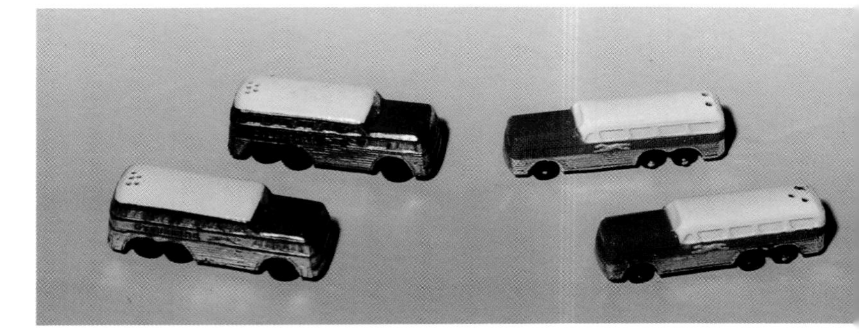

Metal Greyhound buses, those on the left being 3 inches long, those on the right 3-¼ inches. None of them are marked. *Thornburg Collection.*

This pair of cars was obviously made by the same company that made the similarly glazed and labeled locomotives above. They are 4 inches long, are not marked. *Carson Collection.*

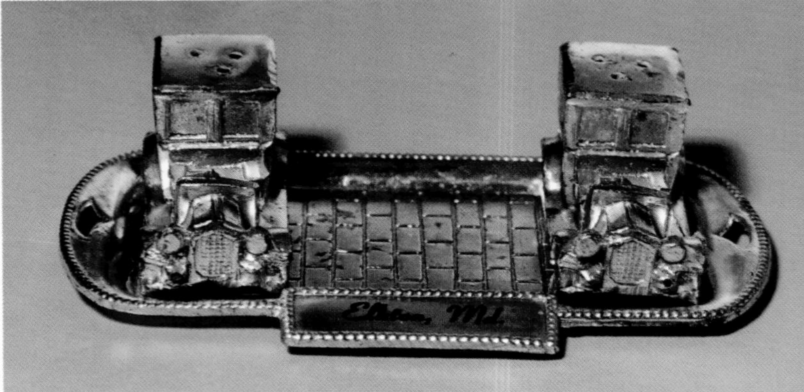

Metal automobiles, originally a souvenir of Elkton, Maryland. The tray is 4-⅛ inches long, marked "Made in Japan." The cars are marked "Japan." *Private Collection.*

More metal. The radiators, which just happened to get mixed in, are 1-½ inches high. They are unmarked. The mark of the cars is shown below. *Carson Collection.*

Mark of the above metal cars.

A popular and widely available plastic set in which the driver and passenger are the salt and peppers. As shown, overall height is 5-¾ inches. The only mark is "Pat. Pend. / Made in USA" on two lines. The steering wheel has a tendency to break off and get lost, so check before you buy. *Carson Collection.*

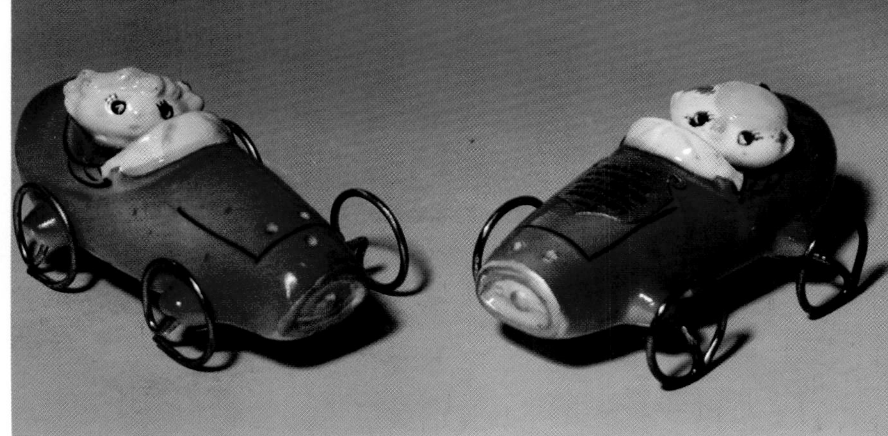

Each of these cars is 3-⅝ inches high. The paper shield on the hood of the one the right reads "Rolling / Racing Car / Salt & Pepper / Authentic / Maserati / Japan" on six lines. One of the cars has a paper label, "A Quality Product Japan," while the other has the Parksmith inkstamp shown below. *Carson Collection.*

Parksmith inkstamp of above Maserati.

This is a complete set, the car being one shaker, the rabbit the other. The car is 4 inches long. It is marked "Made in Japan." As evidenced below, it is apparently one set of a series. *Carson Collection.*

On the left is a pig to match the rabbit above. The horses and wagon in the middle is a wooden set. The ship is 4-½ inches long. It's smokestacks are the shakers. None of these sets are marked. *Carson Collection.*

Combination salt and pepper and toothpick. The metal bike is 4-¾ inches long. The shakers are made of glass. No marks are present on any of the pieces. *Lilly Collection.*

Unmarked Rosemeade sailboats, 3-¼ inches high. *Thornburg Collection.*

Noah's Ark as a condiment set. The ark is 4-¾ inches long. All pieces have Enesco paper labels, in addition to an "E2030" inkstamp on the ark. *Carson Collection.*

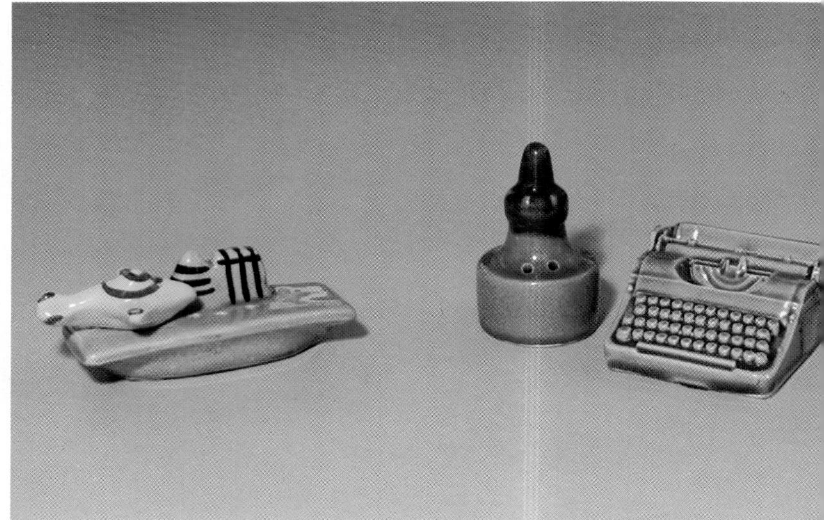

The aircraft carrier is 3-¾ inches long. The set is not marked. The ink bottle stands 2-¾ inches high. The mark of its companion typewriter is shown below. *Carson Collection.*

This gondola is 5-½ inches long, the shakers are 1-⅝ inches high. Marks are shown below. *Lilly Collection.*

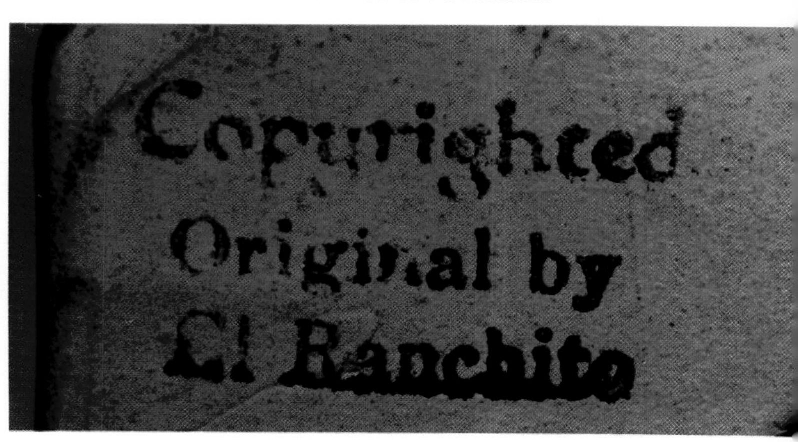

Inkstamp mark of above typewriter.

Marks of the gondola and shakers above. As you can see, one inkstamp shows Occupied Japan, the other simply Japan.

The riverboats are 2 inches high, have "Made in Japan" paper labels. *Carson Collection.*

These unmarked covered wagons are 3-½ inches long. *Courtesy of Cloud's Antique Mall, Castalia, Ohio.*

These unmarked ships stand 3-⅛ inches high. *Carson Collection.*

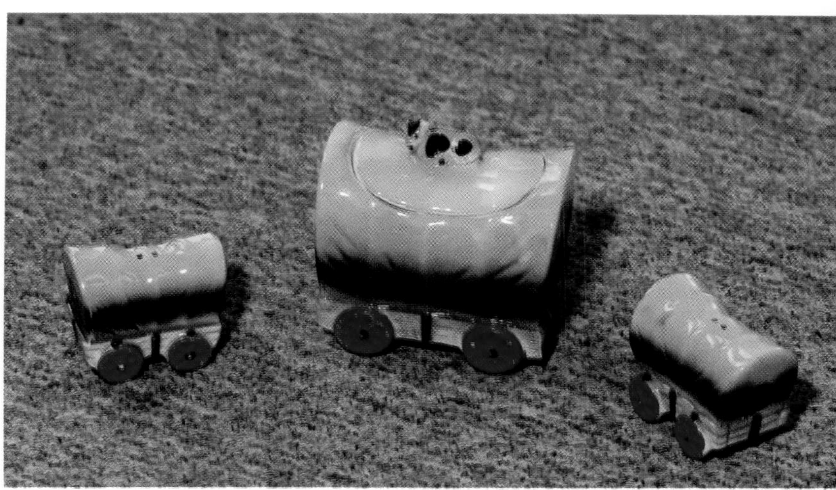

This condiment set wasn't measured but the shakers are about 2-½ inches high, the mustard 4 inches high. The pieces are marked with "Japan" inkstamps. *Private Collection.*

Pieces parts. The ship's wheel tray is 5 inches long. The shakers are 1-½ inches high. All pieces are plastic. The mark on the bottom of the tray is "Made / in / USA" on three lines in raised letters. *Carson Collection.*

On the unmarked set on the left, the salt and pepper shakers combine to form the top of the wagon. The wagon on the right is not marked, but the horse has a "Japan" inkstamp. Measurements were not taken. *Carson Collection.*

The wagon is 2-⅜ inches high, the ox 2 inches high. Both have Ceramic Arts Studio marks. *Thornburg Collection.*

Souvenirs of Knott's Berry Farm in Buena Park, California, the length of these wagons is 2-¾ inches. They have "Made in Japan" paper labels. *Carson Collection.*

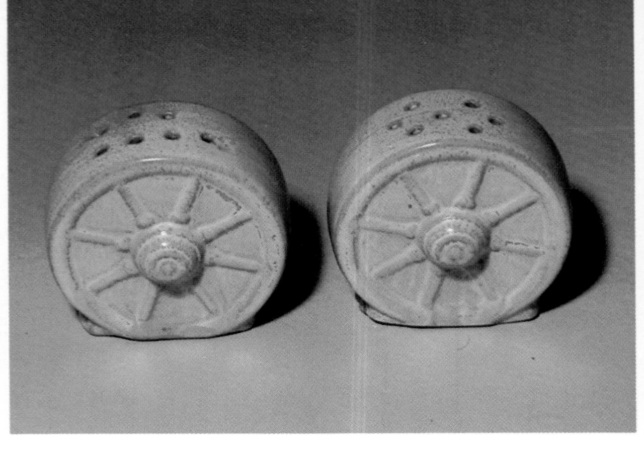

Unmarked wagon wheels made by Frankoma. Height is 2-⅜ inches. *Courtesy of Joyce Hollen.*

The wagons are 4 inches high and unmarked. The stage coaches have "Japan" inkstamps. *Carson Collection.*

Metal buggy and Amishman. The buggy is 2-¾ inches high. There are no marks. *Lilly Collection.*

Height here is 4 inches, mark is "Japan." The shakers form the top of the cart. *Carson Collection.*

Fuel

The term fuel shakers generally refers to plastic gas pumps and tanks. They are extremely popular with salt and pepper shaker collectors, and, unfortunately, with collectors of gasoline and oil company advertising, too. Consequently, you have to get to shows and flea markets earlier than you used to in order to find them. And you can also expect to pay higher prices than you did just a few years ago as the supply dwindles and the demand rises. Some gas pump shakers include local advertising, others don't. Whether or not they do usually doesn't affect value except in the community that the advertising is from.

Standard Oil plastic gas pumps. Sizes are not given on these because all of those shown run about 2-¾ inches in height. This particular set, however, may be a bit taller due to the crown. *Thornburg Collection.*

One of the decals is missing on this set of Zephyr pumps. But on gas pump shakers that are somewhat rare, such as these, it's a good idea to purchase them anyway because you might not see them again. *Thornburg Collection.*

Gas pump shakers have a real nostalgic flavor. Many of the companies represented have either been swallowed up in mergers, or have gone out of business and disappeared. *Thornburg Collection.*

How long has it been since you have used Milemaster or Super 5-D? *Courtesy of Bob Sandoff.*

Sinclair Power-X, another bygone of yesteryear. *Carson Collection.*

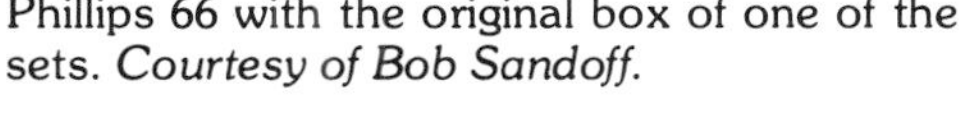

Phillips 66 with the original box of one of the sets. *Courtesy of Bob Sandoff.*

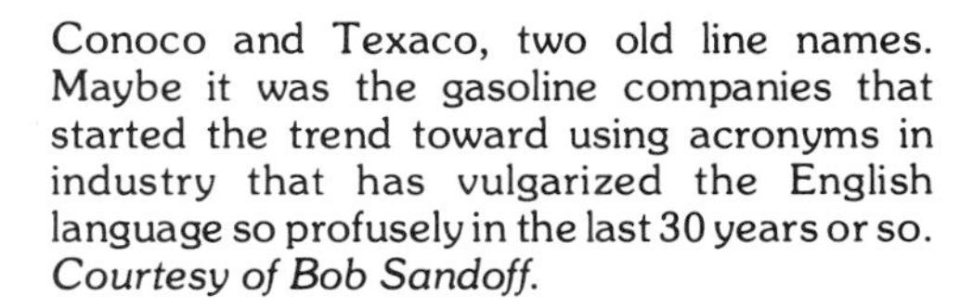

Conoco and Texaco, two old line names. Maybe it was the gasoline companies that started the trend toward using acronyms in industry that has vulgarized the English language so profusely in the last 30 years or so. *Courtesy of Bob Sandoff.*

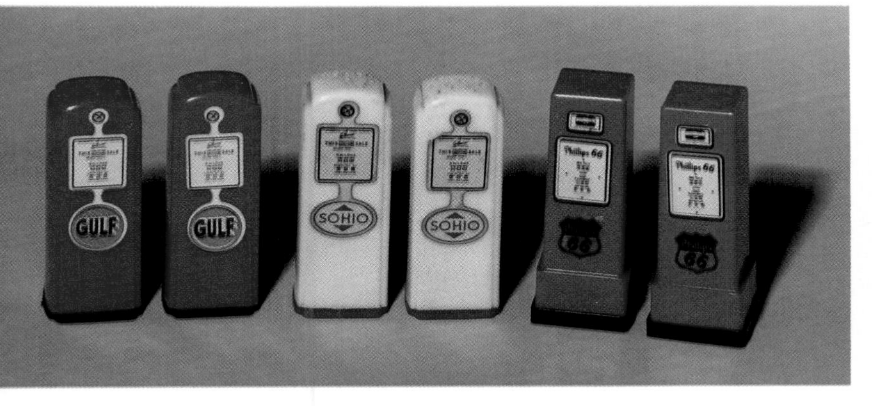

The Phillips pumps are a repeat of the above picture but somewhat significant because they are the only plastic pumps I ran across that had any kind of mark, "AVSCO." The Sohio set represents one of the most recent casualties in the gasoline name game, having been changed to BP a couple years ago. *Thornburg Collection.*

Richfield and Atlantic before the merger, also Shell. *Thornburg Collection.*

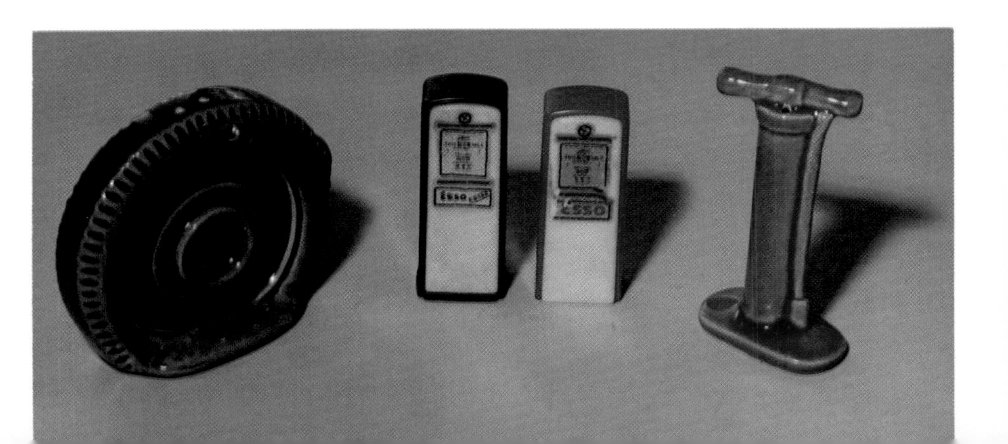

Here the gas pump on the right is missing its bottom. The air pump helps date that set. It has probably been at least 50 years since this type of pump has been associated with anything larger than a bicycle tire. *Carson Collection.*

More recent and made of ceramic, these gas pumps are obviously not an advertising set as there is no name. The shakers are 4-½ inches high, have an impressed mark, "Made in Japan." *Carson Collection.*

Philgas tanks in gray. *Thornburg Collection.*

Liquid petroleum gas tanks, plastic, unmarked, and 2-¾ inches high. *Oravitz Collection.*

Pyrofax gas advertising shakers, plastic, unmarked, and 2-¾ inches high. *Courtesy of Bob Sandoff.*

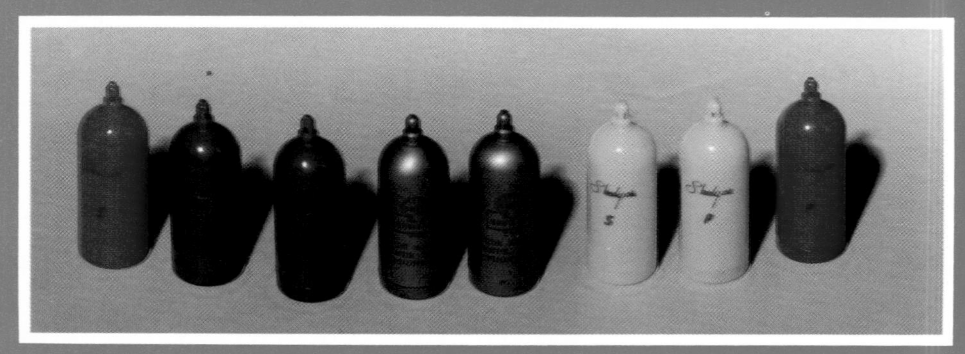

An array of LP tanks, all the same size as above. The silver or gray set has Shell advertising on it, came from a little town in Michigan where the advertiser's telephone number was 51. *Thornburg Collection.*

Plastic Philgas tanks, 1-⅞ inches high, with original box. The mark is shown below. *Private Collection.*

Here is something a little different in the fuel category, a pair of glass oil well derricks. Made by the Oil City Glass Company, they are, like the plastic pumps, 2-¾ inches high. The mark of these shakers is shown below. *Thornburg Collection.*

Mark of the above Philgas tanks. It is the same mark as on the second set of Phillips 66 gasoline pumps, but gives us a bit more information showing the company located in Kansas City, Missouri.

Mark of the above oil derricks. How appropriate for a firm named Oil City Glass Company.

Chapter 9
Structures

From outhouses to birdhouses, from the Empire State Building to the Seattle Space Needle, many structures have been miniaturized as salt and pepper shakers. As a general rule, diligent searching will eventually turn up salt and pepper shakers made in the likeness of any structure that attracts large numbers of tourists.

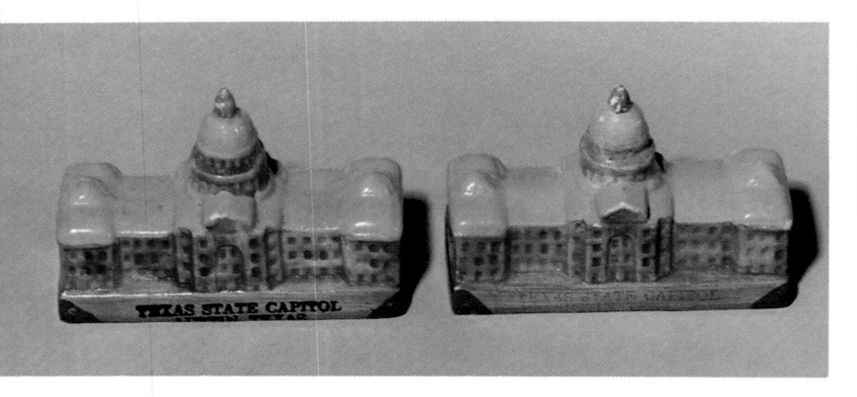

The Texas State Capitol, 2-½ x 3-½ inches. Inkstamp mark reads "Made in Texas." *Thornburg Collection.*

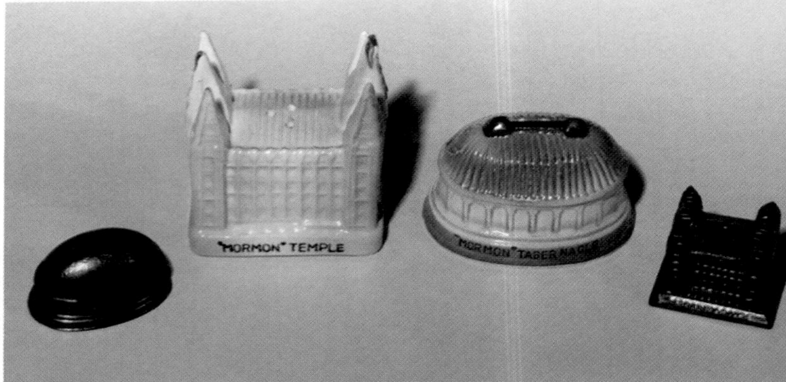

Ceramic and metal renditions of the Mormon Temple and Morman Tabernacle. The ceramic temple is 3-¼ inches high, has a "Made in Japan" paper label. The metal shakers sport "Japan" in raised letters. *Thornburg Collection.*

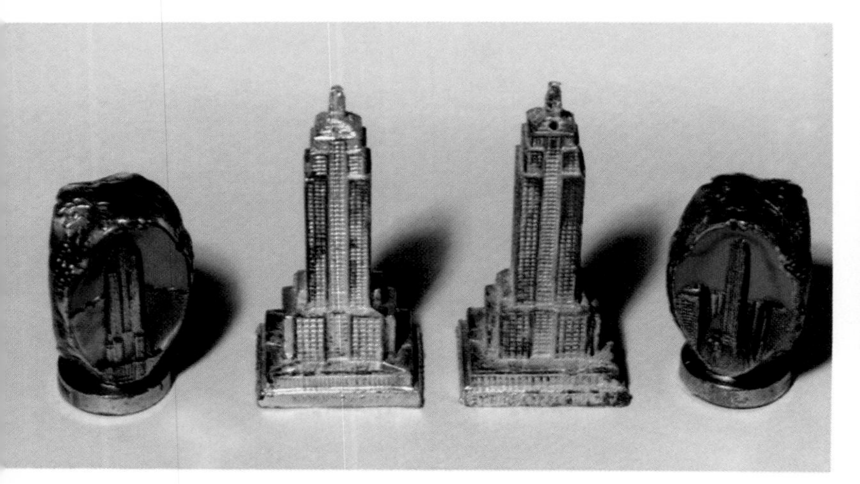

Both these pairs depicting New York's Empire State Building are made of metal. The shakers in the center stand 3-½ inches high. They are unmarked. The pair on the outside is marked "Japan." *Thornburg Collection.*

Metal, 1-⅛ inches high. Mark on the bottoms is "EXCO / Japan" on two lines. Printed around the base on all four sides is "The World's / Only Corn Palace / Mitchell / South Dakota." Another set photographed, identical in size but with the gold either worn off or never applied, had a different mark—"5041 / INCO N.Y.C. / Japan" on three lines. *Carson Collection.*

Here is *Bonanza's* Ponderosa Ranch, 2-⅞ inches high. No mark to designate origin. A paper label reads "Ponderosa Ranch Nevada." *Thornburg Collection.*

Hearst Castle, 3-¼ inches high, unmarked. *Thornburg Collection.*

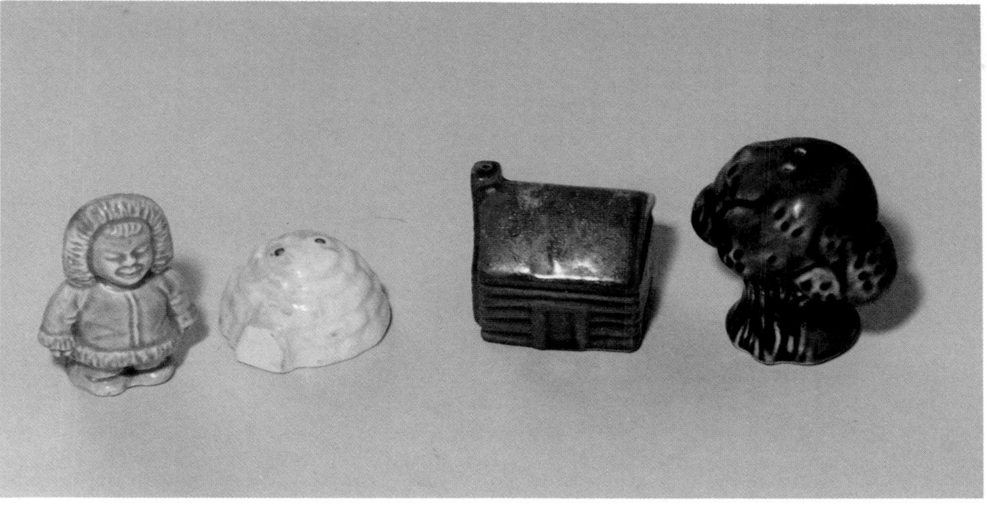

The igloo is 2-¼ inches high, the log cabin 2 inches. None of the four pieces are marked. *Carson Collection.*

Gingerbread houses, 2-⅞ inches high. They are unmarked. There is also a cookie jar that matches them. *Oravitz Collection.*

English cottage condiment set. The tray is 6-¼ inches long, and is marked "Cottage Ware / Reg. no. 645007" on two lines, and "Trice / Bros. / Made in / England" on four lines. The most interesting aspect of this set is that the shaker on the left has one pour hole while the shaker on the right has 15! *Courtesy of Cloud's Antique Mall, Castalia, Ohio.*

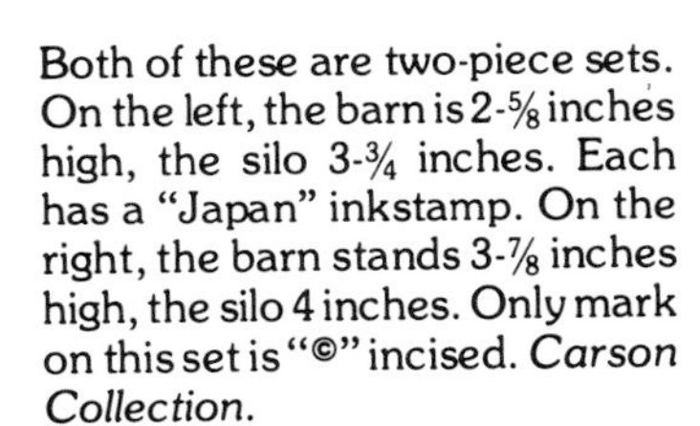

Both of these are two-piece sets. On the left, the barn is 2-⅝ inches high, the silo 3-¾ inches. Each has a "Japan" inkstamp. On the right, the barn stands 3-⅞ inches high, the silo 4 inches. Only mark on this set is "©" incised. *Carson Collection.*

Harvestore silos in two styles. Approximate height of each set is 4-¾ inches. None of the shakers are marked. *Thornburg Collection.*

Another English-style house, 2-⅝ inches high, and unmarked. *Courtesy of Joyce Hollen.*

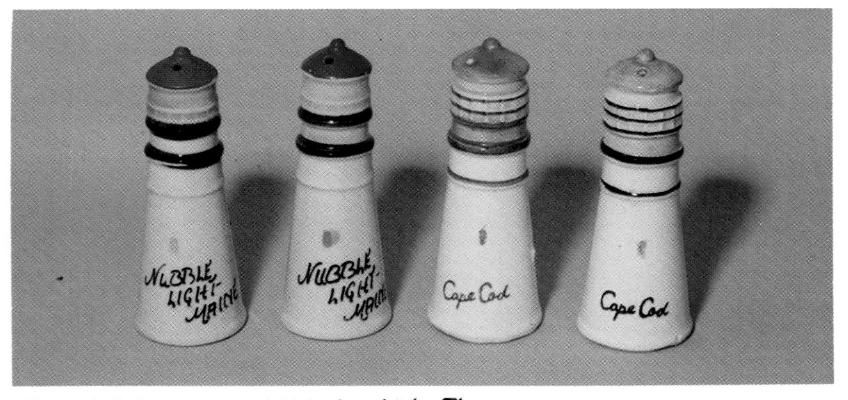

These lighthouses are 4-½ inches high. They have "Japan" paper labels. *Carson Collection.*

The lighthouse of this set stands 3-⅛ inches high. Neither piece is marked. *Carson Collection.*

Cedar outhouses, 2-⅛ inches high. Like most wooden shakers, they are not marked. *Private Collection.*

Oriental birdhouses. Or are they lamps? Either way, they are 3-¼ inches high, and have "Made in Japan" inkstamps. *Carson Collection.*

The cabin and the top of the lighthouse lift off as the shakers. Height is 3-½ inches. There is a "Japan" inkstamp on one of the pieces. *Carson Collection.*

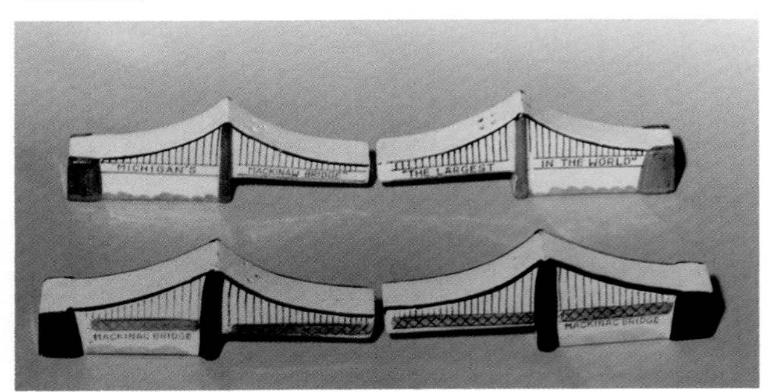

The Mackinac Bridge 5-⅞ inches long, all with "Japan" paper labels. You might not be able to see it in the picture, but the spelling of Mackinac on the rear bridge is incorrect. *Thornburg Collection.*

Another Mackinac Bridge, this one a nodder made by K. Wolfe Studios for the Michigan Shakers Club. Size of the base is 3-⅜ x 4-¼ inches. Bottom is shown below. *Thornburg Collection.*

The taller outhouses are 4-¾ inches high, have "Japan" paper labels. The shorter ones are 2-⅞ inches high, have "Japan" inkstamps. *Carson Collection.*

The outhouses stand 2-½ inches high, the toilets 2 inches. All are marked "Japan." *Carson Collection.*

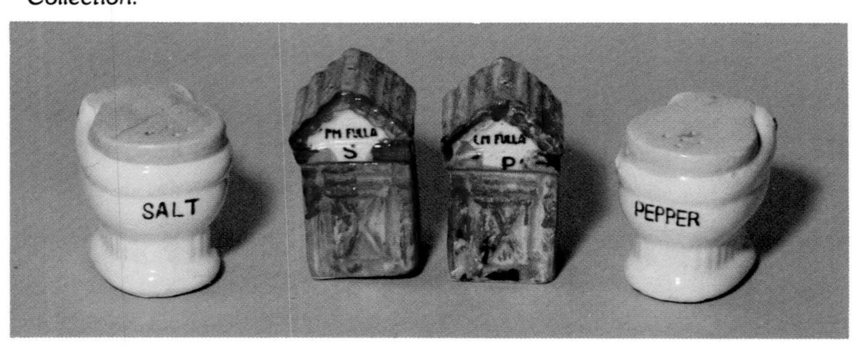

Bottom of above Mackinac Bridge nodders.

The Eisenhower Lock of the St. Lawrence Seaway. The combined length is 4-⅝ inches. Paper label reads "A Quality Product / Japan" on two lines. *Thornburg Collection.*

Back of plastic Mt. Rushmore above. Is the stone a small piece of the mountain? Was it sold that way or added later by the original purchaser? Or was it perhaps added by a dealer to give the lightweight shaker stability and prevent it from blowing off his table at a windy outdoor show? I saw three of these shakers during the course of writing the book, this one with the stone, the other two without it.

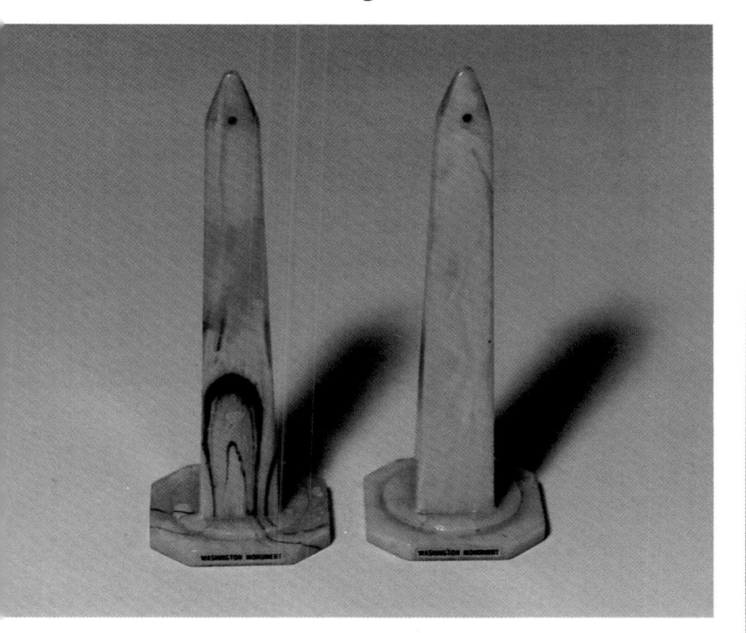

The Washington Monument in Bakelite, 4-¼ inches high, unmarked. *Carson Collection.*

More of the popular Mt. Rushmore. The set in front is metal, 1-¾ inches high. The set directly behind it is plastic, 3-⅛ inches high. The other two sets are ceramic. None of the shakers are marked. *Thornburg Collection.*

This one-piece plastic version of Mt. Rushmore is 1-⅞ x 3-¼ inches. Mark is "© Adrian Forrette 1959. Back is shown at above right. *Thornburg Collection.*

The Seattle Space Needle on the left is 4-¾ inches high, marked "Etc / Seattle USA" on two lines. The set in the center of the back row stands 5-¼ inches high, carries a paper label that says "© Century / Souvenir Co. Inc." on two lines, in addition to a "Japan" paper label. The pair at right rear has a "Made in Japan" paper label. The metal tray set in front is stamped "Japan." *Thornburg Collection.*

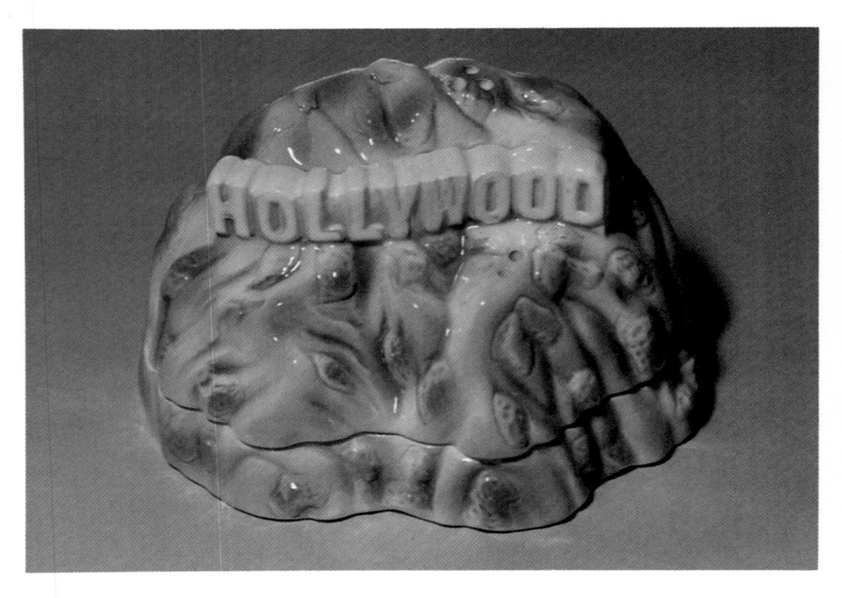

Sarsaparilla Five and Dime Hollywood sign, the base of which is 2-⅞ inches high. The marks of this two-piece set are shown below. *Lilly Collection.*

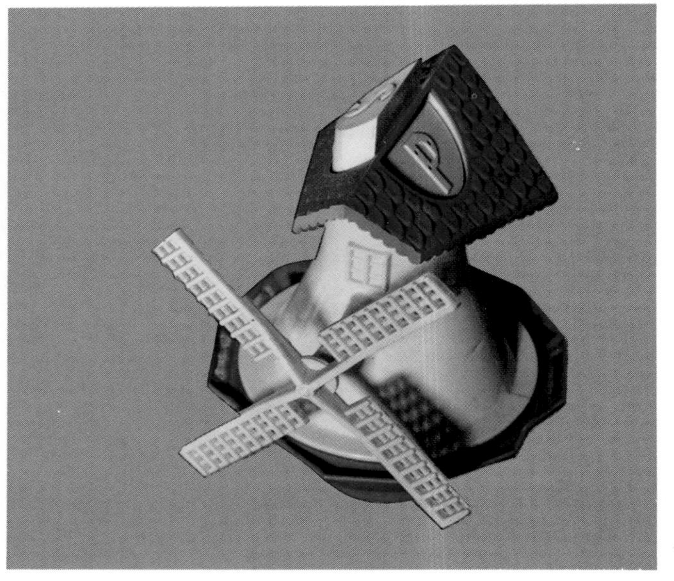

This windmill is made of plastic, stands 5-½ inches high. Turn the blades one direction and the salt shaker rises as shown. Turn them in the other direction, the salt shaker recedes and the pepper shaker rises. Lift off the white part of the building to reveal a sugar bowl beneath it. Mark information was not recorded. *Oravitz Collection.*

Marks and original price sticker of the recent Hollywood sign shaker set above.

Windmill and water mill set, the tray of which is 5 inches long. The set includes a "Japan" paper label. *Carson Collection.*

The windmill set at the left with the yellow rings is 3-⅞ inches high, has a "Made in Occupied Japan" inkstamp. Height of the set at middle rear is 3-¼ inches. It has a crown mark with "Bona / Delfts / Bloup / Holland" on four lines. The set directly in front of it, 2-½ inches high, has a similar mark. The pair on the right is unmarked, 2-¾ inches high. *Courtesy of Joyce Hollen.*

Chapter 10

Appliances and Machines

There are basically two types of shakers in this category, those that were made to be sold, and those that were offered as premiums or giveaways by appliance retailers. Usually those in the latter category are the more desirable, not only because of the advertising but also because the quality is higher.

The stove on the right appears to be the same as that above, but it isn't. This one is ceramic, was sold by Lego. Height is 5-½ inches, and the stove top lifts off for storage of instant coffee, as stated on the door of the oven. The mark and paper label of this set are shown below. The set on the left is redware, 4 inches high, unmarked. The woodburning stove with teakettle is 3-¾ inches high, also unmarked. On this, the stove is one shaker, the teakettle the other. *Courtesy of Joyce Hollen and Bev Urry.*

Inkstamp and paper label of above Lego stove set.

One-piece Tappan stove, plastic, 5 inches high. It is unmarked except for the stoppers which say "Clover / SP-383" on two lines. *Thornburg Collection.*

This plastic stove is 5-¼ inches high. Mark is "Starke Design, Inc. / Bklyn 20, NY" on two lines in raised letters. *Carson Collection.*

The stove on the right is the same as that above but has a gold inkstamp, "Made in Japan." The one on the left measures 3-¾ x 3-¾ inches, is made of redware, and carries a mark that reads "Lugenes / Japan" on two lines. The hot plate and pot have "Japan" paper labels. *Carson Collection.*

A wooden woodburning kitchen stove, 6-¼ inches high, made by Enesco. The coffee pot is 3-¼ inches high. The handle and spout of the pot are made of metal. The stove drawer opens for whatever you might want to keep in it. Paper label of this set is shown below. *Carson Collection.*

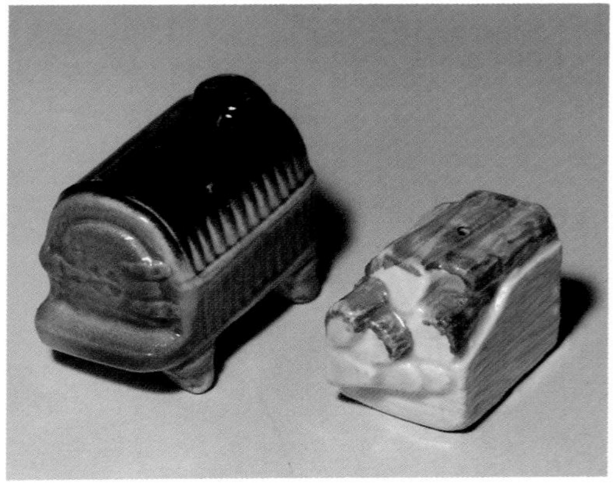

The stove of this go-together set is 2 inches long. Neither piece is marked. *Carson Collection.*

This barbecue grill set is made of glass, plastic and metal. It is 5 inches high as shown. No mark was found on the shakers, the grill, or the box. *Carson Collection.*

Enesco paper label of above wooden stove. This came as a real surprise as Enesco is most often associated with ceramic products. The paper label appears to be from early in the company's history.

The pair on the left, which separates for use, is 3-⅝ inches long as shown. It is made of redware and unmarked. The set is somewhat unusual in that the pepper (note the "S" and "P") is larger than the salt. The set on the right is also unmarked. The stove is 2-¼ inches high, the coffee pot 1-⅝ inches. *Carson Collection.*

Stoves and coffee grinders, both made of metal. Information on these sets was lost, but it is doubtful either is marked, and the stoves are probably no higher than 3-½ inches. *Carson Collection.*

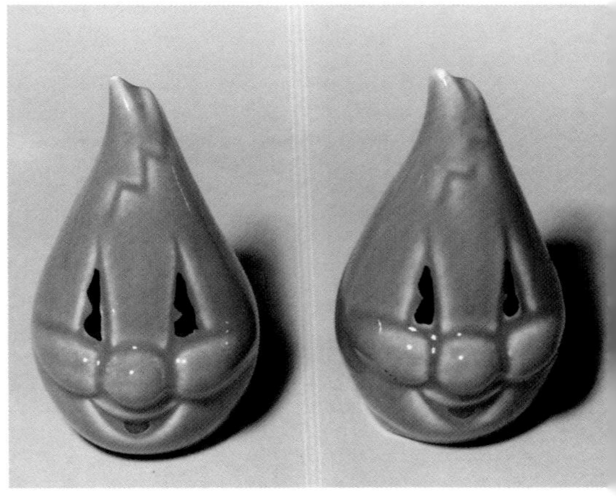

These shakers, each 4 inches high, depict a natural gas flame. Impressed in their backs is "Handy / Flame / ©" on three lines. *Thornburg Collection.*

This stove is 3 x 3-⅛ inches. It is unmarked. As you can see, the shakers, which are the burner covers, lift out. *Carson Collection.*

Then and Now refrigerators with original box. The shaker on the left is 3-½ inches high, the one on the right 4 inches. One (I forget which) is marked "© 1976 / Aluminum / Housewares Co. Inc. / Maryland Heights / 63043 / Hong Kong" on six lines. The other is marked with the same words but different spacing. *Carson Collection.*

Plastic iceboxes, 3-½ inches high. Mark is "Made in Hong Kong / © Boldman-Morgan Inc. / Lego" on three lines. *Carson Collection.*

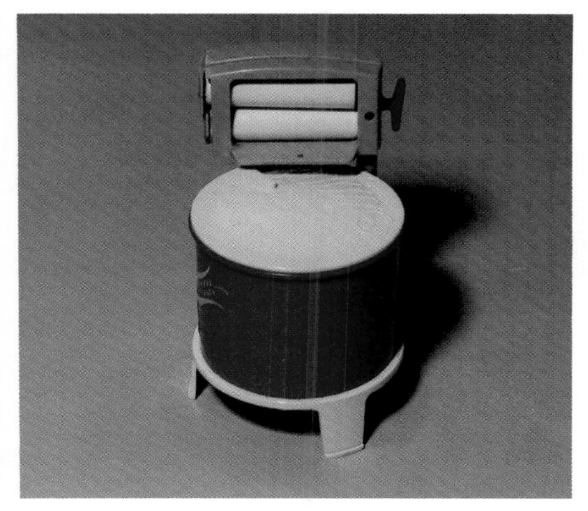

Plastic, 4-½ inches high, marked "Made in U.S.A. Pat. Pend." The rolls of the wringer pull out as the salt and pepper. The top says "Sugar," and removes so that spice may be stored in the tub. *Carson Collection.*

These shakers are made of plastic, stand 2 inches high. They are marked "E.W. Loew Co. Erie, Pa. U.S.A." *Thornburg Collection.*

Gold toaster made of plastic. Height is 4-¼ inches, the bread slices are the shakers. Mark on the bottom of the toaster is "Tucker Products Corp. / Leominster, Mass. / Made in U.S.A." on three lines in raised letters. *Thornburg Collection.*

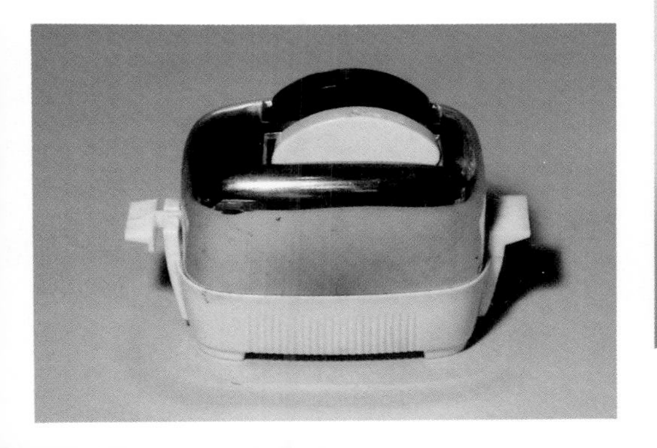

This toaster is made of metal and plastic. It is 3 inches high, and unmarked. *Carson Collection.*

Another plastic toaster, 4-¼ inches wide at the ends of the handles. Only mark is "Pat. Pend. / Made in U.S.A." on two lines in raised letters. *Courtesy of Allen and Michelle Naylor.*

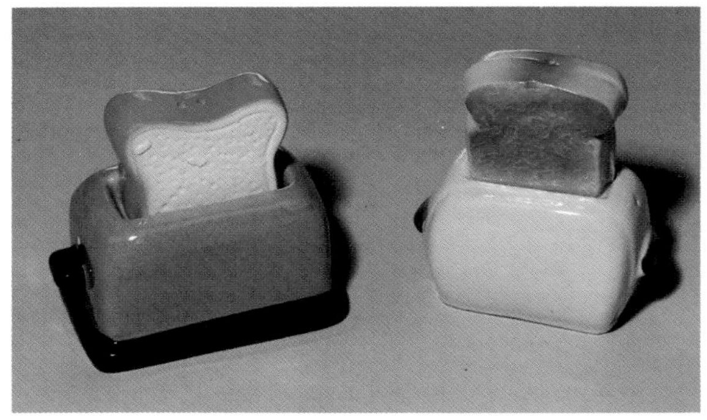

Two sets, each piece of toast and toaster being shakers. The toaster of the set on the left is 1-⅞ inches high, while the piece of toast is 1-⅝ inches high when standing by itself. The set has a Knobler paper label. The set on the right is not marked. *Lilly Collection.*

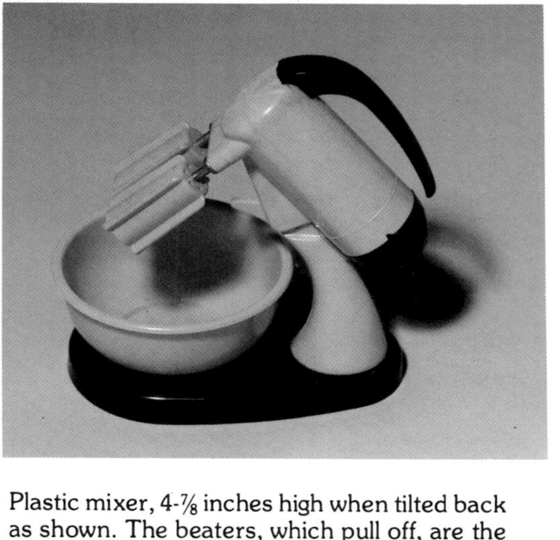

Plastic mixer, 4-⅞ inches high when tilted back as shown. The beaters, which pull off, are the salt and pepper shakers. This set, which is more common in white, is not marked. Quite often I see it for sale minus the handle which, apparently, is subject to breakage. *Carson Collection.*

Blenders made of plastic, 3-¾ inches high and unmarked. To fill these shakers, you remove their tops. *Carson Collection.*

This is a garbage disposal. It is a one-piece that holds salt, pepper and one other seasoning of the owner's choice. It is also plastic, 3-⅞ inches high, and a Kelvinator promotion. The mark is shown. *Thornburg Collection.*

Mark of the one-piece plastic Kelvinator triple action garbage disposal shaker.

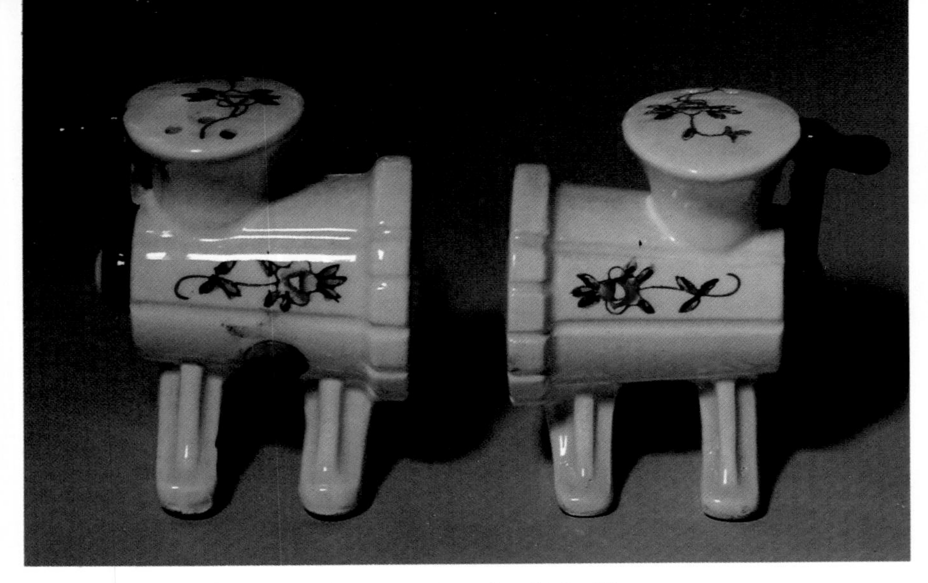

These grinders are 2-¾ inches high. They are not marked. *Carson Collection.*

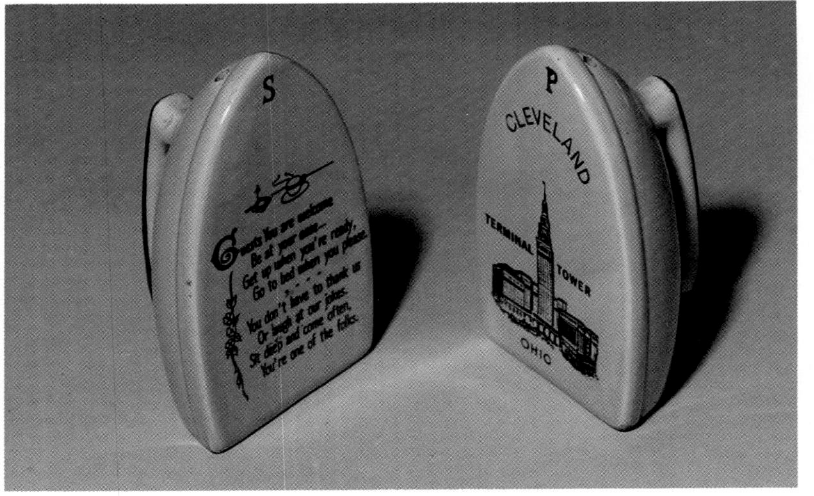

Souvenir irons made with landmarks of various cities on them. Height is 3-¾ inches. This pair has a "Japan" paper label. The poem on the salt shaker is probably too small to see in the picture but is rather nice so you may read it here. "Guests you are welcome / Be at your ease... / Get up when you're ready / Go to bed when you please. You don't have to thank us / Or laugh at our jokes, / Sit deep and come often, / You're one of the folks." This pair of irons belong to Jim and Irene Thornburg, and the verse describes precisely the way they made us feel when we visited them to photograph their shakers. *Thornburg Collection.*

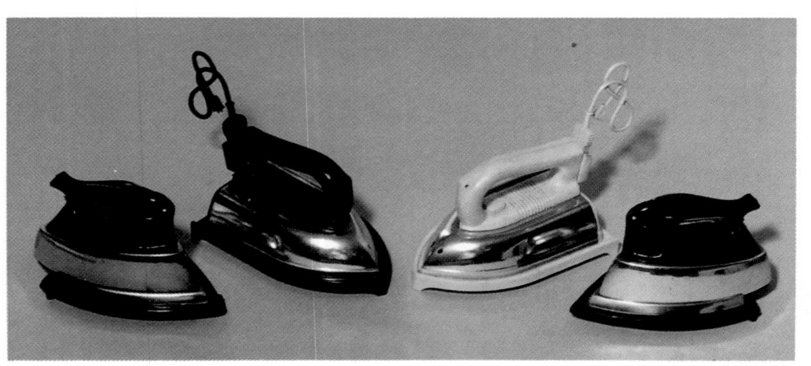

Plastic irons with trivets. The pair on the outside is 1-¾ inches high. None of these shakers are marked. *Carson Collection.*

Plastic sewing machine, 3-½ inches high, unmarked. The drawers are the shakers. *Carson Collection.*

Above sewing machine closed.

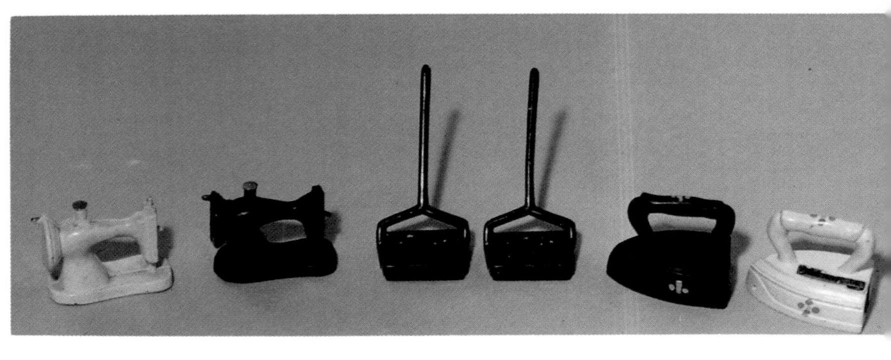

Three metal sets. The irons are 2 inches high. The only marked set here is the sweepers, which say "Japan." *Carson Collection.*

More modern than the sweepers in the previous picture, this plastic pair is 3-¼ inches high. Only mark is the advertising, "Compact Electrolux." *Carson Collection.*

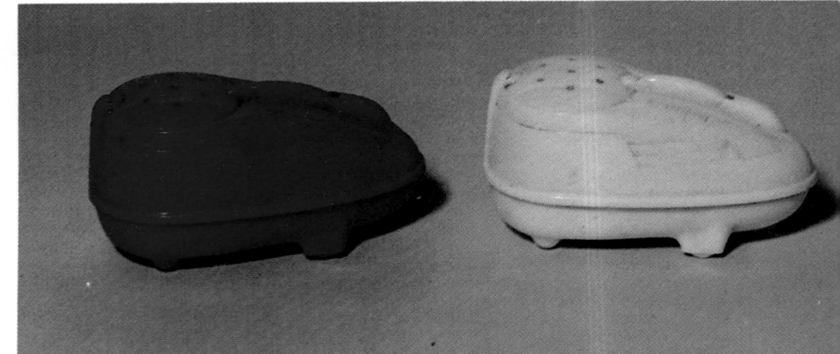

Plastic telephones, unmarked, 2-¼ inches high. *Carson Collection.*

Like the other televisions, but for an advertising campaign. This picture shows how the top slides back and forth to dispense salt or pepper. The mark is a bit different on this one, "The Kobid Corp. / Hong Kong" on two lines. *Carson Collection.*

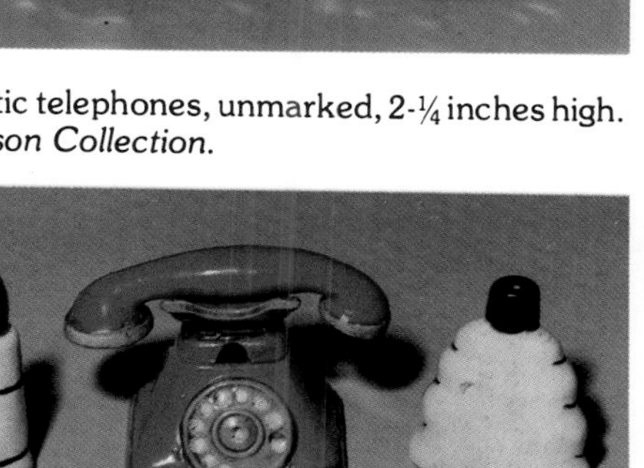

The base of this telephone is the salt, the receiver the pepper. It has a "Made in Japan" inkstamp. For size reference, the lipstick is 2-½ inches high. Neither the lipstick nor perfume is marked. *Courtesy of Joyce Hollen.*

Turn the volume control knob and the shakers (the dark area of the top) pop up. This set is plastic, 3 inches high. "Pat. Pend. / Made in USA" is written *backwards* on the shakers. *Carson Collection.*

The radios are 2 inches high; a matching cookie jar was also made. The record player is 4-¾ inches wide. All four pieces have Vandor paper labels. *Thornburg Collection.*

Here the telephone is 2-⅝ inches high, the directory 1-⅝ inches high. Neither is marked. *Lilly Collection.*

On the right, the same TV as above but in brown, which is the color most often seen. The plastic piano is 3-⅝ inches high, marked "Davis Products / Brooklyn N.Y." on two lines. Press on the keys and the shakers pop up. *Carson Collection.*

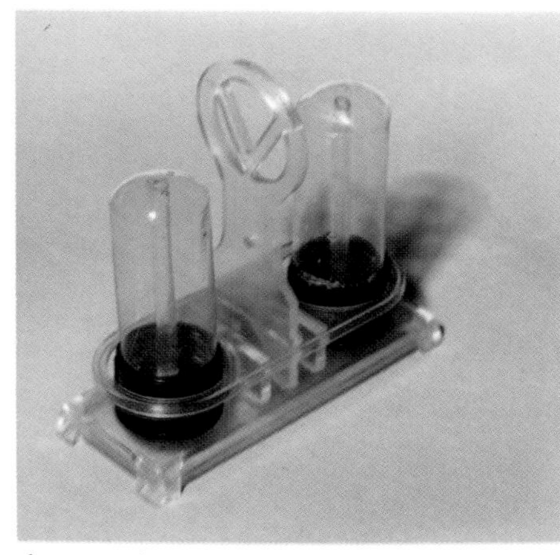

Radio tubes, probably unrecognizable to the youth of today's transistor world, but very much a part of electronics in to the early 1970s. The shakers themselves are 2-¾ inches high. They are marked "Pat. No. 1937505," while the stand is marked "Pat. No. 134265." The first number is for an invention patent corresponding to a date of 1933. The second number is a design patent from 1942. *Thornburg Collection.*

Plastic televisions with water and snow in them. Note that the sizes are different. The snow scene is 2-⅝ x 3-⅝ inches while the beach scenes are 2-¼ x 3-¼ inches. The larger one is marked "Made in Hong Kong / No. 357D" on two lines. The smaller ones are marked similarly, "Made in Hong Kong / No. 357A." *Carson Collection.*

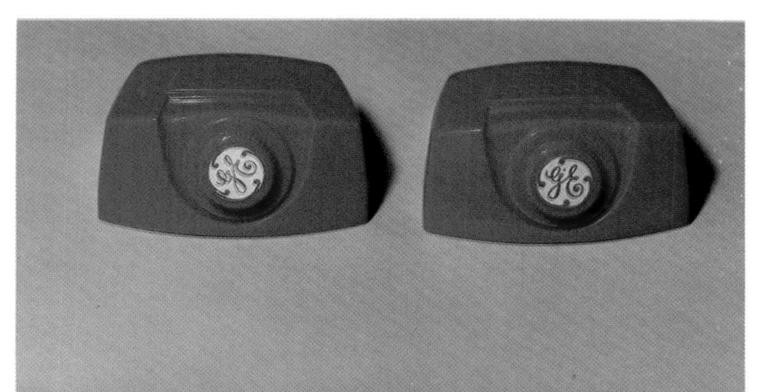

Plastic General Electric blanket controls, ⅞ x 2-⅛ inches. The mark, which gives no clue to origin, is shown below. *Thornburg Collection.*

Raised mark on above electric blanket control shakers.

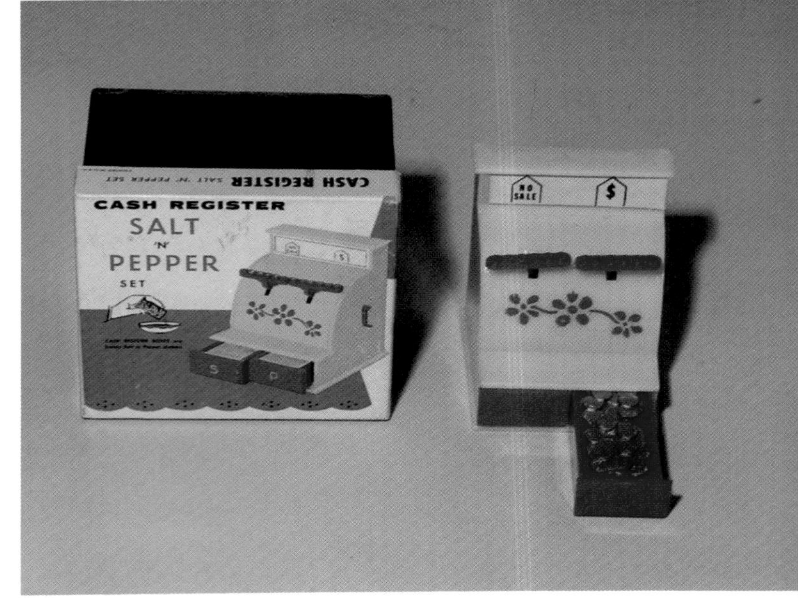

Plastic cash register; push down on the levers and the drawers open to reveal they are the shakers. The only mark is "Made in U.S.A." *Carson Collection.*

Grinder and spinning wheel. Measurements were not recorded. Both pieces have Enesco paper labels. *Carson Collection.*

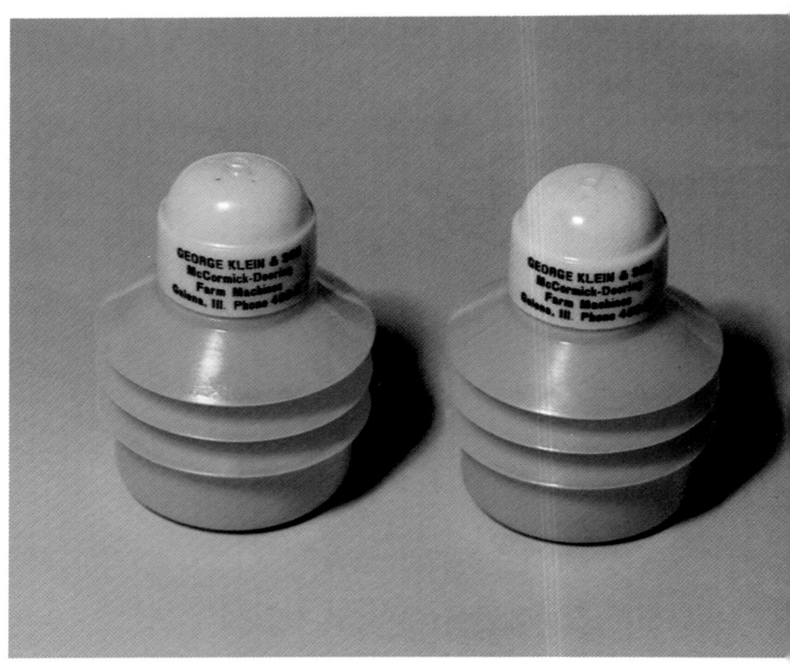

These are plastic, 2-½ inches high, marked "Perma Ware." The printing on this pair may be a bit hard to read. It says, "George Klein & Sons / McCormick-Deering / Farm Machines / Galena, Ill. Phone 408(unreadable numerals)" on four lines. *Thornburg Collection.*

Chapter 11
Containers

This chapter is mostly objects that contain liquid—teapots and coffee pots, beer and pop bottles, mugs and jugs, etc. But there all also a few that contain other things such as snuff, your Christmas gifts, and the daily mail.

Salt and pepper pour out of the spouts of these teapots. Their height is 2-½ inches. They are handpainted and marked "Occupied Japan," which is surprising because the workmanship far surpasses that of most ceramic products bearing that mark. At first glance you might suspect they are English or German porcelain. *Courtesy of Cloud's Antique Mall, Castalia, Ohio.*

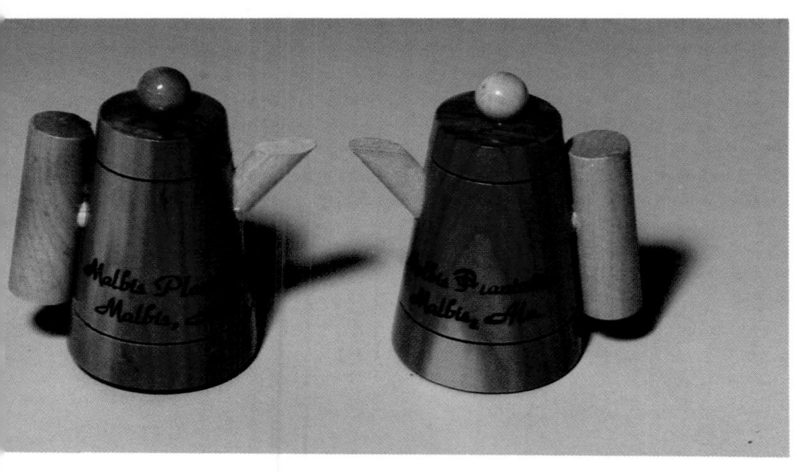

Souvenir cedar coffee pots, 2-½ inches high. They are not marked. *Carson Collection.*

Four Disney shakers. The coffee pots are 2-¼ inches high, have "Japan" paper labels. The mark of the bells is shown below. *Carson Collection.*

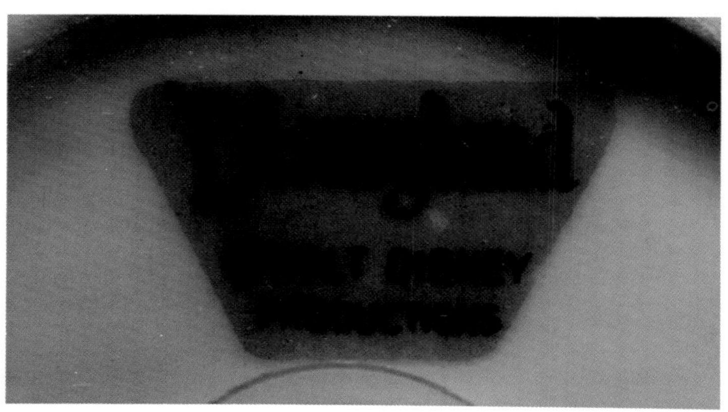

Mark of the above bells.

Unmarked plastic coffee pot, the bottom one shaker, the top the other. Height as shown is 4-¼ inches. *Private Collection.*

Hanging plastic teapot with plastic shakers. The teapot is 6-¼ inches high, the shakers 3 inches high. Mark on the teapot reads "S.P. for / Superlon / Chicago / Ill. / Made in USA / Des. Pat. No. D157497" on six lines. *Courtesy of Stan and Karen Zera.*

Here is another case of Occupied Japan becoming plain Japan once the pieces are removed from the box. As you can see, the box reads "Made in Occupied Japan" but the shakers have only "Japan" inkstamps. The shaker on the left is 2-⅝ inches high. *Lilly Collection.*

The coffee pots are 3-⅝ inches high, have a "Japan" paper label. The tray for the hearts is 6 inches long, the set is unmarked. The clowns stand 2-⅝ inches high, have "Japan" paper labels. *Courtesy of Joyce Hollen.*

Unmarked sitters, 4 inches high on the left, 3-⅛ inches high on the right. *Courtesy of Joyce Hollen.*

The shakers on the outside are 4-⅛ inches high, marked "Handpainted Japan." The pair on the inside have "Japan" inkstamps. *Carson Collection.*

The set in the middle is a repeat of that directly above. On the left are Cinderella coffee pots from Disneyland, 2-⅛ inches high and unmarked. The teapots on the right are 2-½ inches high, marked "Japan." Incidentally, shakers that say Disneyland on them can be no earlier than 1955, the year the Anaheim, California theme park opened. Walt Disney World, near Orlando, Florida opened in 1971, while Tokyo Disneyland debuted in 1983. *Courtesy of Joyce Hollen.*

This picture is a montage of materials. The shakers on the left are made of metal, 2-⅝ inches high and unmarked. In the center is a metal hotplate, 3-⅝ inches across, with glass shakers on top. None of the pieces are marked. The coffee pots on the right are plastic, 3-⅛ inches high, also unmarked. *Carson Collection.*

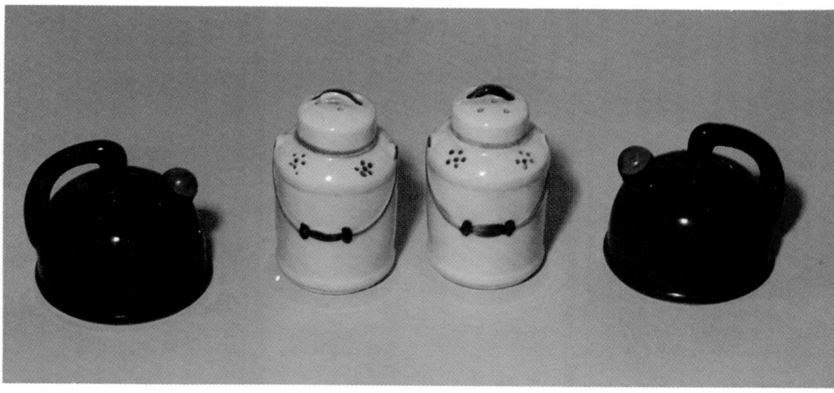

The teakettles are 2-¾ inches high, unmarked. The milk cans are unmarked, too, but they are Shawnee. They stand 3-¼ inches high. *Carson Collection.*

Redware teapots with African faces and metal handles. They are 4-¼ inches high to the tops of the handles. They are not marked. *Courtesy of Joyce Hollen.*

These milk cans are McCoy. There is also a cookie jar in the same shape, and several other matching pieces with identical glaze and design. The decal is called Spice Delight. Height of the shakers is 3-¾ inches. They are marked "USA." *Lilly Collection.*

More teapots and coffee pots with faces. The pair on the left measure 3-¾ inches. Like the pair on the right, they are not marked. The set in the center has A Quality Product paper label. *Thornburg Collection.*

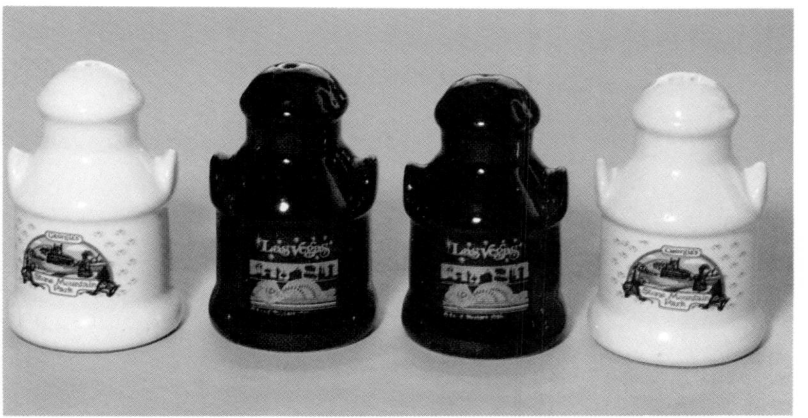

These were obviously made by the same company, but only the black pair is marked. It has a paper label that reads "By: Karol / Western / Los Angeles California / Made in / Taiwan" on five lines. Measurements were not taken but they run between 3 and 4 inches. *Carson Collection.*

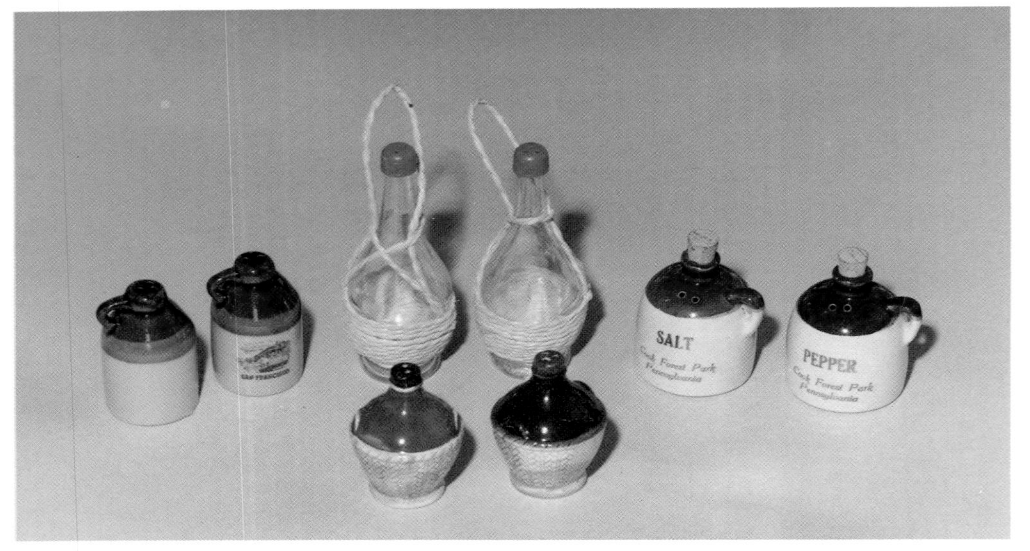

The glass bottles are 5-½ inches high, unmarked. The jugs on the right, souvenirs of Cook Forest, Pennsylvania, stand 3 inches high, are marked "Handcrafted / 22K Gold /

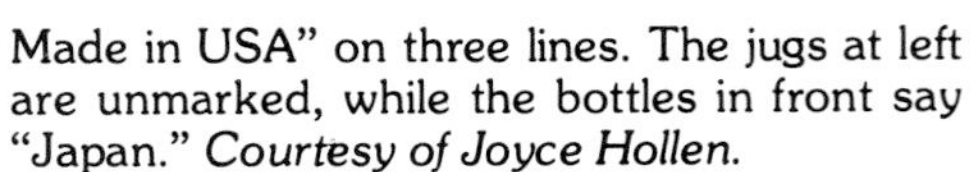

Made in USA" on three lines. The jugs at left are unmarked, while the bottles in front say "Japan." *Courtesy of Joyce Hollen.*

I guess corks are not containers but this seems like the most appropriate place to put them. The white ones are 2-⅝ inches high, marked "Porcelain du France" with a gold inkstamp. The brown ones are unmarked. You might have a hard time finding them at your local garage sale as they were sent from France to Irene and Jim Thornburg as a thank you gift for crewing during a Michigan hot air balloon rally. They are the only such set I ran into, so it seems probable they were not exported on a commercial basis. *Thornburg Collection.*

Ceramic Mason jars. Height on the left is 3-½ inches, height on the right is 3-¾ inches. Marks, or lack of them, were not recorded. *Thornburg Collection.*

These Coca-Cola bottles are 3-¾ inches high. No mark information was recorded. *Thornburg Collection.*

Advertising whiskey bottles, made of glass, 5-⅛ inches high and unmarked. Holder is metal. *Carson Collection.*

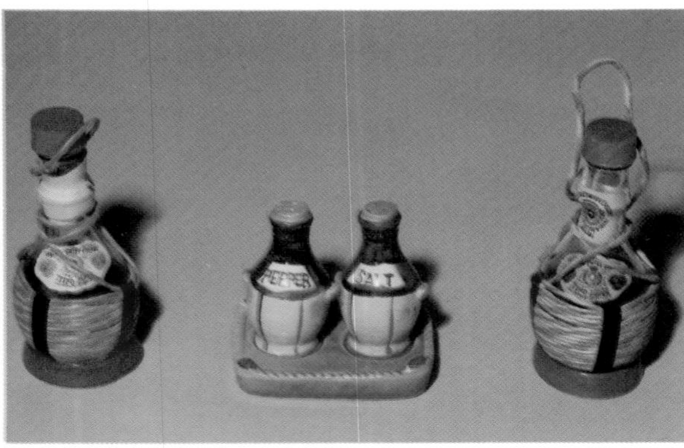

The glass Chianti bottles are 4-¾ inches high. They are not marked except for the Italian Swiss Colony advertising labels. The shakers of the tray set in the center are 2-¾ inches high, the tray is 3-¼ inches long. All three pieces have "Japan" inkstamps. *Lilly Collection.*

All glass, no pertinent marks. The Golden Age bottles are 5-½ inches high. The canning jars in front have the Ball logo. *Carson Collection.*

The seven-up bottle is 4-⅛ inches high. This may be a true pair, may not. Both pieces have Enesco paper labels. *Courtesy of Jazz'e Junque Shop, Chicago, Illinois.*

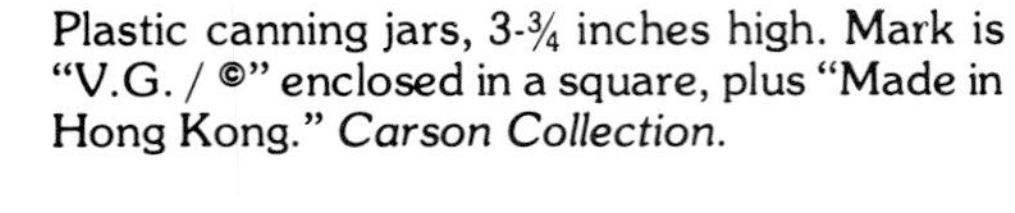

Plastic canning jars, 3-¾ inches high. Mark is "V.G. / ©" enclosed in a square, plus "Made in Hong Kong." *Carson Collection.*

The Sealtest bottles are 3-¼ inches high, the Seven-Up bottles 4 inches high. None are marked. *Carson Collection.*

Coors Golden Export Lager Beer. Height of these bottles is 3-½ inches. They are not marked. *Thornburg Collection.*

Beer bottles in two different sizes, 4 and 4-⅜ inches. The pair on the right is marked on their bottoms in raised letters, "Muth & Son Buffalo NY" in a circle with "Pat. Pend." inside. Mark of the pair on the left is "Muth / Pat. Pend. / Buffalo" on three lines. *Courtesy of Allen and Michelle Naylor.*

These are the 4-inch variety. Marks were not checked. *Carson Collection.*

The pair of Budweiser bottles on the right are 4 inches high. Oddly, they are quite a bit heavier than the similar pair on the left. As a general rule with beverage bottle salt and pepper shakers, the ones with metal caps, such as those in the middle of this picture, are older and more difficult to find. *Carson Collection.*

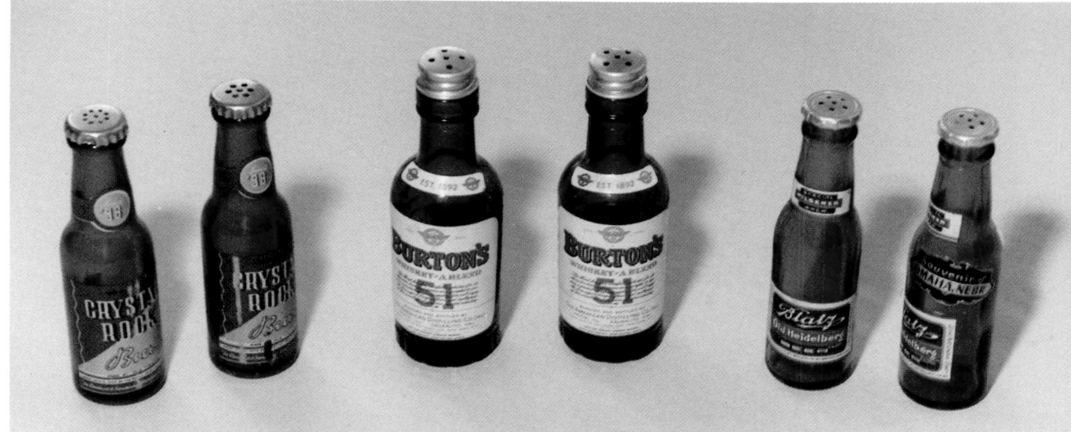

The Crystal Rock bottles are 4-5/16 inches high, Burton's Whiskey 4-⅝ inches, and Blatz Old Heidelberg 4-⅜ inches. Mark information was not recorded. *Carson Collection.*

The bottles in the middle are 4-½ inches high, a souvenir of Cleveland, Ohio. Both sets of bottles are glass. The cans are cardboard. *Carson Collection.*

Here is an interesting situation, a pair of bottles advertising Knox Glass, Inc., of Knox, Pennsylvania but made by Muth. They are marked "Muth / Pat. Pend. / Buffalo" on three lines. The size of this set was not recorded. *Lilly Collection.*

New Era potato chip cans, 3-¼ inches high, made of metal. Mark is shown below. *Thornburg Collection.*

These are plastic, marked "Hong Kong" on the bottom. The salt is 3-¼ inches high. *Carson Collection.*

These are tall, 6-⅝ inches. That is because they are not salt and pepper shakers but real beer bottles to which plastic caps have been added. *Carson Collection.*

Embossed Decoware mark of New Era potato chip cans above.

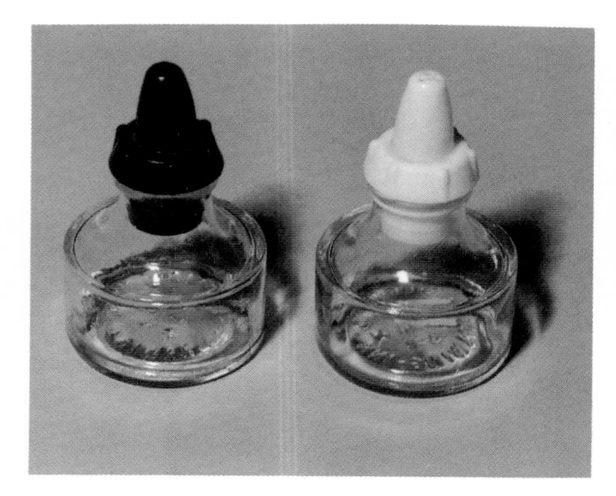

Higgins ink bottles, glass with plastic tops, 2-¾ inches high. Marks indicate they were made by Hazel Atlas. *Thornburg Collection.*

Still in the original packaging, these Budweiser bottles are plastic and 4 inches high. Purchased recently, the price sticker says $3.99. *Lilly Collection.*

Metal Campbell soup cans, 4 inches high, with original box. No marks. *Thornburg Collection.*

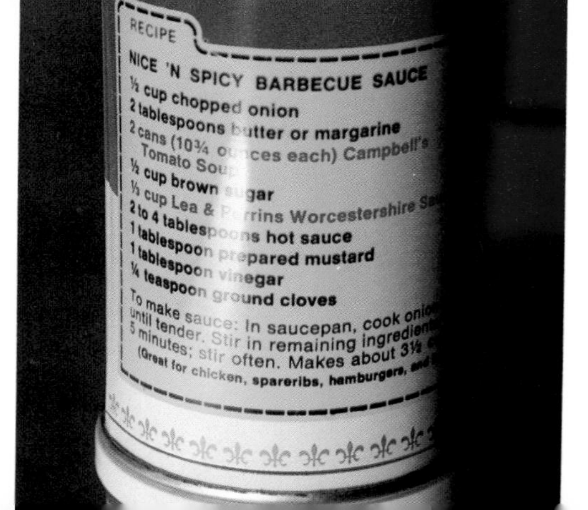

RECIPE

NICE 'N SPICY BARBECUE SAUCE

½ cup chopped onion
2 tablespoons butter or margarine
2 cans (10¾ ounces each) Campbell's Tomato Soup
½ cup brown sugar
½ cup Lea & Perrins Worcestershire Sauce
2 to 4 tablespoons hot sauce
1 tablespoon prepared mustard
1 tablespoon vinegar
¼ teaspoon ground cloves

To make sauce: In saucepan, cook onion until tender. Stir in remaining ingredients. 5 minutes; stir often. Makes about 3½ (Great for chicken, spareribs, hamburgers, and

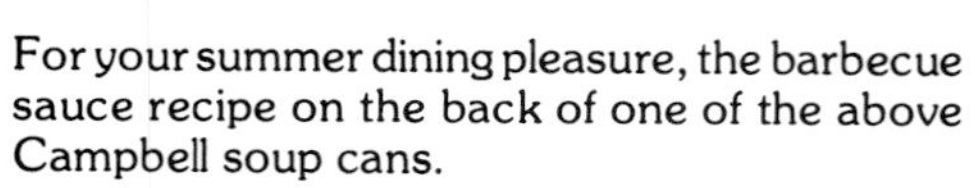

For your summer dining pleasure, the barbecue sauce recipe on the back of one of the above Campbell soup cans.

Plastic snuff bottles made to imitate cinnabar. Height is 3-¼ inches. They are marked, "Made in Hong Kong" and "No. 003" with No. and 003 widely separated. *Carson Collection.*

Cedar bottles and pots. Bottles are 3-¾ inches high, pots are 3-⅛ inches high, not including the metal handles. *Carson Collection.*

Metal minis. The carrier is 1-⅝ inches wide, the smaller of the shakers just ⅝ inches high. *Thornburg Collection.*

Don't know exactly what these are. Seems they should have handles. Height is 3-¼ inches, the tops are plastic. Mark is shown below. *Courtesy of Joyce Hollen.*

Mark of above shakers without handles.

These appear to be the same as those above but are far from it, mainly because they are made of milk glass instead of ceramic. Height is 3-¼ inches. Only mark is on the plastic tops, "Pat. No. 4783" which doesn't make any sense since that number was issued more than 100 years ago whether it signifies an invention patent or a design patent. Perhaps the number signifies a foreign patent. *Carson Collection.*

Souvenirs of Canada, 4-⅛ inches high. They have cellophane labels that say "Japan." Courtesy of Cloud's Antique Mall, Castalia, Ohio.

Steins, 2-⅞ inches high, "Japan" inkstamp. *Carson Collection.*

The handled containers in the center (you may call them what you want) are 3-¾ inches high. They were a souvenir of Deer Forest in Coloma, Michigan. The pair on the left is 2-¾ inches high, the pair on the right 2-½ inches high. *Carson Collection.*

The pair on the left stands 4-¼ inches high, has a "Japan" paper label. The pair on the right is 2-½ inches high, has "Japan" inkstamps. *Courtesy of Joyce Hollen.*

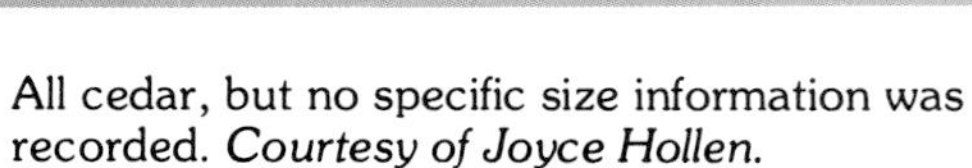

All cedar, but no specific size information was recorded. *Courtesy of Joyce Hollen.*

The plastic cups are 4 inches high. They are marked "Burroughs Mfg. Corp. / Los Angeles, Calif. USA / 308 / Burrite" on four lines. The inside pair measure 2-⅞ inches in height. They are also made of plastic but not marked. *Private Collection.*

The mugs are 2-¼ inches high, marked "Made in Japan / Takahshi / S.F. / 94103" on four lines. *Courtesy of Joyce Hollen.*

Plastic holders, wood shakers, metal tops. Height is 2-¾ inches, no marks. *Carson Collection.*

The brown mugs are 2-¾ inches high, have "Made in Japan" paper labels. The blue ones stand 3 inches high. Their paper labels are shown below. *Carson Collection.*

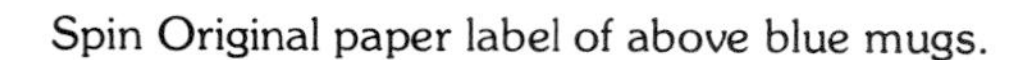

Spin Original paper label of above blue mugs.

Looks the same as above but a little different story. The blue shakers are plastic, marked "Made in Hong Kong." The boxed set is the same as above with the exception that it is marked "Japan." Both sets are 2-¾ inches high. *Courtesy of Joyce Hollen.*

Glass shakers, wood handles, metal bands, plastic tops. Height is 3-⅜ inches. Mark is "Artmark Made in Taiwan." *Carson Collection.*

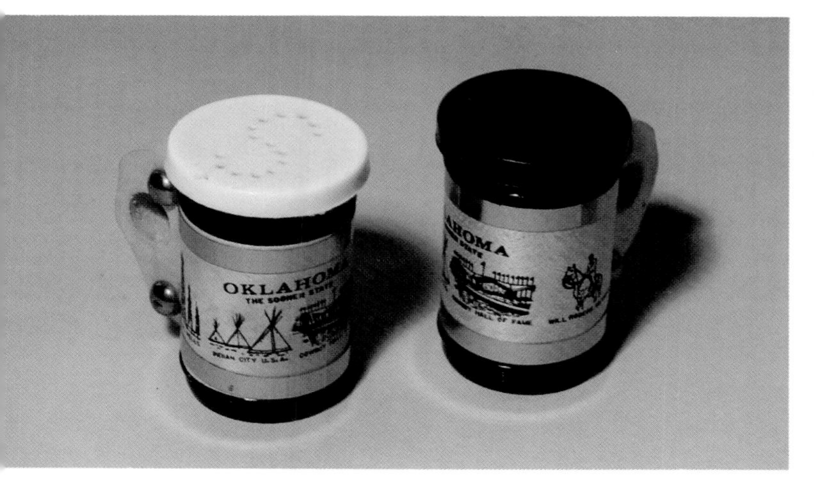

These are smaller than above but made of the same materials. Paper label says "Taiwan." *Carson Collection.*

The tan pair is 2-⅜ inches high, the brown pair is ⅛ inch higher. Paper labels are shown below. *Carson Collection.*

Kelvin's paper label of the tan shakers above.

Thrifco paper label of the brown pair.

McCoy salt and pepper and creamer. The shakers are 3-¼ inches high, the creamer is 5 inches high. None of the pieces are marked. *Lilly Collection.*

Shawnee jugs with handpainted heartland tulips. Height is 4-¼ inches. They are not marked. *Smith Collection.*

These are unmarked Purinton Pottery, 2-½ inches high. *Carson Collection.*

Pepper mill and salt shaker, made of Italian alabaster. Size was not recorded but they are about 4-½ inches high, as I recall. *Courtesy of Stan and Karen Zera.*

Plastic lunchbox with vacuum bottle shakers. I got in a bit of a hurry here, wrote down 2-⅞ inches but failed to note whether it was the box or the shakers. The lid of this lunchbox had broken off. The box is shown below with the lid on it. *Courtesy of Allen and Michelle Naylor.*

The gifts stand 3-⅜ inches high, are marked "Inarco." The decal on the front of the pair in the middle says Santa Workshop North Pole New York. The pair on the right are marked "Ron / Gordon / Designs, Inc. © Taiwan" on three lines. *Carson Collection.*

The shaker on the left is 2-⅛ inches high. Its paper label is shown below. *Thornburg Collection.*

Paper label of above gift shaker.

The present is 1-½ inches long, the card is 1-½ inches high. Neither is marked. *Oravitz Collection.*

Above lunchbox with lid attached. The bottom is marked "Starke Design Inc. / Bklyn 20, N.Y." in raised letters on two lines.

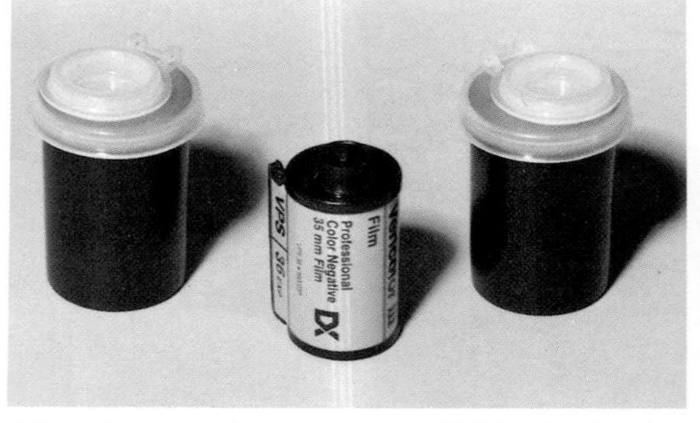

After photographing more than 2000 sets of salt and pepper shakers, to me this pair is the most familiar of all. It is Kodak film containers fitted with special lids so they maybe used as shakers. *Thornburg Collection.*

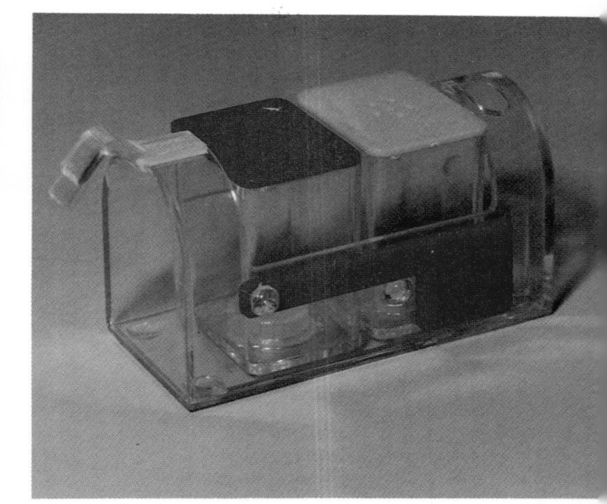

Plastic mailbox with a hole for toothpicks at the rear. The shakers, of course, are the pieces with the yellow and red tops. The box is between 4-½ and 4-¾ inches long. It is marked "Made in Hong Kong." *Carson Collection.*

Chapter 12
Miscellaneous Figural Subjects

This chapter is larger than I would have liked it to be. It includes all those subjects for which there were not enough pictures to require separate chapters. Because it is so large and varied, and to facilitate your use of it, the individual subjects are listed in the Table of Contents.

Athletics and Games

This section very closely reflects weekend television sports programing—baseball, football, boxing, golf and bowling—with a few lesser known games sprinkled in.

The unmarked boxing gloves are 2-¾ inches high. *Carson Collection.*

The boxing gloves are the same height as those in the previous picture, 2-¾ inches, but do not appear to be the same because of the laces. The ball and glove on the right was apparently a souvenir of a Pittsburgh Pirates baseball game. None of these sets are marked. *Carson Collection.*

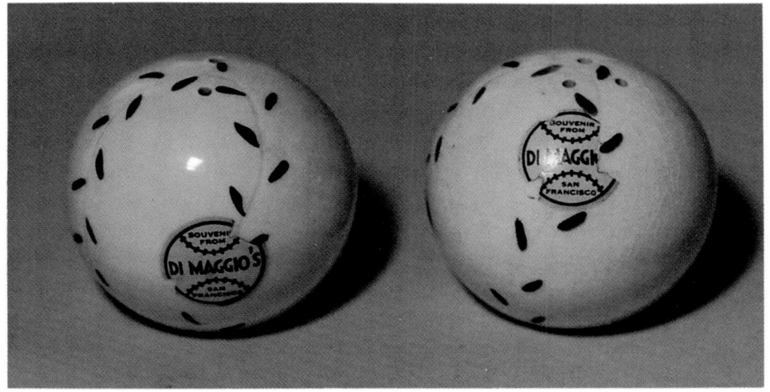

Here is a souvenir set of shakers from (Joe) DiMaggio's restaurant in San Francisco. The shakers are 2 inches high, unmarked. *Carson Collection.*

The ball and glove might be the same as above. The glove is 3-⅛ inches long. The football is 3 inches long. None of the pieces are marked. *Carson Collection.*

Metal golf balls here. The base, also metal, is 4-¼ inches long. Note the club at the front of the base. *Courtesy of Allen and Michelle Naylor.*

These footballs are 3-½ inches long. They are not marked. The golf balls and holder are made of wood. The balls are 2-⅝ inches in diameter. "Made in Japan" is on the bottom of the base. *Carson Collection.*

Wooden bowling ball and pin, 2-¼ and 5-⅞ inches high. "Japan" is impressed. *Lilly Collection.*

This is a plastic condiment set, the ball being the mustard. The bowling pins are recessed on the bottom so they may fit over upright prongs in the base that hold them in place. The pins are 3-½ inches high. The only mark is "Made in USA" in raised letters on the base. *Lilly Collection.*

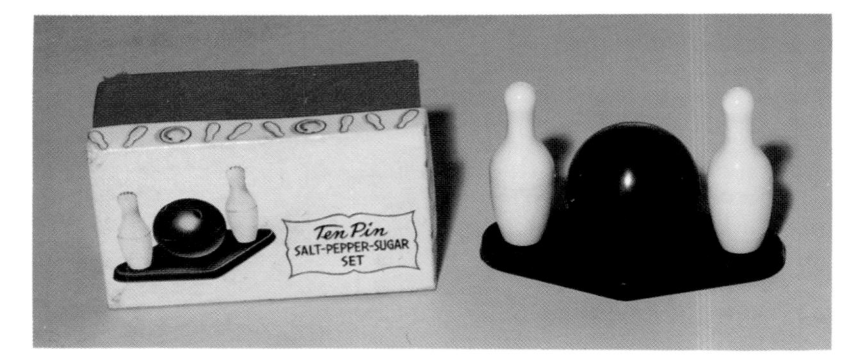

The footballs in the rear are 2-½ x 4 inches. The helmet in front is 2-⅛ inches high. None of these shakers are marked. *Lilly Collection.*

Height here is 2-⅝ inches. Each shaker has a "Japan" inkstamp. *Carson Collection.*

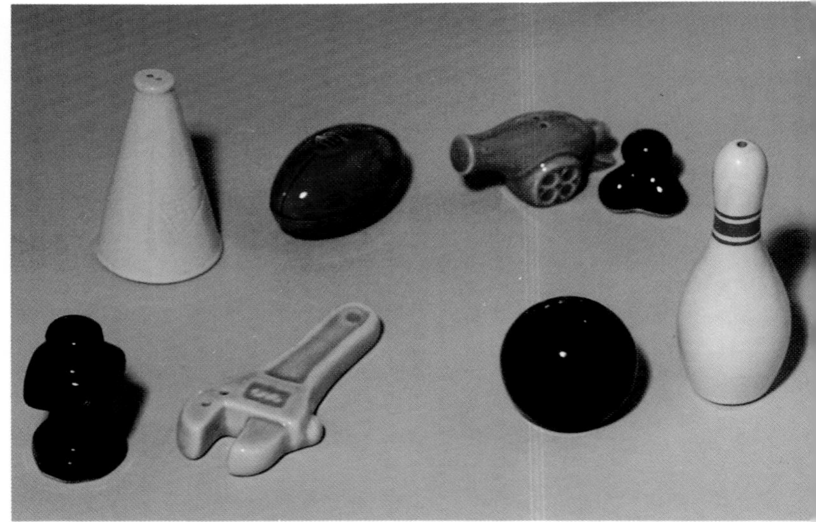

The bowling pin is 4-½ inches high, the megaphone is 3-⅜ inches. None of the pieces in this photo are marked with the exception of the megaphone, which has a Bernard Studio paper label. *Carson Collection.*

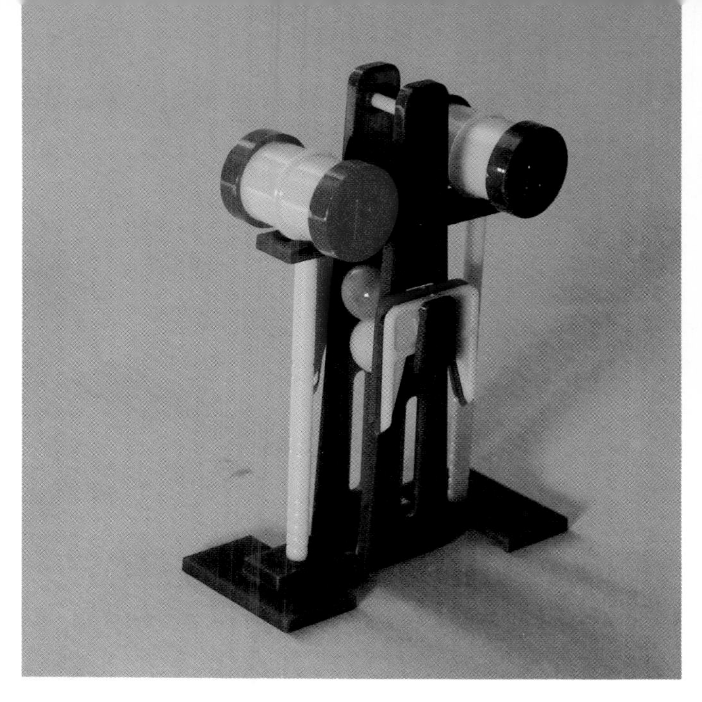

The stand of this plastic croquet set is 5 inches high. The only mark is "Pat. Pend. / Made in USA" on two lines on the bottom. *Courtesy of Allen and Michelle Naylor.*

The King and Queen of diamonds, 2-½ inches high, and unmarked. *Carson Collection.*

Here are spades and clubs, now all we need is a heart. The wooden shakers are 4-½ inches high. The set is marked with a "Japan" inkstamp on cork. *Lilly Collection.*

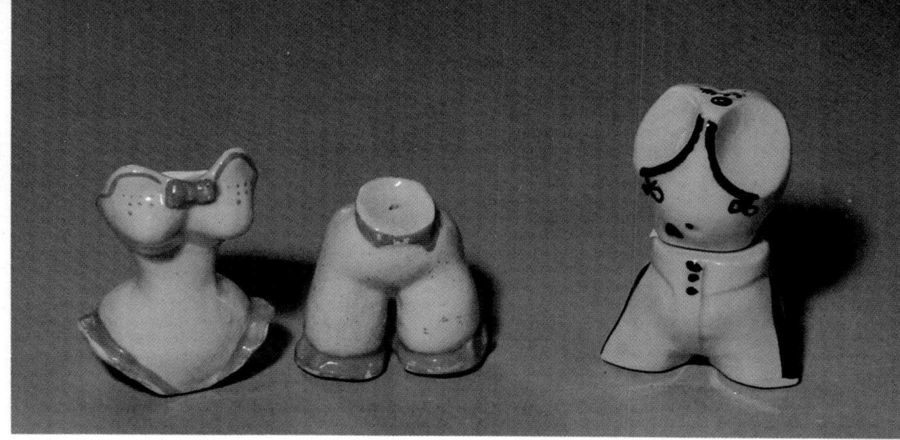

The best way to build anything, including a section in a book, is to start with a good foundation so that's what we will do here. Both of these sets are sitters. The shaker at far left is 3-¼ inches high. None of the pieces are marked. *Carson Collection.*

Clothing, Apparel and Accessories

Shoes and hats seem to be the big vote-getters with collectors in this area. Nearly every general collection I have seen has at least one pair of each.

These shoes are 3-½ inches long, have a "Japan" inkstamp. *Lilly Collection.*

The metal boots are 2 inches high. The tray under the teapots is 4-⅝ inches long. All pieces are marked "Made in Japan." *Carson Collection.*

Plastic wooden shoes, a souvenir of Holland, Michigan. They are unmarked, 2-⅞ inches long. *Carson Collection.*

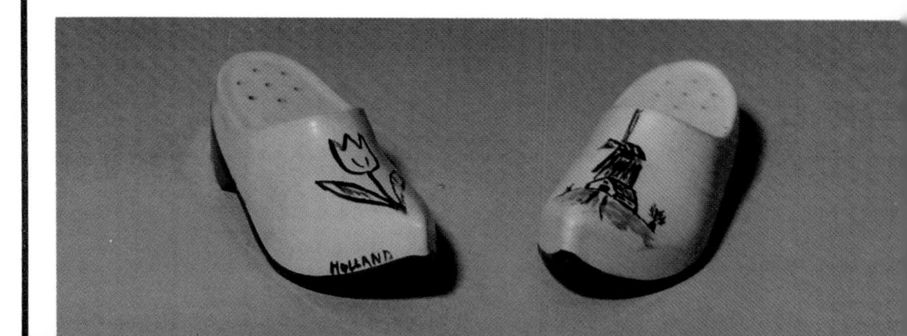

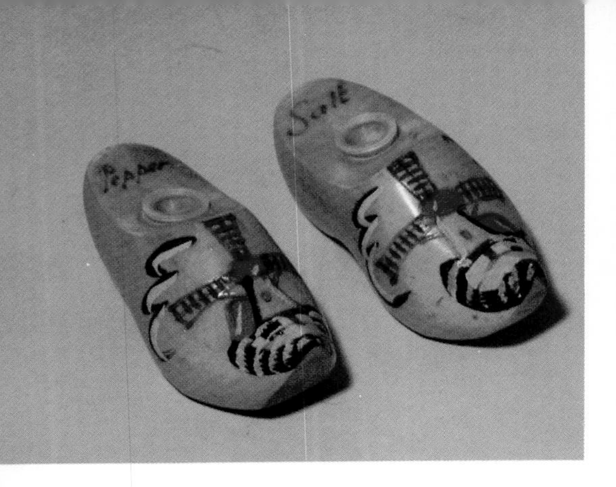

Here is the real thing, wooden shoes. Length is 4-¼ inches, marks are lacking. *Carson Collection.*

From Holland it's off to the Orient for some Japanese shoes. They are 2 inches high, unmarked. *Carson Collection.*

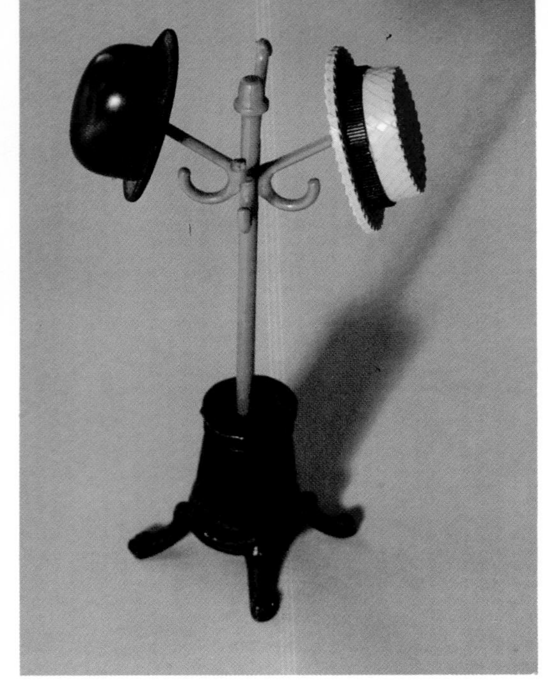

Sometimes this set is seen with two hats, sometimes four. The stand, which is weighted for stability, is 6-½ inches high. It is not marked. *Carson Collection.*

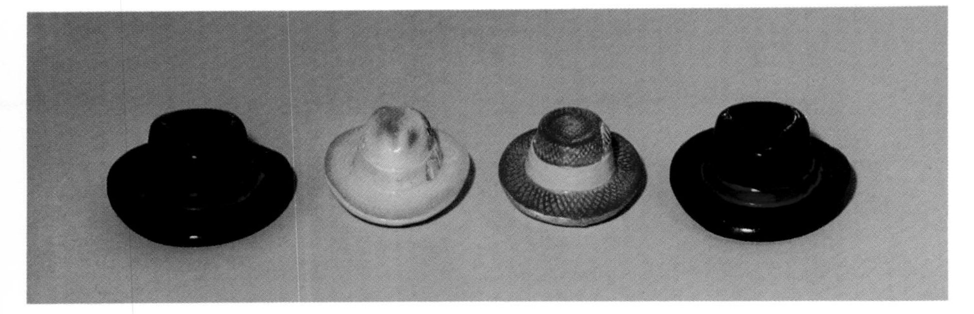

My notes say that the pair of hats on the outside are green, but you couldn't prove it by the picture (another of my indecent exposures, so to speak). Size and mark information was not recorded. *Carson Collection.*

General Douglas MacArthur's hat and pipe. The hat is 2 inches high. Neither piece is marked. *Carson Collection.*

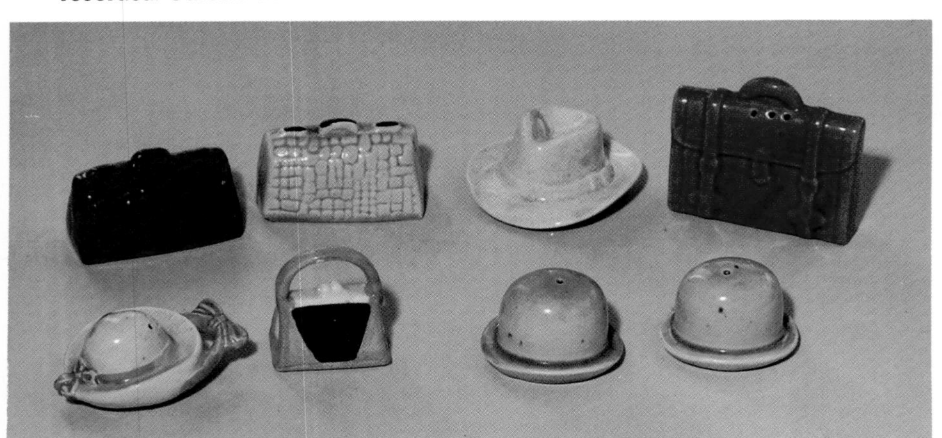

The purse in the front row is 2-7/16 inches high. Both it and its mate are marked "Japan." The other pieces are not marked. *Carson Collection.*

For size reference, the palm tree is 3-½ inches high. It, the date, and the man and lady's hats are unmarked. The pipe and holder is a Trevewood set, "Trevewood" impressed. *Carson Collection.*

The plastic umbrellas are 3-¼ inches long, while the umbrella stand is 6-¾ inches high. Note the S and P on the handles of the shakers. Like the hall tree above, the stand is weighted for stability. The base is marked "Made in USA Pat. Pending." *Carson Collection.*

Eating Utensils

Just four pictures here, but you get the idea. The three identical spoons and forks shown in the first two pictures are by far the most prolific sets. They are all over the place. Whoever made them did a nice job, which obviously accounts for their popularity.

As shown below, this spoon and fork was made in several different color combinations. Both members are 5-⅛ inches high, and marked "Japan." *Carson Collection.*

The shakers in the center and at right are marked as those above. The set on the left is unmarked. *Carson Collection.*

This set is 4-¼ inches high, marked "Japan." *Carson Collection.*

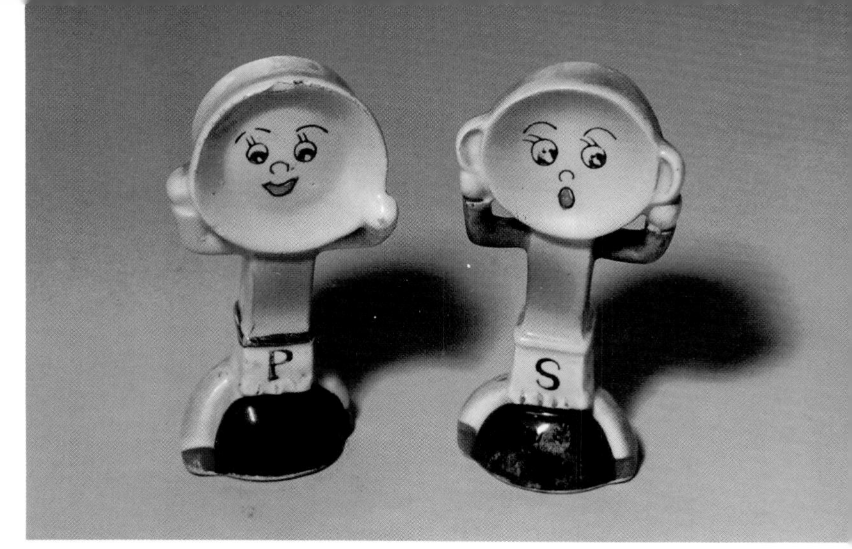

The height of these is 3-¾ inches. They are not marked. *Carson Collection.*

Food

It has always seemed to me that God played a trick on us in the food department. The better it tastes, meaning the more likely we are to eat it, the worse it is for us. Shown below is everything you need for a short happy life—cholesterol in the eggs, cheese and ice cream, caffeine in the coffee and chocolate, and too many calories in just about everything. And to top it all off every one of these "foods" holds salt, which the medical community now says is not good for us either!

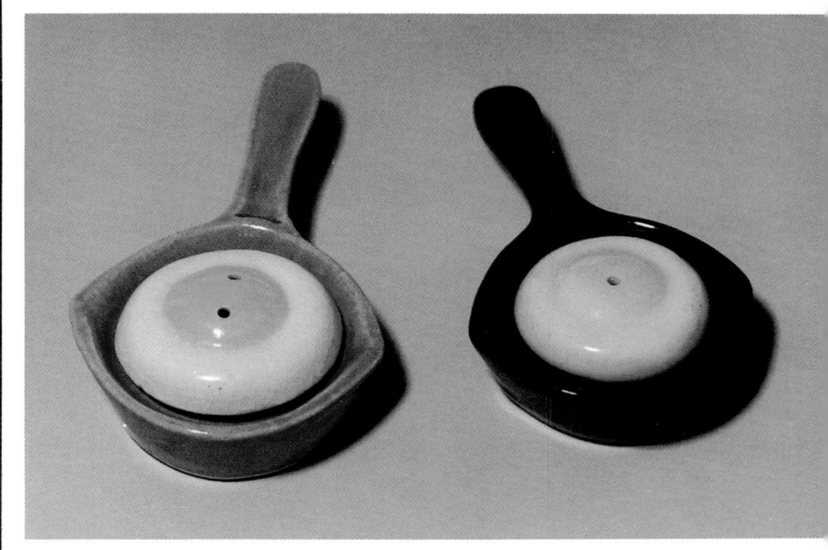

This is two sets, with each the egg being one shaker, the skillet the other. And they are different sets, apparently, since the skillet on the left is 4-½ inches long, the skillet on the right 4-¼ inches. None of the pieces are marked. *Carson Collection.*

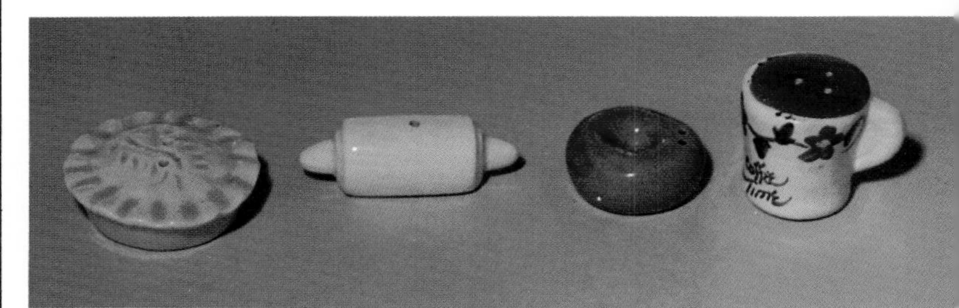

These are miniature sets. The rolling pin is 1-¾ inches long, the coffee cup 1-⅝ inches high. There are no marks. *Carson Collection.*

This pair of shakers represents Walt Disney World's nutrition show cereal sisters. Oats is 2-¾ inches high. Each is marked "© 1981 / Disney / Japan" on three lines. *Thornburg Collection.*

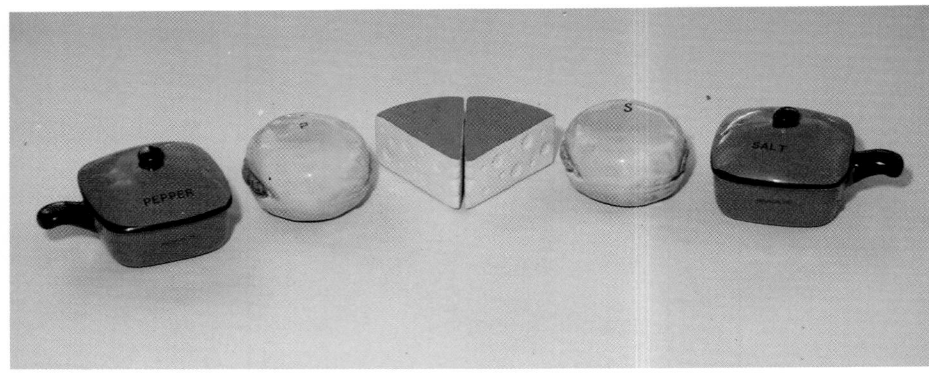

The shakers on the ends are unmarked. They are 3-¾ inches long. The hamburgers are 1-⅝ inches high, have Our Own Import paper labels, shown below. The cheese in the middle is marked with a paper label that says "Made in Taiwan / Republic of China" on two lines, and an inkstamp, "4S-10Z," on one line. *Carson Collection.*

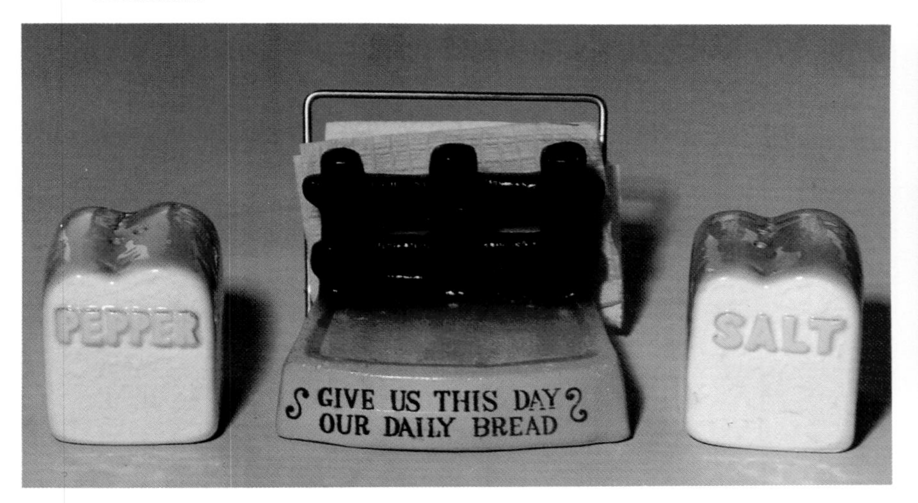

Paper label of above hamburger shakers.

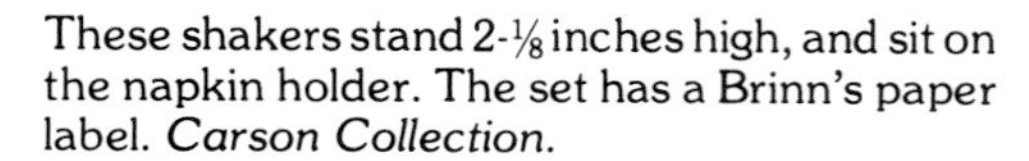

These shakers stand 2-⅛ inches high, and sit on the napkin holder. The set has a Brinn's paper label. *Carson Collection.*

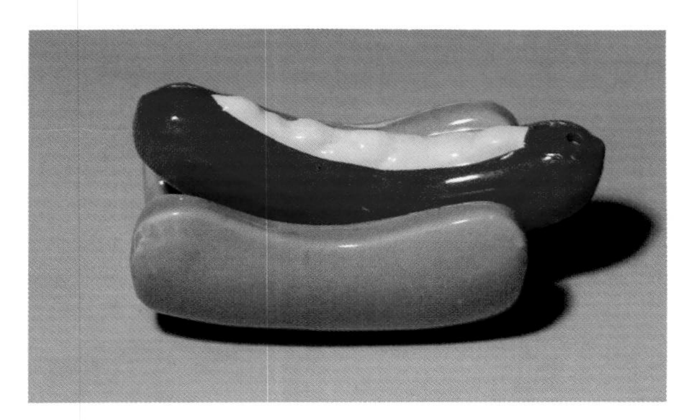

The hotdog is 3-½ inches long, the bun 2-⅞ inches. Paper label reads "Knobler / Taiwan" on two lines. *Lilly Collection.*

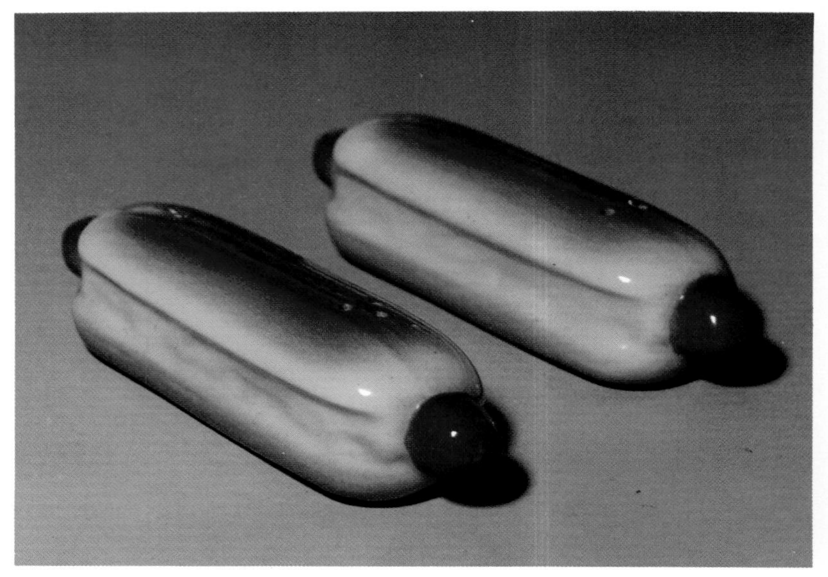

A pair of hotdogs, 4-¼ inches long. They are marked "Bernard Studios, Inc. / Mfgrs of Ceramics / Fullerton, Calif" on three lines. *Carson Collection.*

The hamburgers are 1-½ inches high, the skillets are 4 inches long. None of the pieces are marked. *Courtesy of Bev Urry.*

The ice cream cones are 3-$\frac{3}{4}$ inches high. The cones contain the words "Saf-T Cup." None of the shakers are marked. *Carson Collection.*

The coffee cup and donut is a repeat of above. The cookbook is 2-$\frac{1}{8}$ inches high, the sundae, or banana split if you prefer, is 4 inches long. None of these shakers are marked. *Courtesy of Bev Urry.*

The steaming turkey is a repeat. It measures 2-$\frac{3}{4}$ inches in length, has a "Japan" paper label. The fish and skillet are unmarked. *Carson Collection.*

Artmark chocolate rabbits, 4-$\frac{1}{4}$ inches high, with a "Taiwan" paper label. *Lilly Collection.*

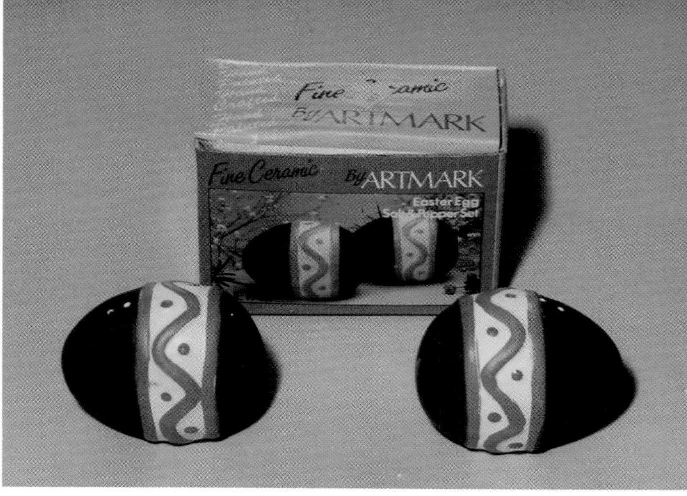

The original price for these Artmark Easter eggs, which are quite recent if not current, shows $1.99. The unmarked eggs are 2-$\frac{1}{4}$ inches high, 3 inches long. *Lilly Collection.*

Candy cane shakers, 4-$\frac{5}{8}$ inches high, with "Japan" paper labels. *Lilly Collection.*

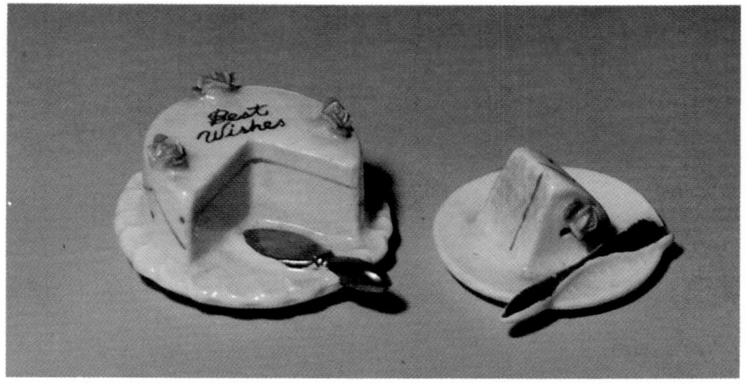

Here is a very detailed miniature set. The cake is just under an inch high, while the slice is 1-$\frac{5}{8}$ inches wide. Original price tag shows it was sold at Fred Harvey's for $0.98. *Lilly Collection.*

Plastic ice cream sodas, 3-½ inches high to the top of the straws. They are unmarked. *Carson Collection.*

Black walnut shakers actually made of black walnuts. They are 1-⅜ inches high, marked only by the cracks and crevices of their growth habit. *Lilly Collection.*

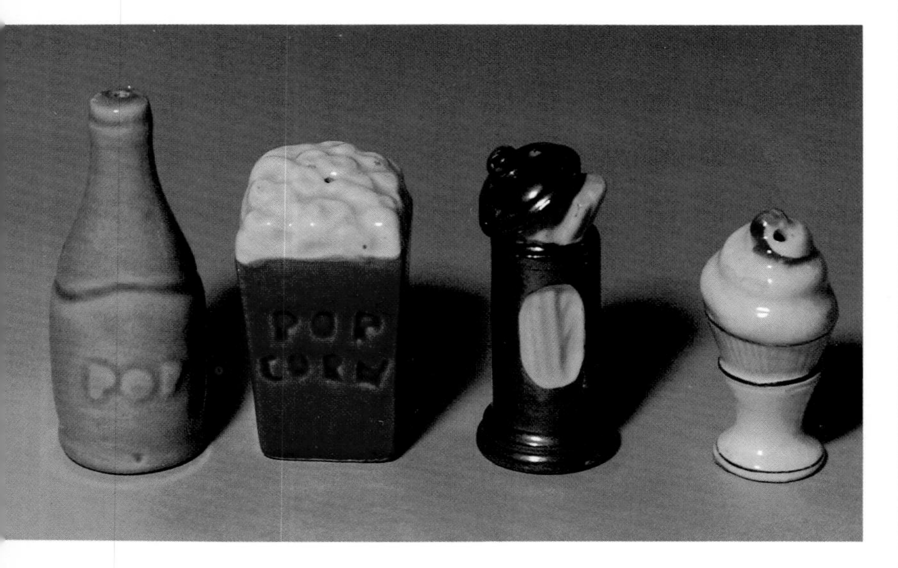

Pop and popcorn, ice cream soda and straws. The pop bottle is 3-¼ inches high. The set is unmarked. The ice cream soda has a Napco paper label along with a "2167" inkstamp. *Carson Collection.*

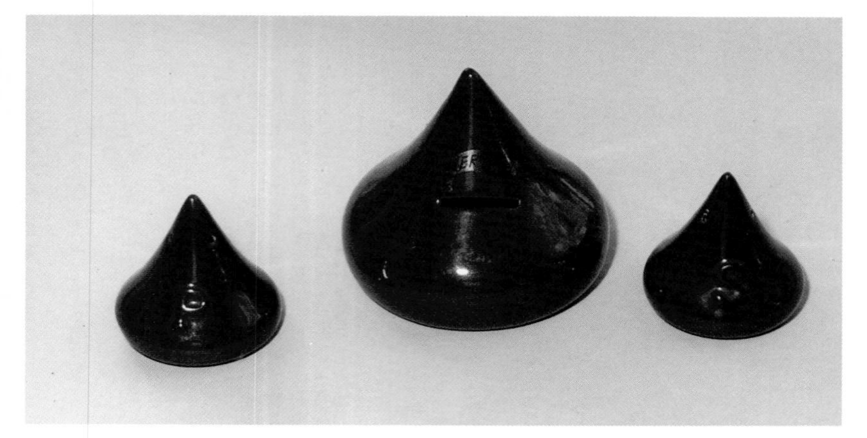

Hershey's chocolate kisses, in both shaker and bank form. The bank is 4-½ inches high. The pieces are not marked. *Carson Collection.*

LIGHTING

Lighting subjects include candles, lanterns, light bulbs—anything we use, or have used in the past, to provide artifical sunshine. It seems this would be a broad field, but when you begin looking around you just do not find that many shakers with a lighting theme.

These candles are 3-½ inches high. No other information was recorded. *Carson Collection.*

This pair of sitters stands 6 inches high as shown. Neither piece is marked. *Courtesy of Joyce Hollen.*

Shakespeare would write, "A pair or not a pair, that is the question." Whatever they are, the candle is 3 inches high, has a "Made in Japan" paper label with a "5572" inkstamp. *Carson Collection.*

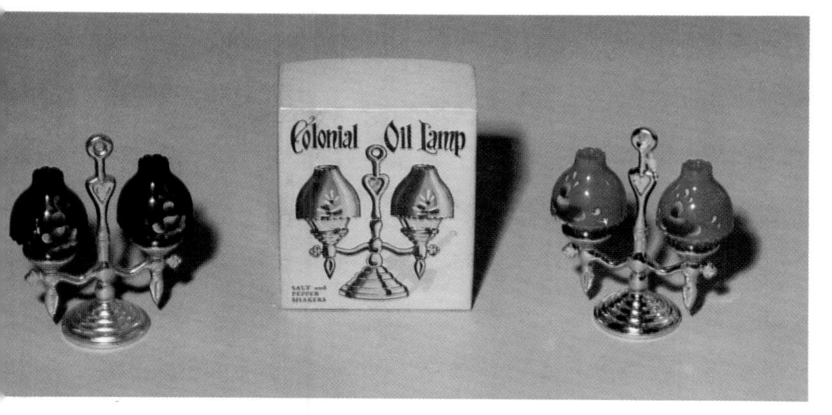

The stands of these plastic sets are 4-½ inches high, the shakers themselves being 1-¾ inches. The stands are marked, "Pat. Pend. / Made in U.S.A." on two lines on the bottom. *Lilly Collection.*

The set in the middle is simply another color of the above set. The lamps on the outside are made with metal bases and plastic tops. They are 3 inches high. Mark information was not recorded. *Courtesy of Joyce Hollen.*

I assume these are Oriental lamps. They are 3-¼ inches high, have a "Japan" inkstamp. *Courtesy of Cloud's Antique Mall, Castalia, Ohio.*

Wooden street lamps, unmarked, 3-⅞ inches high. These are a cut above many other wooden shakers as the finials are actually threaded, not just wedged in. *Carson Collection.*

These lamps stand 4-⅜ inches high, have Artmark paper labels that indicate they were made in Japan. *Courtesy of Cloud's Antique Mall, Castalia, Ohio.*

The bases of the sets at left and center are metal. The rooster is plastic. The base in the middle is 5-⅞ inches high, the shakers are 3-¾ inches long. None of the pieces are marked. *Carson Collection.*

This is a lantern candy container on which the manufacturer included holes in the top so it could be used for a shaker. It is made of glass and metal. Raised lettering on the globe indicates it was made by the Victory Glass Company of Jeanette, Pennsylvania. *Courtesy of Cloud's Antique Mall, Castalia, Ohio.*

Glass light bulbs, unmarked, 5-⅜ inches high. The caps appear to be the same as used on real light bulbs, and the paper label of the shaker on the right warns that the bulbs are for ornament only. *Courtesy of Allen and Michelle Naylor.*

Music

It wasn't too many generations ago that if you wanted to hear music, you had to go to town on Saturday night. Now with radio, television, and piped in music in offices and retail establishments, many of us can say there is seldom a time when we don't hear it. This ongoing relationship with melodious sound shows up often in salt and pepper shaker collections.

The center piece of the unmarked set on the left is a napkin holder. It is 3-⅜ inches high. The pair on the right, 3-⅛ inches high, carries a Knobler paper label. The mark of the set in the center is shown below. *Thornburg Collection.*

Inkstamp mark of the musical notes, above center.

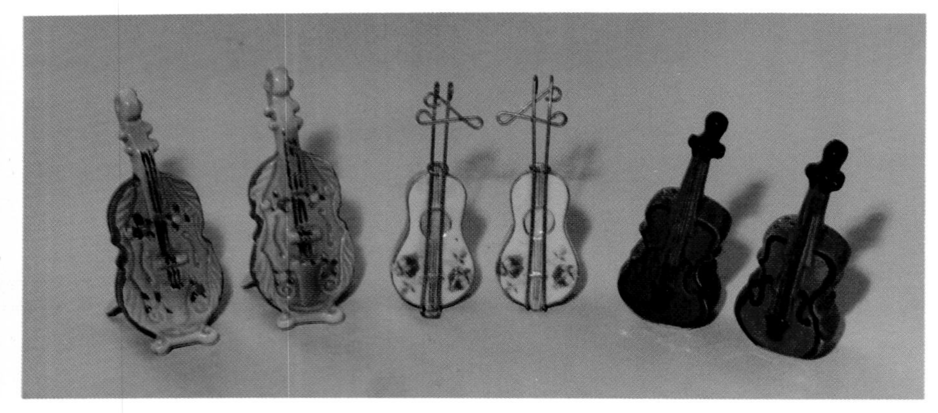

Heights left to right are 4-½, 4-¼ and 4 inches. The pair on the left has a "Japan" inkstamp. The paper label of the pair in the center reads "Relco / Hand Decorated / Japan" on three lines. The pair on the right has a Fern paper label. *Carson Collection.*

Height of these one-piece guitar shakers is 4-⅝ inches. They are made of plastic, "Pat. Pend." in raised letters on their backs. The music stands are a combination of plastic and metal. *Carson Collection.*

These shakers are 3-½ inches high, are unmarked except for the original owner's mark shown below. *Carson Collection.*

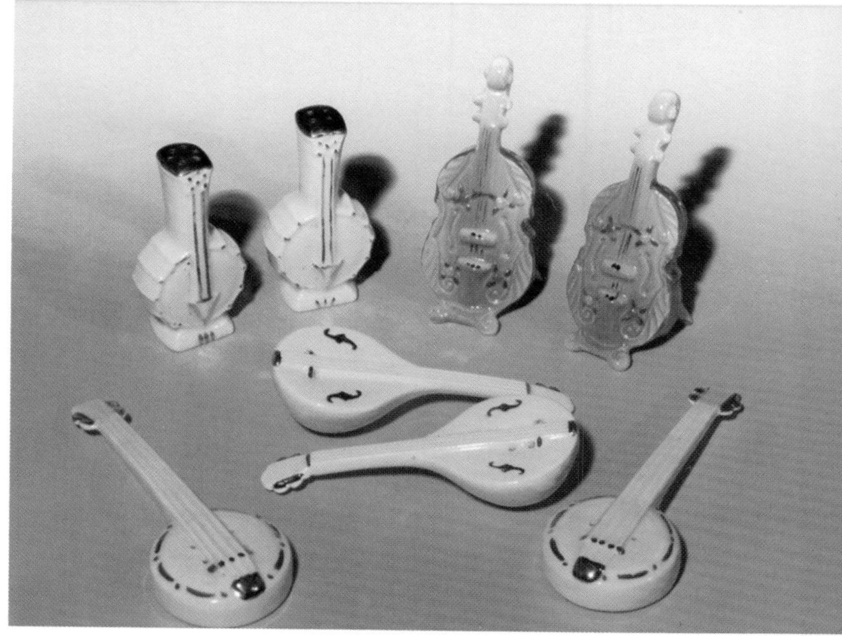

None of these sets are marked. The bases are 4-⅛ inches high, the banjos in the foreground, 5-¼ inches long. *Thornburg Collection.*

To some, one of the more interesting aspects of collecting shakers is knowing a bit of the history behind individual sets. As shown here, this pair was apparently picked up in Puerto Rico while the purchaser was vacationing on a cruise in March, 1982. Markings like these are also a great aid in dating shakers.

Instruments with faces, 4-¼ inches high, and unmarked. Note that this pair is not identical, one being a guitar, the other a violin. *Thornburg Collection.*

Violins and hands. The shakers are 4-½ inches high. Their inkstamp mark reads "Japan." *Courtesy of Cloud's Antique Mall, Castalia, Ohio.*

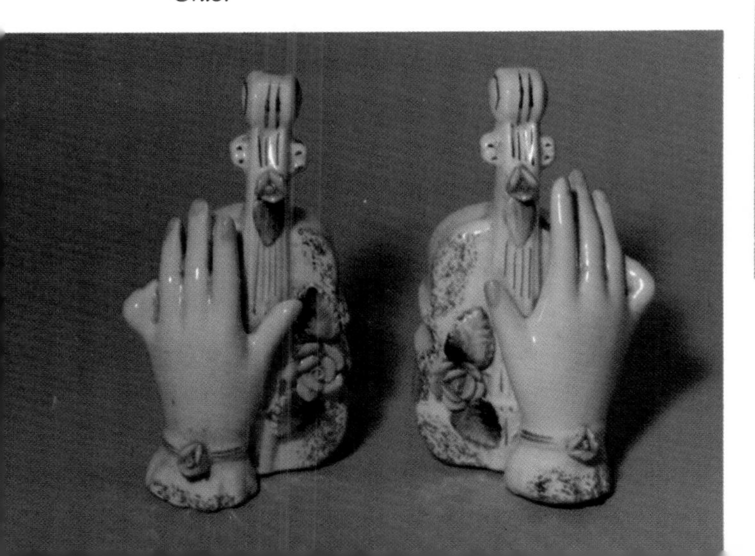

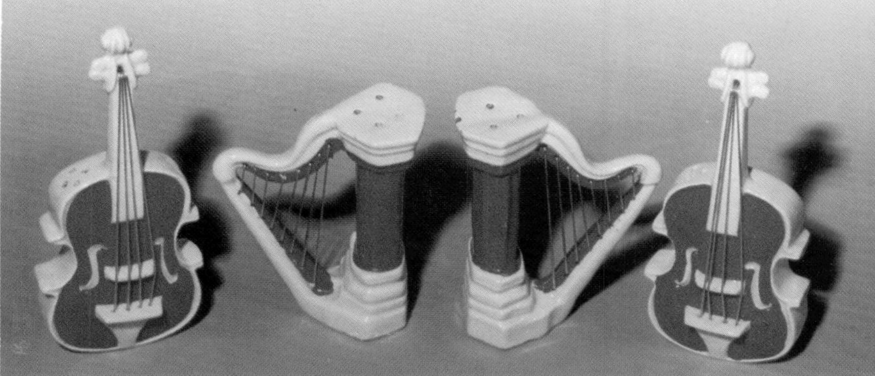

The cellos are 5-¾ inches high. Both sets have "Japan" inkstamps. Also made but missing from the picture was a set of mandolins. *Thornburg Collection.*

Height of the cello is 3-⅜ inches, width of the music book is 2-⅝ inches. Neither piece is marked. *Lilly Collection.*

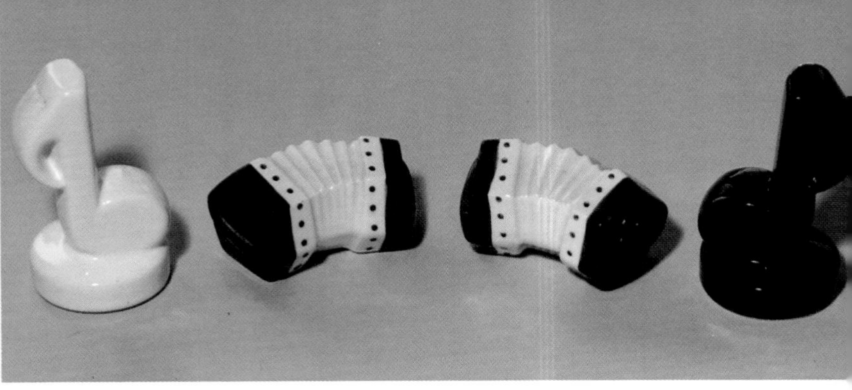

The notes are 3-⅛ inches high and marked "Sasparilla Deco Design." The concertinas measure 2-⅞ inches in length, have "Japan" inkstamps. *Carson Collection.*

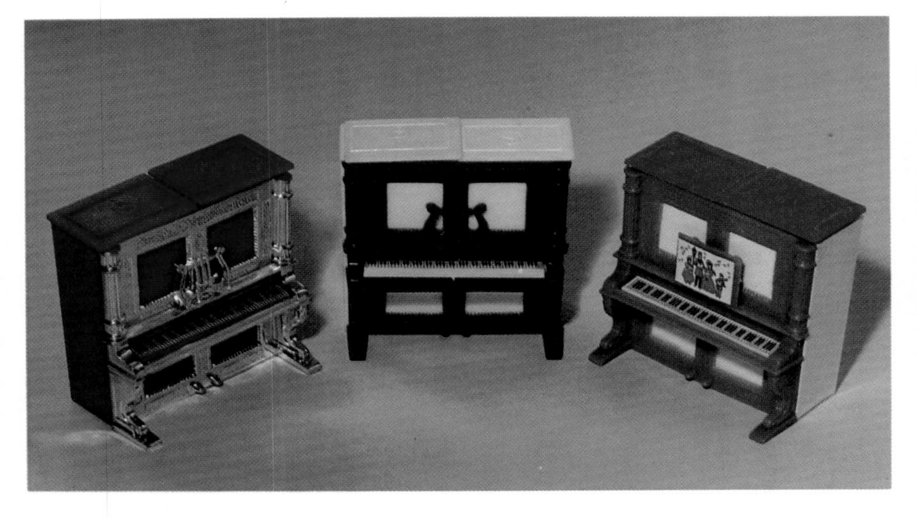

All three pianos are plastic. All are 3-½ inches high. But only one, the green, red and white, is marked. Its mark is "Davis Products / Brooklyn N.Y. / Pat. Pend. Made in USA." When you depress the keys of these pianos, the shakers rise. *Carson Collection.*

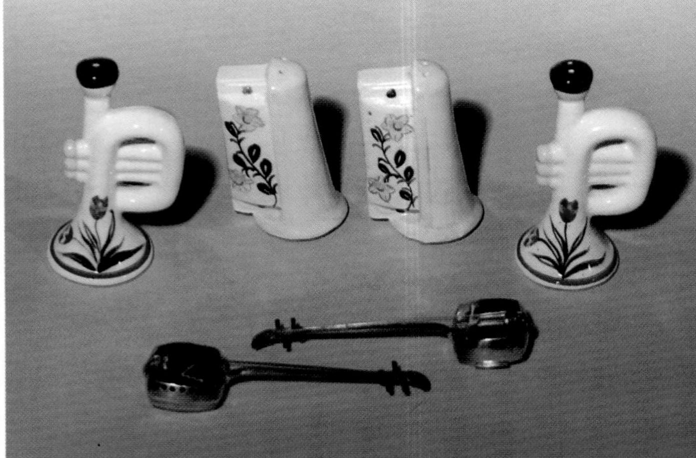

The pair on the outside is 3-⅝ inches high, has "Japan" inkstamps. The pair on the inside is marked similarly. Perhaps the pair in front is most interesting. They are metal, 4-⅛ inches long, and marked "Sterling .950" on the handles. The mark is unusual in that true sterling is .925 fine silver. They have metal strings, and sliding doors for filling that are shown below. *Thornburg Collection.*

Three go-with pianos with their benches or stools. The piano on the left is 2 x 3-⅛ inches. Its stool is 1-⅛ inches high. Both have Five and Dime paper labels. The shakers in the center have "© Nec" incised, while the ones on the right sport "Japan" inkstamps. *Thornburg Collection.*

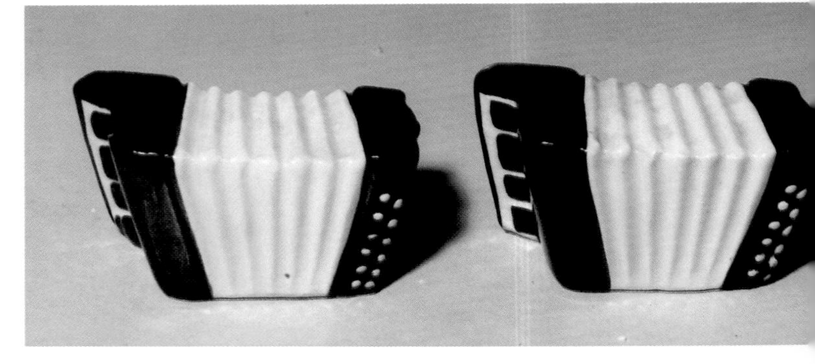

Accordians, 2-½ x 2-¾ inches. They have "Japan" inkstamps. *Lilly Collection.*

Sliding doors of the above metal shakers.

These harmonicas measure 3-¼ inches in length. They have "Japan" inkstamps. *Carson Collection.*

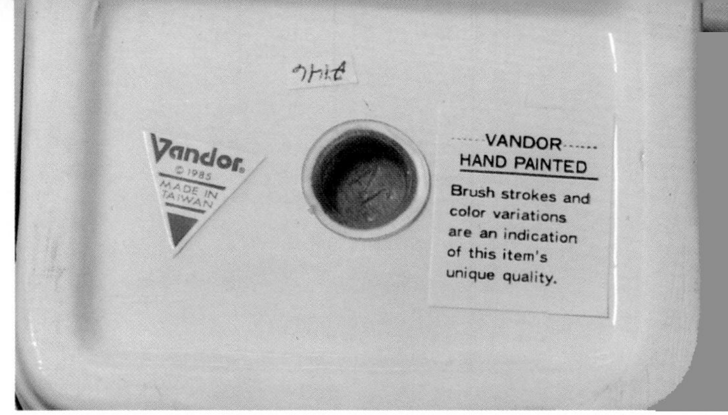

Bottom of Vandor jukebox. I enjoyed the disclaimer so much that I just had to show it. Of course, I feel the same way about this book. Any errors in spelling, grammar, or mistakes in identification are an indication of its unique quality.

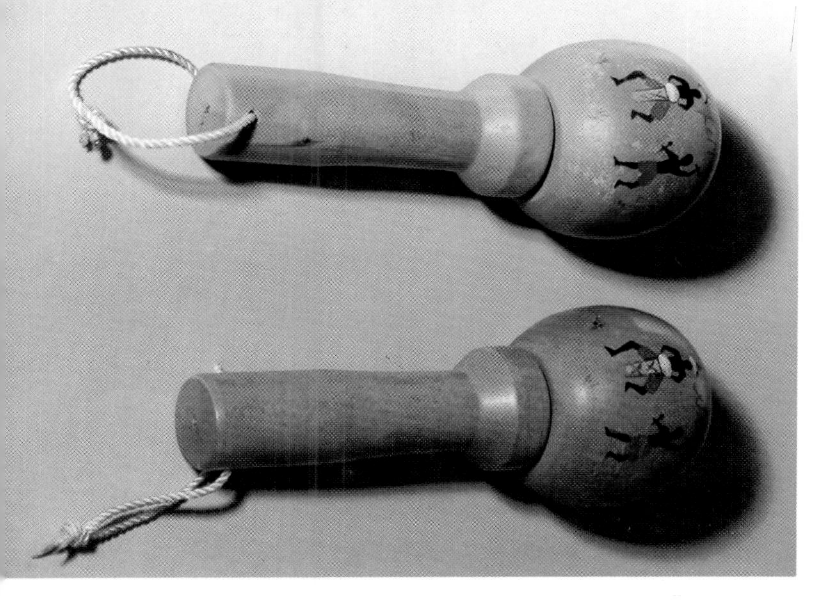

Here's a pair of shakers that are really shakers. They are made of wood, 6-¼ inches long, and unmarked. Each has a flat spot on the back to prevent it from rolling around on the table. *Thornburg Collection.*

The jukebox set on the left was made by Vandor. Its paper label is shown below. The pair on the right, which sits one on top of the other, is 3-⅞ inches high as shown. It has a Five & Dime paper label. *Thornburg Collection.*

Potluck

Here we go, the catch-all section of the catch-all chapter. Clocks and watches, scarecrows, insulators and gravestones are just a few of the subjects found here.

These bells are 5-¼ inches high, unmarked. They really ring. Only the handles hold the salt and pepper leaving the bottom free to function as a chime to call people to dinner. *Courtesy of Trina Cloud.*

The eyepieces of these binoculars are the salt and pepper. They lift out. The set is unmarked. Measurements were not taken. *Courtesy of Cloud's Antique Mall, Castalia, Ohio.*

The Liberty Bell, made of metal, 1-¾ inches high, and unmarked. *Carson Collection.*

A couple go-togethers. The case of the binoculars is 3 inches high. The set is not marked. The telephone is marked "Japan." *Carson Collection.*

The books are 2 inches high. The paper label reads "Bone / China / Japan" on three lines. *Carson Collection.*

These wooden alarm clocks are 3-⅛ inches high. Inkstamps on their backs indicate they were once a souvenir of the New York Thruway. *Zera Collection.*

I didn't know where to put this rare Regal China pair, so I'm slipping them in here. They are Tweedle Dum and Tweedle Dee from *Alice in Wonderland*. Height is 4-⅝ inches. Their impressed mark reads "Walt Disney / Production / Company" on three lines. *Private Collection.*

Here is another pair of where-do-you-put-thems, Al Capp's Smoos. The one on the right is 4 inches high. Each has a "Made in Japan" paper label along with an "8294" inkstamp. *Thornburg Collection.*

Alarm clocks, 3-¼ inches high, with a "Japan" inkstamp. *Carson Collection.*

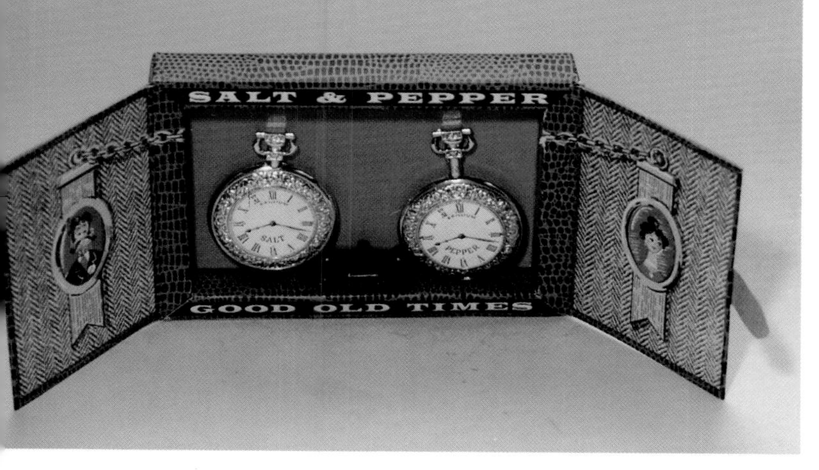

Watch salt and peppers in their original box. The shakers are 3 inches high. The back of the box is shown below. *Courtesy of Jazz'e Junque Shop, Chicago, Illinois.*

Wood chopping blocks made of cherry, oak and walnut. They were not measured. They have Enesco paper labels. *Carson Collection.*

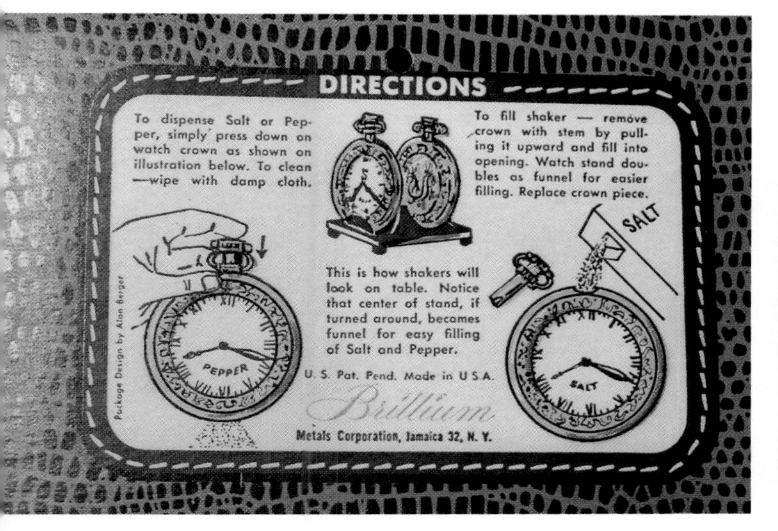

Back of the box of the above watch shakers.

The couch and chair are unmarked. They are 2-¾ inches long and 1-¾ inches high, respectively. The jockey is 3 inches high, the horse 2-¾ inches long. They have "Japan" inkstamps. The toothbrush measures 4-⅝ inches. Mark information on the toothbrush and toothpaste was not recorded. *Carson Collection.*

The chairs are 3-⅛ inches high. Both sets have "Made in Japan" paper labels. *Carson Collection.*

The toilet of the condiment set in the middle is 2-⅜ inches high. Its mark indicates it was made in Japan. The thundermugs on the ends are not marked. *Courtesy of Joyce Hollen.*

The table of this set is 5 inches long. It has a "Made in Japan" paper label. *Carson Collection.*

The base that holds the pipes is 3-⅝ inches long, while the pipes measure 2-½ inches. The base has a "Made in Japan" inkstamp. *Lilly Collection.*

Both these sets are wood. Neither is marked. The bullets stand 2-¾ inches high, the grave markers 2-¼ inches. *Carson Collection.*

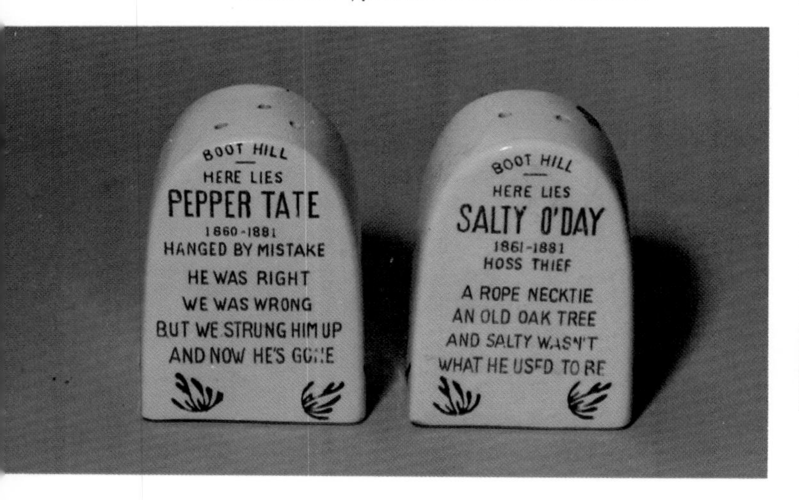

Another pair of gravestones, these having much more originality than the "Here Lies Salty" and "Here Lies Peppy" on the wooden examples above. Height is 3-¼ inches. There are no marks. *Carson Collection.*

These insulators are 1-¾ and 2 inches high. They are unmarked. *Courtesy of Joyce Hollen.*

The comb is 3-½ inches long, the brush 4-⅜ inches. Mark information was not recorded for either set. *Carson Collection.*

A couple suitcases, the one on the left 2-¼ inches high. They each have a "Japan" paper label. *Carson Collection.*

The set on the left is a two-piece. The ones in the middle are 4-½ inches high. None of the shakers in this picture are marked. *Carson Collection.*

The length of these pans is 4-¾ inches. Mark is "Brillium / Pat. Pend." on two lines. *Carson Collection.*

Scarecrows and grandpa and grandma. Gramps is 2-½ inches high, unmarked. The scarecrows stand 4-½ inches high. They have "Japan" paper labels. *Carson Collection.*

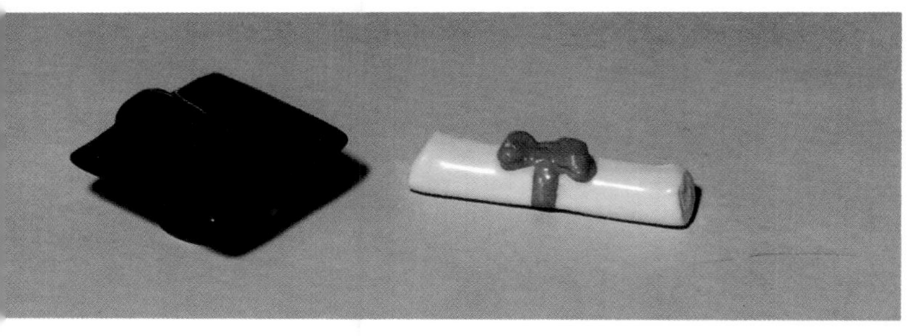

The cap is ¾ inch high, the scroll 2-⅝ inches long. Neither piece is marked. *Lilly Collection.*

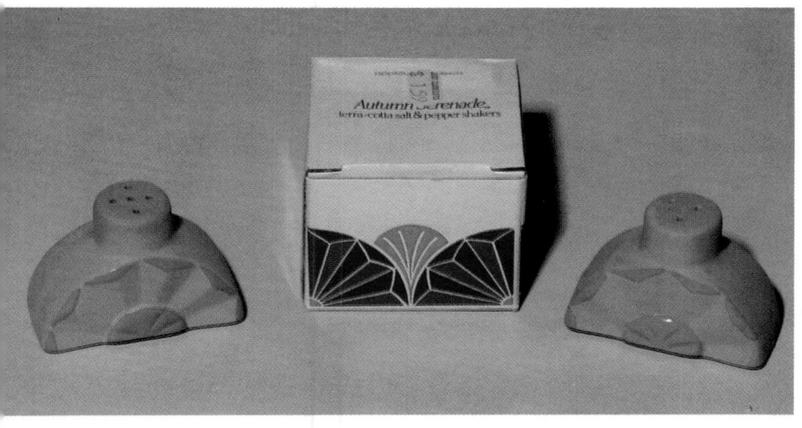

The box calls this set Autumn Serenade. It was recently purchased in a store as a closeout for $1.59. Height was not recorded, but as I recall they are about three inches. Paper label is shown below. *Lilly Collection.*

Mark of Autumn Serenade set shown above.

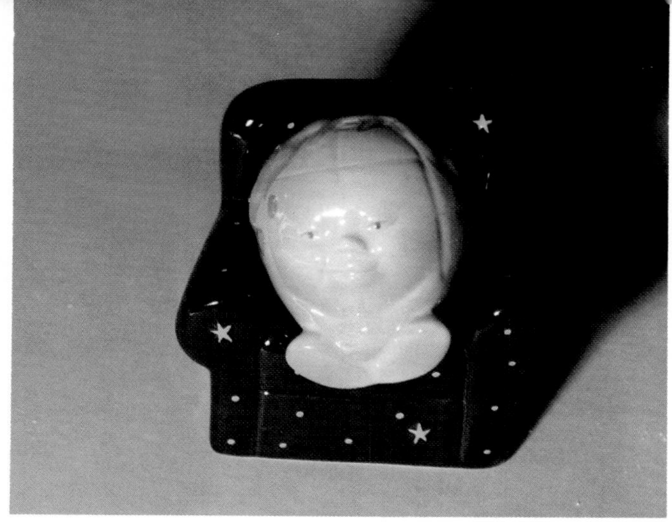

This is called Earth/Easy Chair, as shown on the picture of the box below. It a popular recent set by Clay Art. The Earth is 2-½ inches high, the chair 2-¾ inches. Paper label is shown below. *Courtesy of Lesia Dobriansky.*

Paper label of Earth/Easy Chair.

Box of above set showing a date of 1990 and a price of $14.99.

Mountains? Buttes? Rosemeade? Reliable information is 2-½ inches high and unmarked. *Carson Collection.*

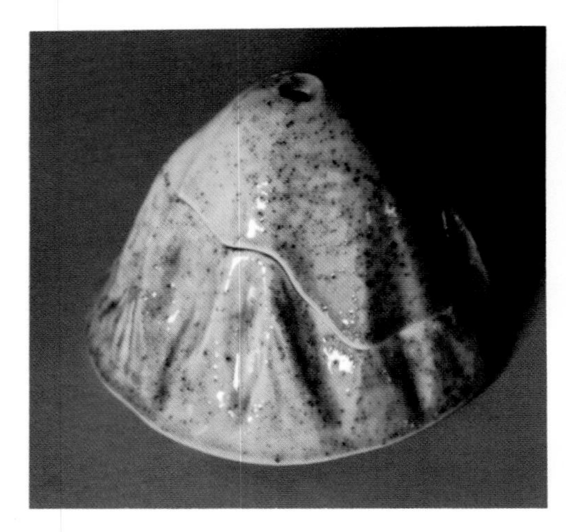

This is a rather interesting two-piece set in that it shows Mt. St. Helens before and after it erupted in 1980, and is made in part from volcanic ash from that eruption. As shown it measures 3-¼ inches high. *Thornburg Collection.*

Mt. St. Helens after it erupted.

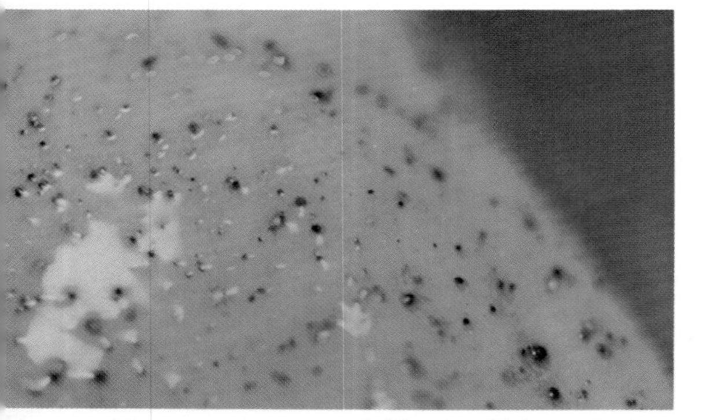

Extreme close-up of Mt. St. Helens shaker showing minute pieces of volcanic ash in the glaze. (Color is due to incorrect exposure.)

Impressed mark of Mt. St. Helens shaker.

Plymouth Rock, 3-⅛ inches high, unmarked. *Courtesy of Joyce Hollen.*

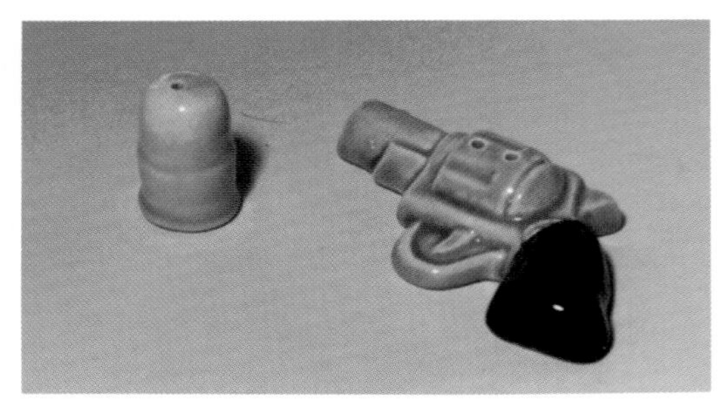

The gun is 2¼ inches long, the bullet 1-⅜ inches high. The pair is not marked. *Carson Collection.*

The shotgun shells are plastic, 2-⅜ inches high, and unmarked. *Carson Collection.*

Metal is the material of these shakers. They are 2-½ inches high, unmarked, but the box says "Made in Hong Kong JSNY." *Thornburg Collection.*

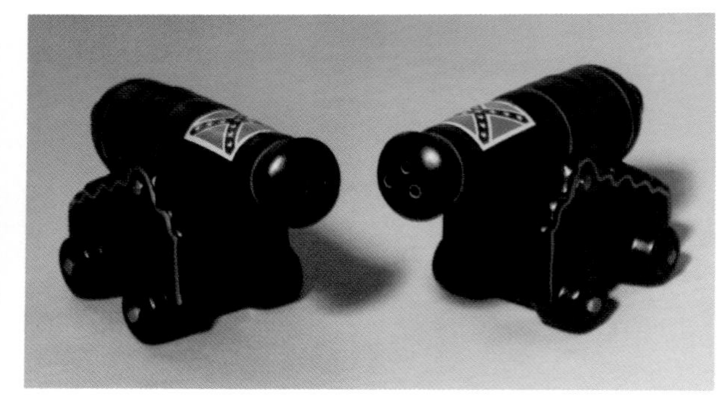

Metal pistols, 4-¾ inches long. They have "Made in Japan" paper labels. *Carson Collection.*

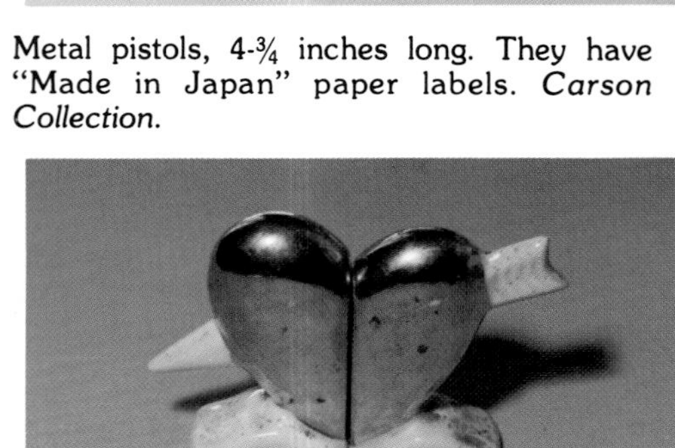

This broken heart is unmarked. It is 3 inches high. *Carson Collection.*

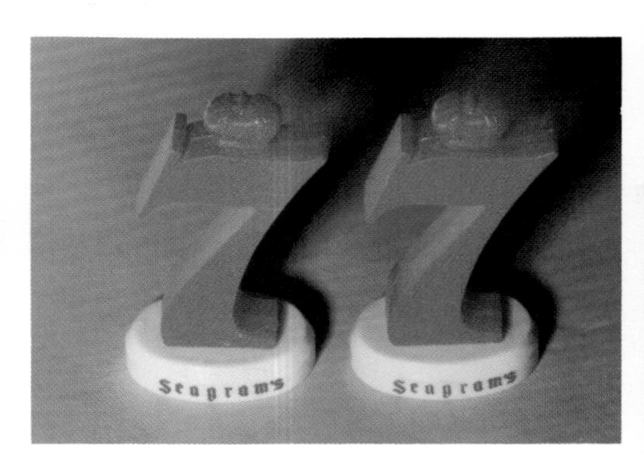

These plastic shakers are 3-½ inches high. They are not marked. *Thornburg Collection.*

These cannons are 4-⅜ inches long. They are unmarked. *Carson Collection.*

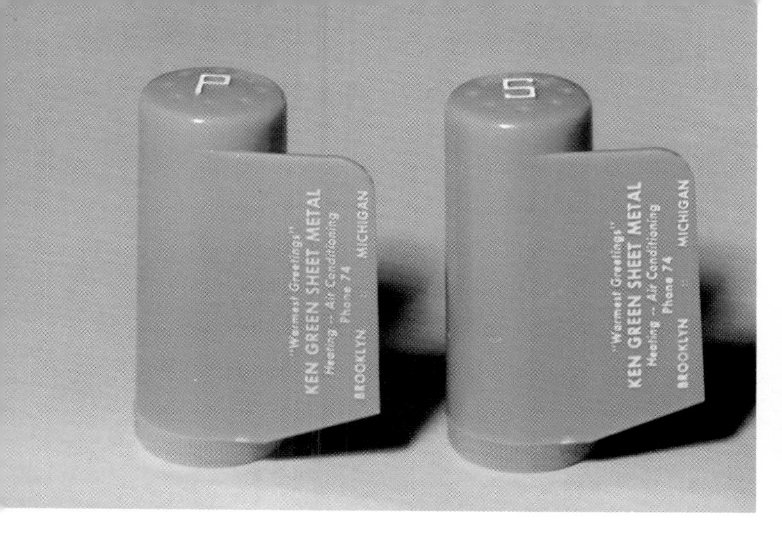

Early plastic shakers advertising a sheet metal company. They are 2-⅜ inches high, unmarked. *Thornburg Collection.*

Close-up of the advertising on the above sheet metal shakers. Note the telephone number.

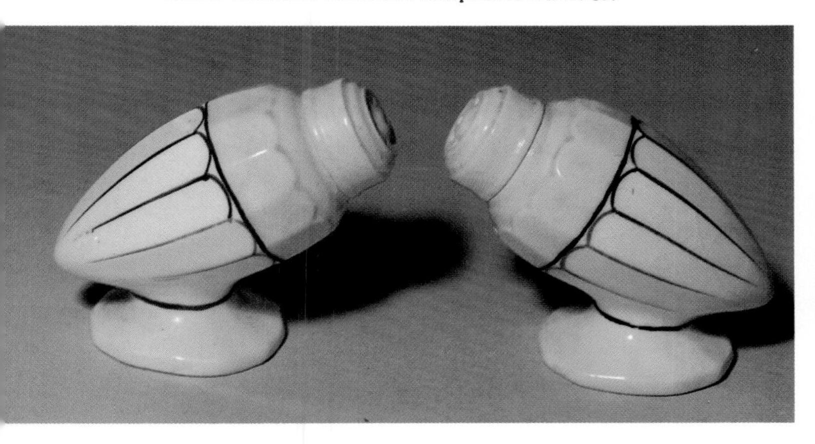

These ceramic shakers have threaded lids, something that cannot be accomplished on a cheaply made product. They are 2-½ inches high. An inkstamp mark says "Made in Japan." *Carson Collection.*

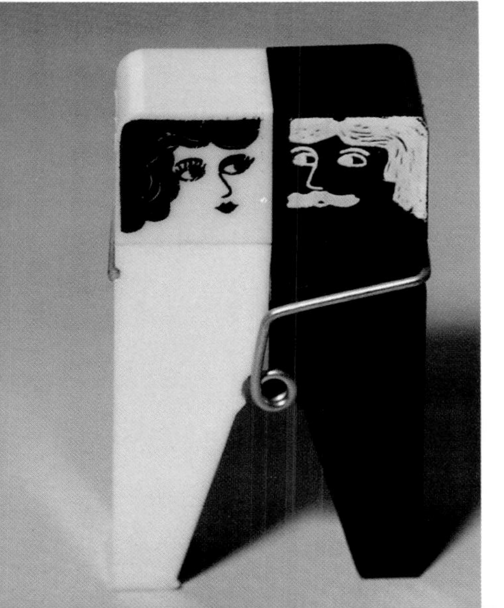

Plastic and metal, 3-¾ inches high. There is no mark. *Carson Collection.*

Snow People

Shakers representing snowmen and women make wonderful additions to Christmas displays. Most are decorated with the Christmas colors red and green.

The snowman on the left is 3-⅛ inches high. His counterpart on the right stands 3 inches. Both sets are plastic, neither is marked. I have seen the set on the right with a red tray shaped like a ribbon bow. *Carson Collection.*

Sandy Srp miniatures, the man being 1-¼ inches high. *Carson Collection.*

The pair on the left is 3 inches high. It has a Lefton paper label, as does the pair on the right. The pair in the center is marked "Japan." *Carson Collection.*

This is a sitter, the snowman is the salt, the hat is the pepper. It has a "Made in Taiwan" paper label, and was recently purchased at a discount store for $1.99. *Oravitz Collection.*

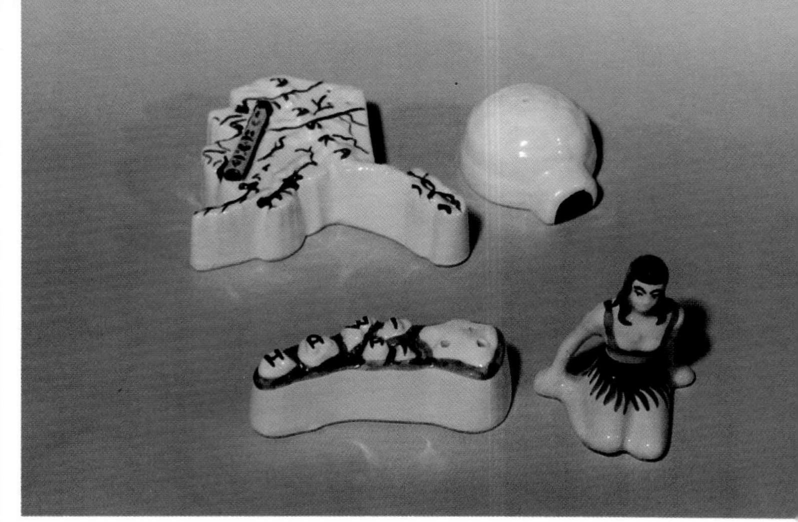

The 49th and 50th states, obviously from the later set. *Thornburg Collection.*

States

Completing a set of Parkcraft states is high on the list of many collectors' priorities. Parkcraft made two sets of states, an early one that did not include Alaska and Hawaii, a later one that did. Some of the accompanying pieces were changed for the second set. See Appendix III for a complete list. KlayKraft (Milford) also made a states set, but in theirs both the salt and the pepper were in the shape of the state.

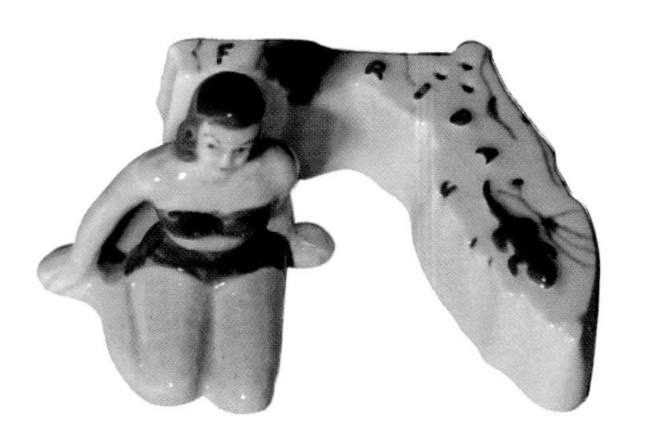

The bathing beauty with Florida is from the first set. The second set had a fish to go with the state. *Thornburg Collection.*

All of the Parkcraft states are similar in size so I didn't measure each one. To give you an idea, Ohio is 2-⅝ inches across. *Thornburg Collection.*

Note that the Parkcraft states were not made to scale. Here Texas is no larger than Missouri, Kansas or Colorado. *Thornburg Collection.*

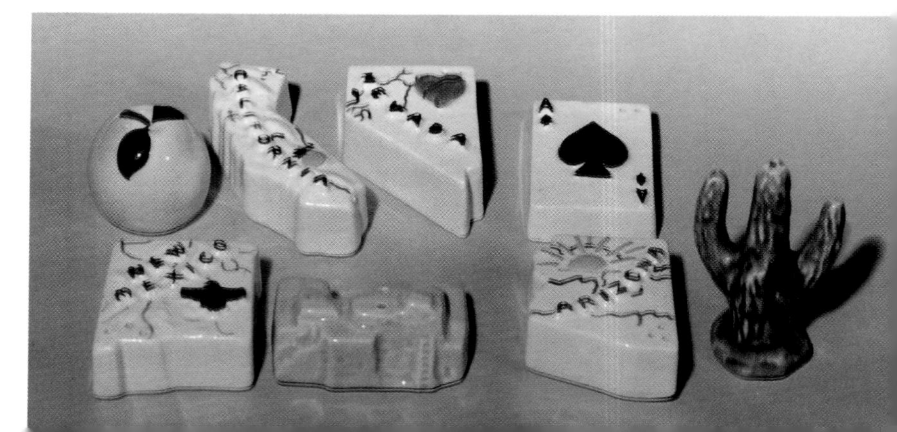

If this set was being made today, California's orange might lose out to something representing Hollywood, the Silicon Valley, Disneyland, or numerous other subjects. *Thornburg Collection.*

As stated above, Bob Ahrold solicited the opinions of state governors concerning what object would best represent their state. Obviously well schooled in politics, the Oregon governor apparently decided not to offend either of his state's most important athletic teams at the time—the Oregon Ducks and the Oregon State Beavers. *Thornburg Collection.*

How many readers would associate North Dakota with an oil well. (I personally would prefer a pair of Rosemeade shakers!) *Thornburg Collection.*

Take a close look at New Jersey and you will see the same girl that appears with Florida above. This is wrong. It should be Miss America. *Thornburg Collection.*

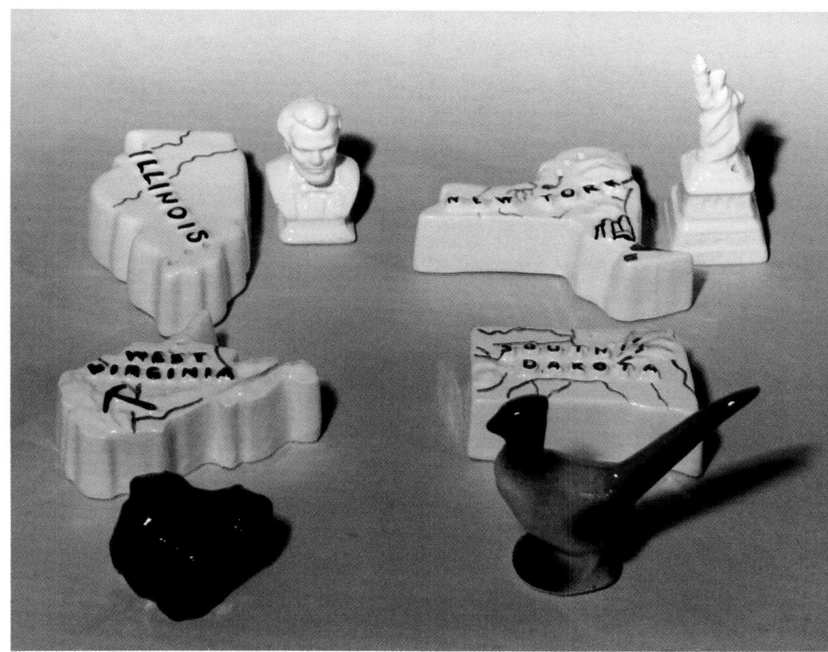

When the Parkcraft states were made, South Dakota probably had the densest pheasant population anywhere in the world. Then, due to a few bad winters and modern farming methods, it plummeted to the point that today's younger collectors might not recognize the symbolism. But now, through the efforts of a conservation group called Pheasants Forever, the numbers are again rising and indications are that the former glory days of South Dakota's pheasants may return. So perhaps the symbolism is not so outdated after all. *Thornburg Collection.*

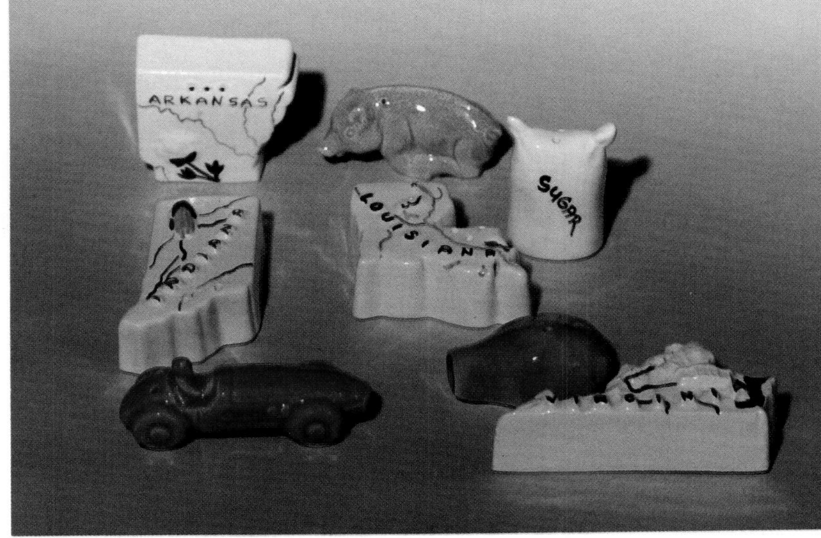

Not that it means anything, but Arkansas is my least favorite of all the Parkcraft states, not because of Arkansas itself but because of the Razorback. When I was drafted into the army in the fall of 1965, my company commander in Basic Training was a graduate of the University of Arkansas. The very night we arrived he made an intimidating speech, saying that each Saturday the Razorbacks lost we would have an extra two hours of physical training and dismounted drill the following Monday. Fortunately, Arkansas posted a 10-1 record that year, losing only to LSU in the Cotton Bowl. But I still don't like the Razorbacks. *Thornburg Collection.*

The North and South Carolina go-withs might not be instantly recognizable. They are a pack of cigarettes and a cotton boll. *Thornburg Collection.*

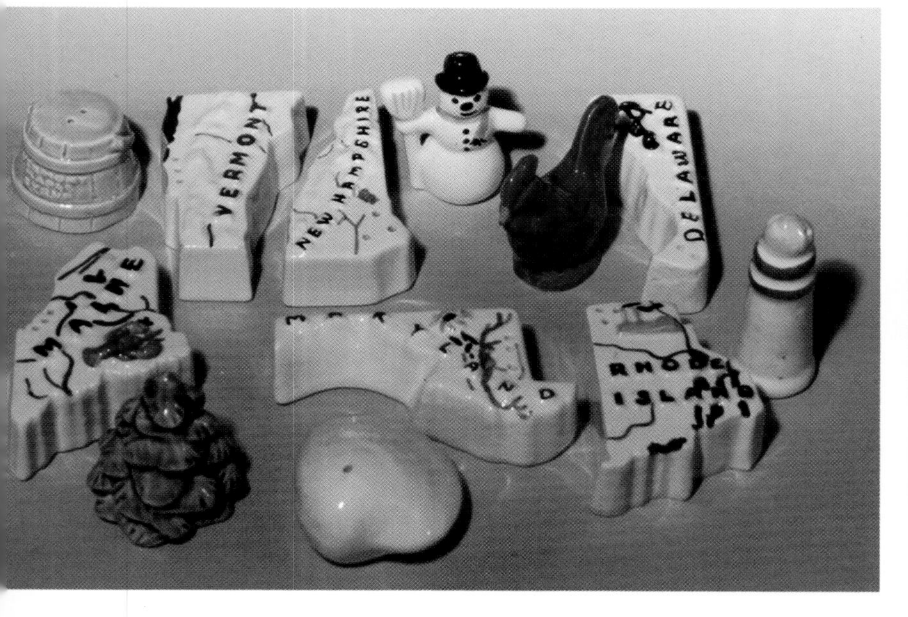

Here the Rhode Island red rooster obviously goes with that state, while the lighthouse goes with Delaware. The symbol for Maryland is an oyster. *Thornburg Collection.*

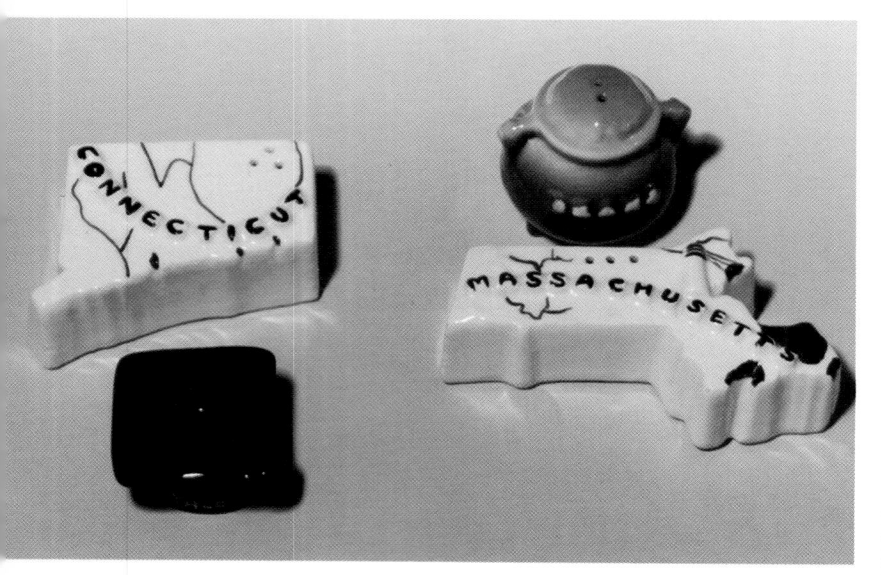

Look closely at the mortar board with Connecticut and you will see the word Yale at the base. *Thornburg Collection.*

Mark of Parkcraft Missouri above.

Mark of Parkcraft Ohio above. (Color is due to improper exposure.)

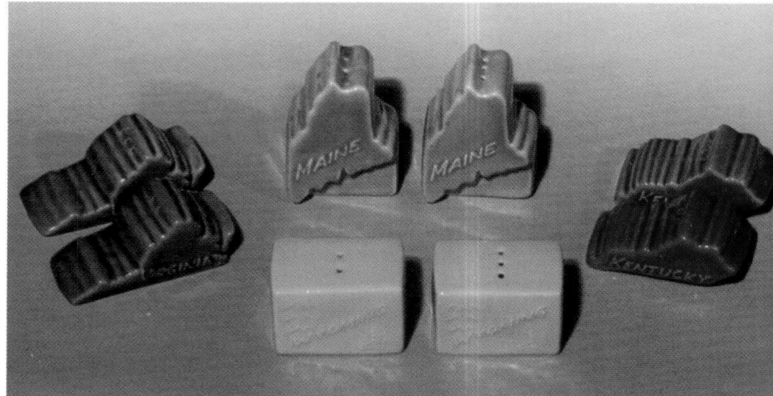

Four states from the Milford Pottery series. Maine is 2-⅜ inches high. These also came in white, tan, and possibly other colors. Marks are shown below. Milford also made sugars and creamers to match the shakers, perhaps other pieces, too. *Thornburg Collection.*

Mark of Milford Wyoming above.

Mark of Milford Kentucky above.

Tools and Utensils

A good place for these shakers is way on the back of the shelf, practically hidden from view. If you see them too often, they will remind you of all the work you should be doing instead of cruising garage sales and attending flea markets in search of more shakers.

The hatchet is 3-½ inches long, the stump 1-⅛ inches high. Neither is marked. *Lilly Collection.*

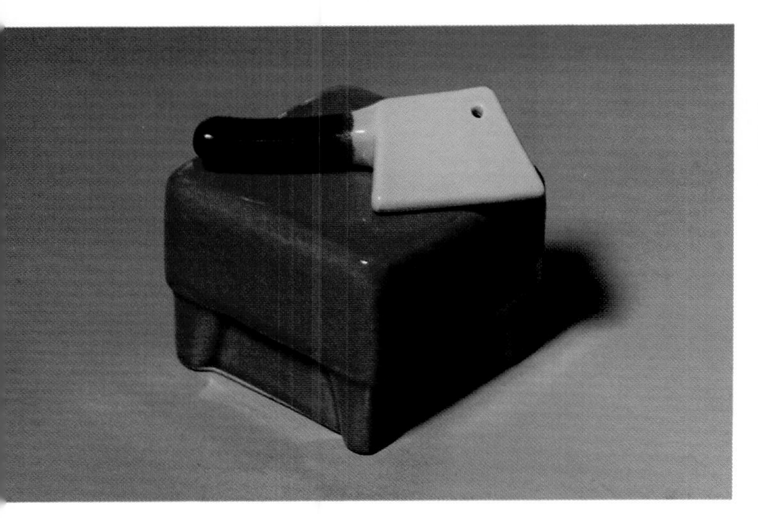

This set is not marked, either. The chopping block is 1-⅝ inches high. *Carson Collection.*

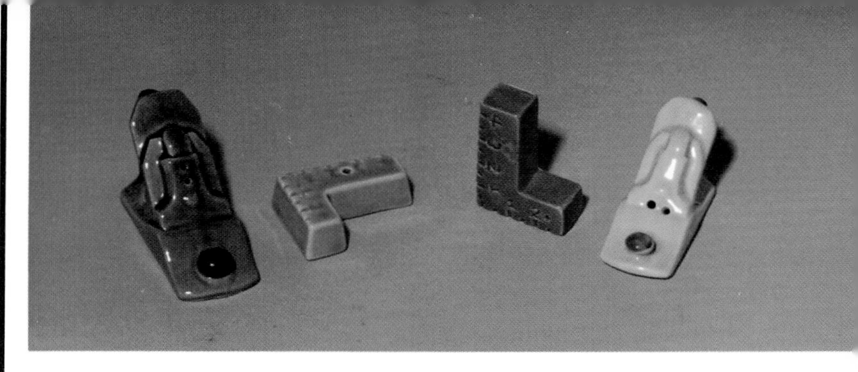

The planes appear to be slightly different while the squares look the same. The squares are 2-⅛ inches high when standing as the one on the right is. Mark information was not recorded. *Carson Collection.*

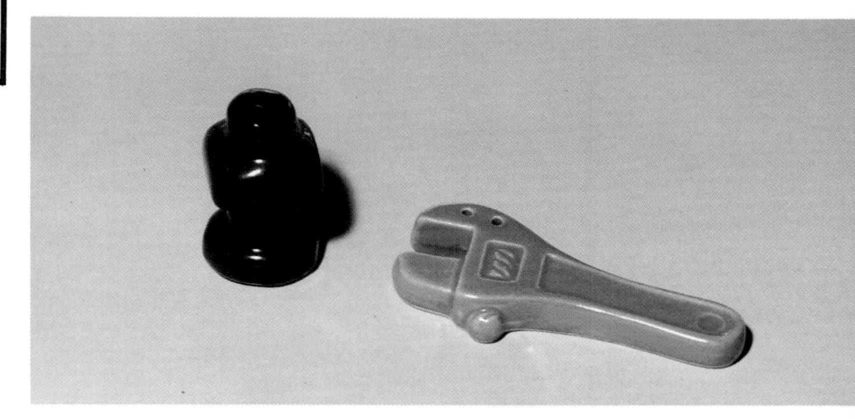

The adjustable wrench is 4-¼ inches long, the nut and bolt 2-⅜ inches high. The set is unmarked. *Lilly Collection.*

Unmarked bellows, 3-¾ inches high. *Carson Collection.*

Another picture that was taken prior to establishing the format of the book. The anvil and hammer are made of redware, marked "Japan." The bowling ball and pin are unmarked. The pin is 4-⅜ inches high. The bread and butter were made by Bernard Studios, a Bernard paper label shown on the next page. *Courtesy of Joyce Hollen.*

Paper label of Bernard Studios.

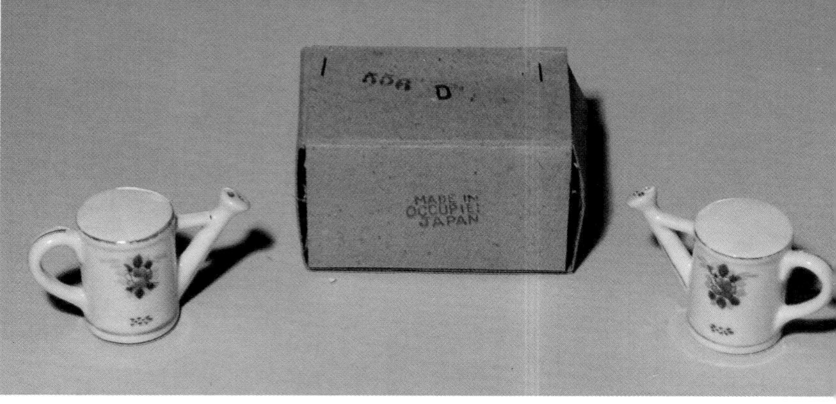

These watering cans are 1-¾ inches high. They are marked "Japan," while the box they came in is marked "Made in Occupied Japan."

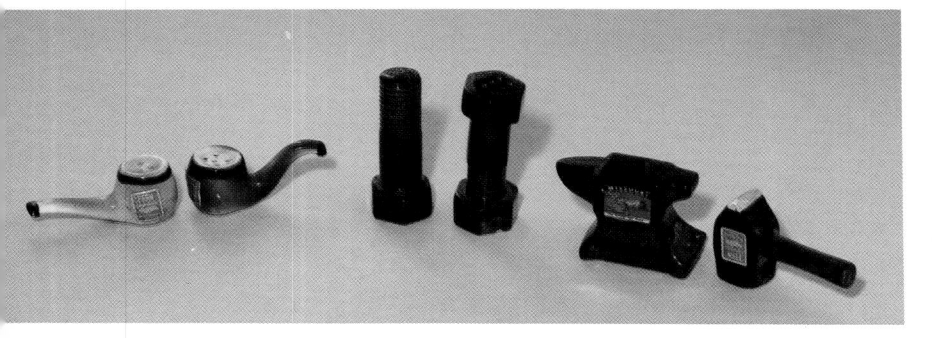

The pipes are 3-⅛ inches long, the nuts and bolts 3-⅜ inches high. The hammer is 2 inches long. The pipes and the hammer have Fern paper labels. *Carson Collection.*

Plastic watering can and flower pot. Height of each is 1-⅞ inches. Neither is marked. *Private Collection.*

All of these are metal; no other information was recorded. *Carson Collection.*

Sprinkling cans, 2 inches high, with a paper label that says "Japan." *Carson Collection.*

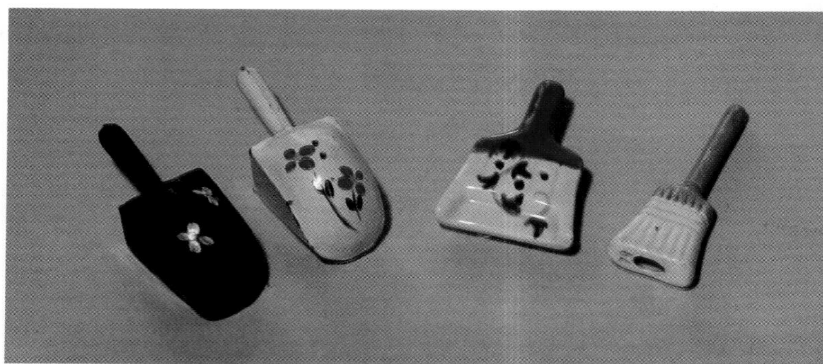

The set on the left is metal, the set on the right ceramic. The ceramic dustpan is 3 inches long. None of the pieces are marked. *Carson Collection.*

These three sets are Shawnee. The wheelbarrows are 2 inches high, the milk cans 3-⅜ inches, and the watering cans 2 inches. None are marked except for the paper label on the one milk can. *Carson Collection.*

Section III: Non-Figural Salt and Pepper Shakers

Non-figural Shakers

While the main thrust of this book is on figural salt and pepper shakers, non-figurals have a way of creeping into every collection. Condiment sets are probably the most common, but there are also souvenir shakers, nondescript plastic sets, dinnerware accessories and others.

Chapter 13
Non-Figurals

Here are but a few of the numerous non-figural sets we found in various collections. As you will see in the price guide, values run form practically nothing to amounts that can quickly flatten a semi-healthy wallet.

Condiment Sets

Most of the non-figural condiments you see were made in Japan, but a few were made in Germany, and perhaps other countries. Condiment sets were often made to complement dinnerware, as in the case of the Noritake Azalea set shown below.

This is a Noritake Azalea pattern condiment set. The shakers are 2-½ inches high, the tray 7-⅝ inches long. The mark is shown below. *Courtesy of Virginia Sell.*

Mark of the tray of the above condiment set.

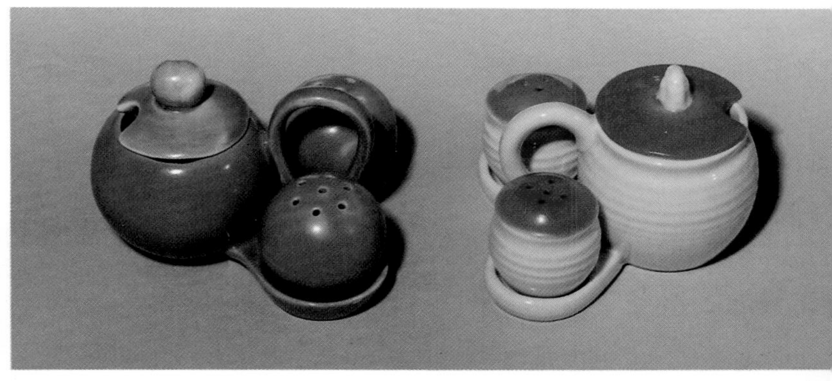

The two sets shown here were made by Goebel. The mustard on the left is 3-¼ inches high. The set on the right has "6905" impressed. The mark of the tray on the left is shown below. *Thornburg Collection.*

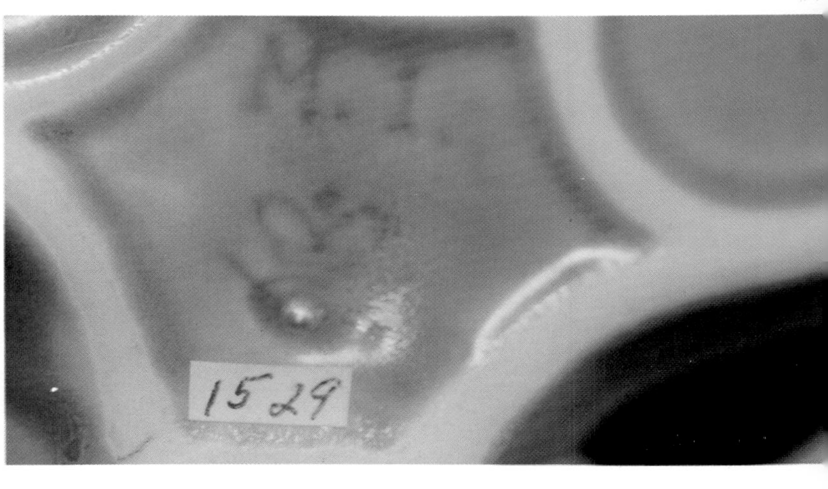

Mark of the above Goebel condiment set.

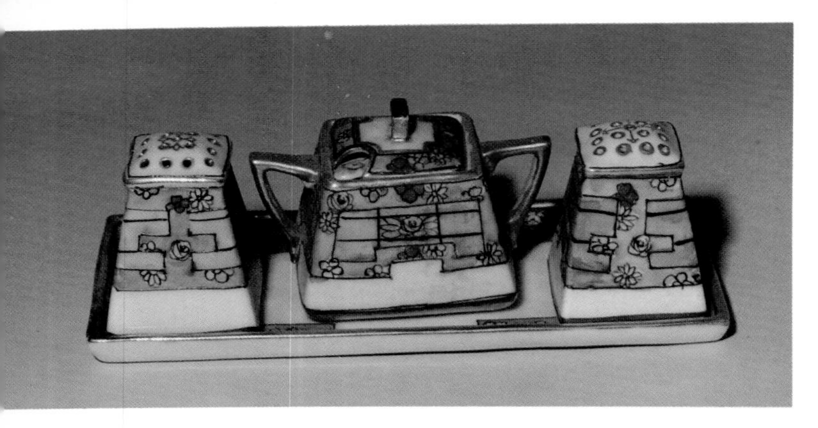

The mark of this set is "Hand Painted Nippon." The tray is 7-½ inches long, the shakers 2-½ inches high. *Courtesy of Joyce Hollen.*

Another Japanese condiment set. This one includes a toothpick holder. The tray measures 5 inches in diameter, the shakers stand 2-½ inches high. The mark is Japanese. *Courtesy of Cloud's Antique Mall, Castalia, Ohio.*

Here's one that goes a bit further by having small cruets for vinegar and oil. The shakers stand 2-½ inches high. The mark of the set is shown below. *Courtesy of Cloud's Antique Mall, Castalia, Ohio.*

Even the United States Army is not beyond using a condiment set once in awhile. This one is made of aluminum. The salt and pepper are weighted. Measurements were not taken. *Courtesy of Cloud's Antique Mall, Castalia, Ohio.*

Mark of the above condiment set with cruets.

This looks like there is a lid missing but I'm not so sure because of the gold finish on the rim of the mustard, which would quickly wear off if a lid was used. The shakers are 2-⅞ inches high. Note that the one on the right is used as a shaker, while the one on the left has a removable cap so you may spoon from it. Strange set. The holder is silverplate, the inkstamp on the ceramic pieces is "B5754." *Courtesy of Cloud's Antique Mall, Castalia, Ohio.*

China Painting

The china painting craze in the United States lasted from about 1875 to the beginning of World War I. During that time painting sessions were often substituted for the afternoon teas commonly attended by upper class ladies. China painting is still being done today, but on a smaller scale with the work generally being sold at craft shows.

Height of this one-piece is 2-¼ inches, length is 3-¾ inches. It is not marked. *Lilly Collection.*

Same whiteware as above, different painting. Another difference is that this one is marked a couple times, "Germany" in green, "Couts" in red. *Carson Collection.*

The set on the right was probably painted by a crafter way back when. It is unmarked, 2-¼ inches high. The picture shows the front of one shaker, the back of the other. The set on the left is 3 inches high, has a Noritake mark. *Courtesy of Cloud's Antique Mall, Castalia, Ohio.*

Both of these shakers have been broken and glued, but when I saw them at a flea market for 50 cents I just couldn't pass up the mushroom motif. They are 2-⅞ inches high, unmarked.

Known American and Foreign Potteries

Some of the shakers shown below, such as the Fiesta, were made to complement dinnerware. Others, the light blue Rosemeade, for example, were apparently made to reach that corner of the market that wanted shakers that would appear attractive but not cutsie with a semi-formal table setting.

The Homer Laughlin Company made Fiesta salt and pepper shakers in 11 different colors from the late 1930s into the 1970s. Starting at the top and going clockwise, the colors are: chartreuse, yellow, turquoise, ivory, cobalt, rose, red, green, gray, and dark green. The plate is medium green, the only color not shown as a shaker. Fiesta shakers are 2-⅝ inches high, and nearly always unmarked. *Courtesy of Cindy Schneider.*

Rosemeade, 2 inches high, also unmarked. *Thornburg Collection.*

This is a pair of Harlequin shakers, also made by Homer Laughlin and sold exclusively by F.W. Woolworth stores. Height is 3-½ inches. Harlequin shakers were made in most of the colors of Fiesta. *Courtesy of Cindy Schneider.*

These shakers were made by the Purinton Pottery Company. There was also a matching cookie jar. They are unmarked, 4-½ inches high. *Courtesy of Stan and Karen Zera.*

These look like Purinton, too. But they are Shawnee. Height is 3-¼ inches. They are not marked but can easily be identified by their large Shawnee fill holes. *Smith Collection.*

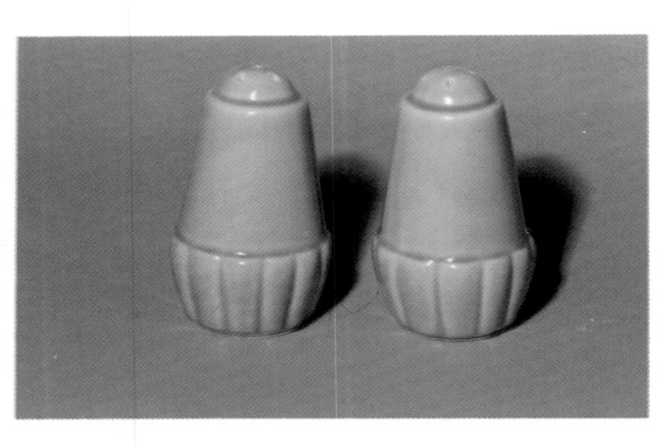

Lu Ray shakers, made by Taylor, Smith and Taylor. They are 3-¾ inches high, marked "USA," and were made is a variety of pastel colors. *Courtesy of Heidi Calhoun.*

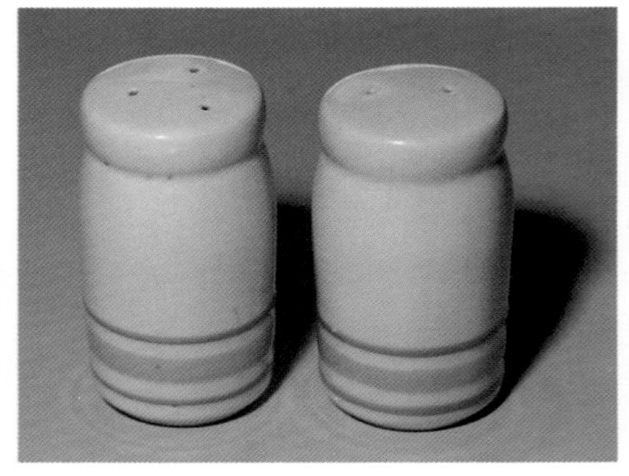

McCoy, 3-⅝ inches high, unmarked. *Lilly Collection..*

Wedgwood blue jasperware, 4-⅛ inches high. The impressed mark is "Wedgwood / Made in England" on two lines, and in another area of the bottom, "1953." *Carson Collection.*

This set, 3-½ inches high, is somewhat of a mystery. They have the same large size fill holes as Shawnee shakers, which suggest they were made by that company. But the uncleaned mold seam on the left side of the shaker on the right is not indicative of Shawnee's generally fine work. On the other hand, I've sold two pair of these sets at antique shows in the past year, and both purchasers were Shawnee collectors who spotted them from a distance and seemed **very happy to get them. They are not shown in either of the Shawnee books currently in print.**

This set is by Royal Winton, another English pottery. The shakers are 2-½ inches high. The corks of this set have a rather curious addition—metal around the tops, apparently to make them easier to grip. The mark of this set is shown below. *Carson Collection.*

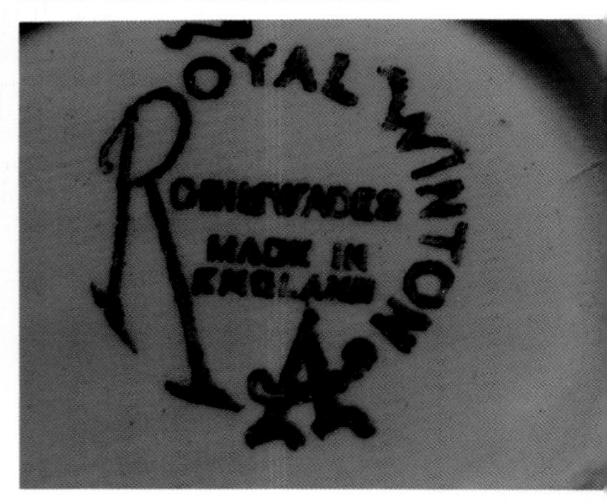

Mark of the above Royal Winton shaker set.

Here is the Noritake Azalea pattern again, as shown in the condiment set at the beginning of the chapter. The pair on the inside stand 2-¾ inches, have a "Made in Japan" inkstamp. The pair on the outside go 3 inches. Their inkstamp mark is shown below. *Courtesy of Virginia Sell.*

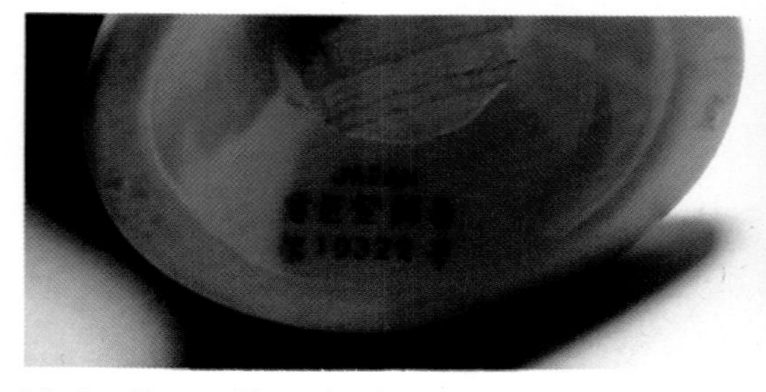

Mark of larger Noritake Azalea shakers shown above.

Miscellaneous Non-Figurals

Shown below are some miscellaneous ceramic shakers, souvenir shakers, plastic sets in geometrical shapes, and a few advertising sets.

The height of these shakers is 2-½ inches. They are unmarked, but the tray is marked "Delft Holland." As you can see if you look close, the tray also has a picture of a windmill. *Courtesy of Joyce Hollen.*

The shakers on the left are 2-¾ inches high, have a "Made in Japan" inkstamp. The pair in the middle stand 3 inches, have an "Imperial Nippon" inkstamp. Height at right is 3-¼ inches. Inkstamp reads "Japan." *Courtesy of Cloud's Antique Mall, Castalia, Ohio.*

For size reference, the sugar shaker is 3-⅝ inches high. All pieces are marked "Nippon" except the pairs at front center and rear right, which are unmarked. *Courtesy of Joyce Hollen.*

The boat holding these shakers is 4-¼ inches high. It is marked "Nippon," while the shakers are marked "Japan." *Carson Collection.*

These appear to be Chinese, but are not marked. Height is 2-¾ inches. *Courtesy of Cloud's Antique Mall, Castalia, Ohio.*

Another Chinese-looking pair but also unmarked. They are 2-⅜ inches high. *Courtesy of Joyce Hollen.*

Both of these sets are made of plastic. The Red Lobster shakers are 3-⅝ inches high. Neither pair is marked. *Carson Collection.*

Plastic kitchen shakers 2-¼, 2-¾ and 2-⅞ inches high going left to right. Only the pair on the right is marked, "Dapol Plastics, Inc." in raised letters.

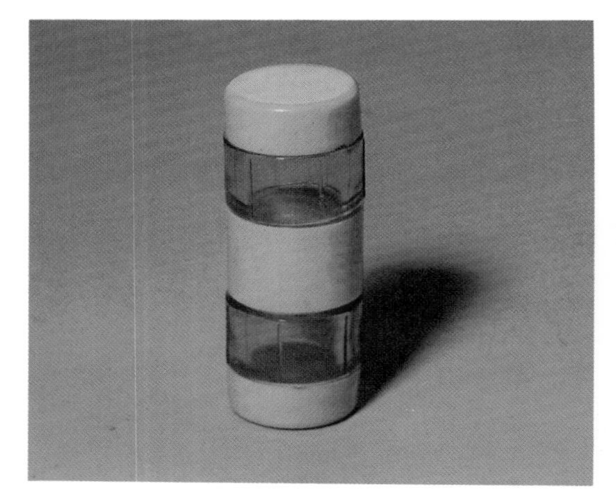

Purse or pocket one-piece, 2-¼ inches high, unmarked. They also come in red, green, and other colors.

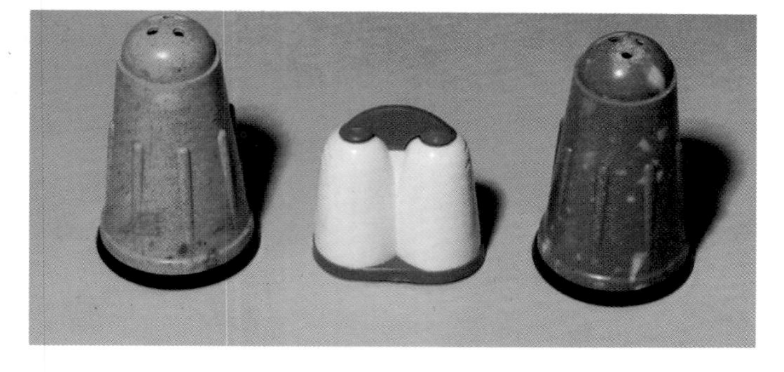

The only mark on these plastic shakers doesn't say much, "Pat. Pend. Made in USA." It's on the inside set, which is 1-¾ inches high. The set on the outside is 2-½ inches high. *Lilly Collection.*

The plastic holder is 5-¼ inches high, the shakers themselves 4 inches. Mark is "Made in China." *Lilly Collection.*

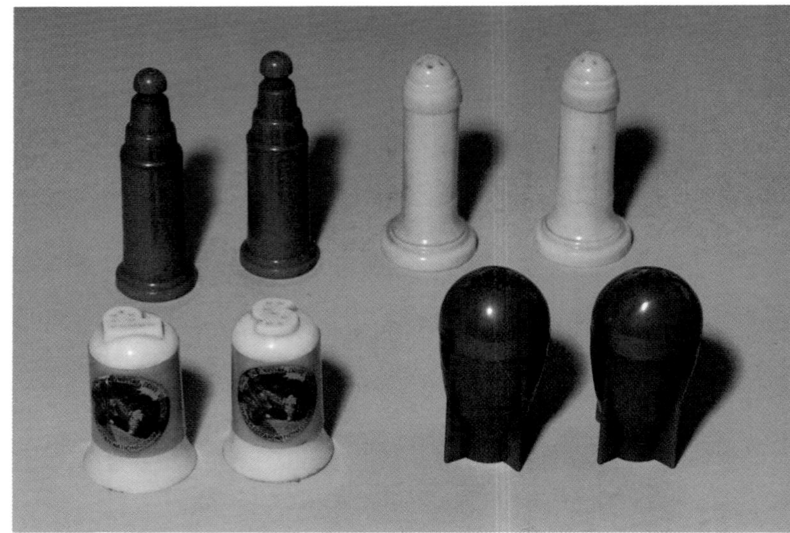

All plastic. The green pair at left rear are pushbuttons. They are 3-¾ inches high. All shakers in this picture are unmarked. *Carson Collection.*

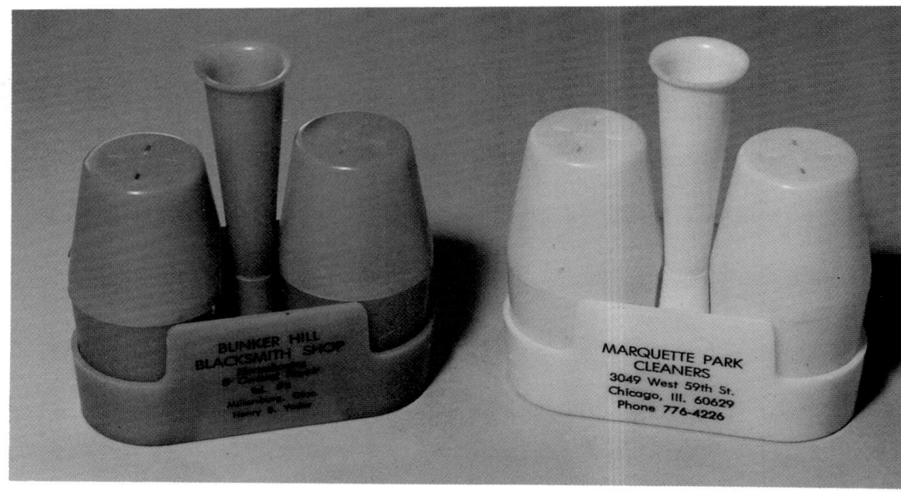

The trays of these plastic advertising sets are 3-¾ inches high. They are not marked. How ironic that the set on the left uses a modern material to advertise an ancient trade. *Carson Collection.*

I don't know how many of these plastic shakers were made but the total would probably have to be written exponentially as you see them from just about every state, city and tourist attraction in the Union. They are 2 inches high, have "Made in Hong Kong" in raised letters on their bottoms.

Same as above except the ones on the right have a little plastic or metal figure attached instead of a decal. *Carson Collection.*

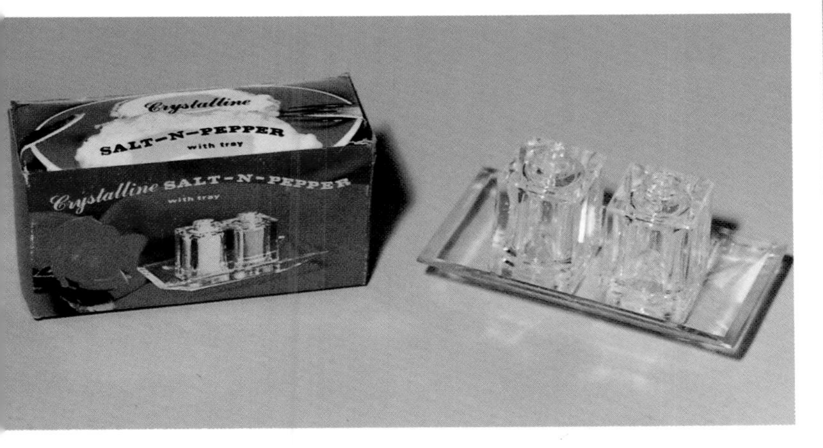

And here is the box in case you ever wondered what it looked like. *Lilly Collection..*

Handpainted plastic but unmarked. Height is 2-¼ inches. *Courtesy of Joyce Hollen.*

Same as above, different color and picture. *Carson Collection.*

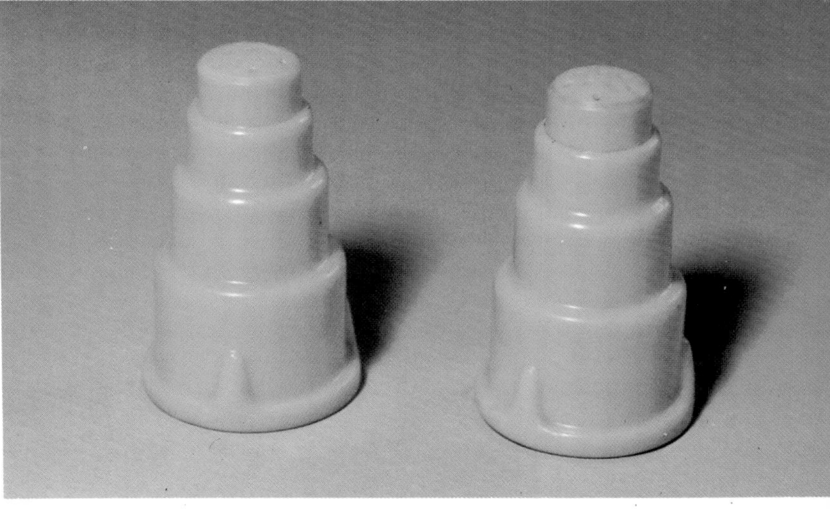

This plastic pair is 2-½ inches high. Only mark is a raised "4" on the bottom. *Carson Collection.*

Bakelite, 1-¾ inches high, unmarked. *Courtesy of Allen and Michelle Naylor.*

This pair is plastic in a somewhat Deco style, 2-¼ inches high, and unmarked. *Lilly Collection.*

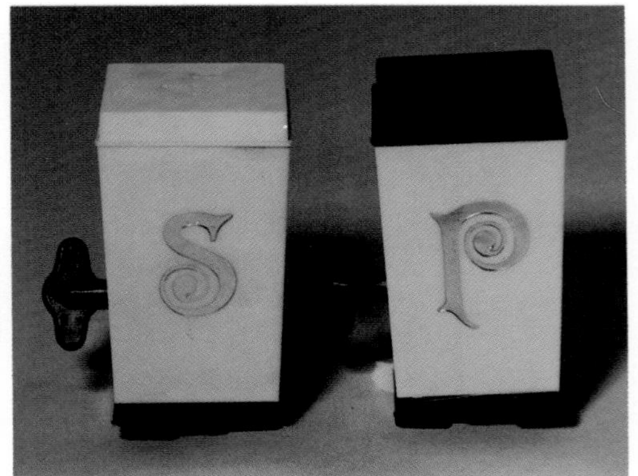

Plastic walkers. Wind them up and they will move across the table. They are 3 inches high, are marked "© 1968 Dan-Dee Imports / New York, NY" on two lines. *Carson Collection.*

The tab sticking out on the left side of the top of this plastic one-piece is a bar that slides back and forth to let out salt or pepper. The shaker is 3-⅜ inches high. The label visible on the front says "Made of / Styron" on two lines. *Carson Collection.*

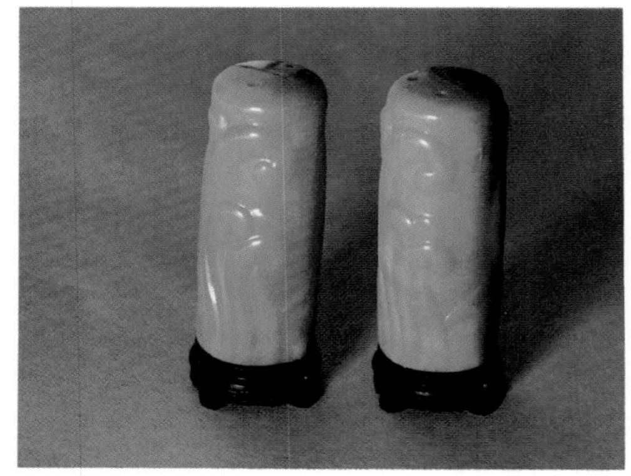

I guess these shakers are actually figurals, but what do they represent? They are plastic, 2-¼ inches high. The mark is "Superlon / Products / 4 / Chicago 12, USA" on four lines within a circle. *Carson Collection.*

The brass shakers on the left are marked "Annettes." They are 1-½ inches high. The one-piece plastic shaker in the middle was an advertisement for The Farmers Elevator Company in Beaverdam, Ohio. It was not measured. The Ace Lunch Box advertising shakers on the right are made of aluminum, stand 1-½ inches high. *Carson Collection.*

A one-piece plastic arch. 2-⅜ inches high. It is unmarked. *Lilly Collection.*

Plastic, and as common as algae in August. The shakers are 2 inches high, marked, "St. Labre Indian School / Ashland, Montana / 59003" on three lines on the bottom.

Plastic, 2-½ inches high, and unmarked. Both these establishments were in York, Pennsylvania. Lincolnway is still in business. *Thornburg Collection.*

These shakers with comical sayings on them are 1-⅞ x 3-¾ inches. They are inkstamped "Japan." *Carson Collection.*

These are all 2-⅞ inches high. All have "Made in USA" impressed in their bottoms. *Carson Collection.*

Same as above, different sayings. *Carson Collection.*

The flags are 2-¼ x 4 inches. They have a Parkcraft inkstamp. *Thornburg Collection.*

The shakers are made of metal, the basket plastic. Shakers are 2-½ inches high, marked "Made in Hong Kong," both on the neck under the lid, and on the bottom. *Carson Collection.*

Five-inch high shakers of the U.S. Navy. Apparently silverplated at one time, they are marked "Astor Shaker Set Pat. Applied for" around the bottom. *Carson Collection.*

These shakers are 1-⅝ inches high, the tray is 3 inches long. The tray is marked "H. Kato / Made in / Occupied / Japan" on four lines, while the shakers are marked "Japan.". The box they came in was also marked Occupied Japan. *Lilly Collection.*

This handpainted pair is 3 inches high. Only mark is an inkstamped "S90." *Carson Collection.*

Chrome tops with plastic bases (look close and you will see the base of the salt shaker), this pair is 3 inches high. It is not marked. *Carson Collection.*

A brass pair from which the bottoms unscrew to fill. Height is 2-½ inches. There is no mark. *Courtesy of Cloud's Antique Mall, Castalia, Ohio.*

The pair on the left is made of small bits of shells glued to glass, probably a souvenir of a seaside resort area. Height is 3-¾ inches, there are no marks. The set on the right is 2-¾ inches high. A paper label reads "Japan Handptd." *Carson Collection.*

Aluminum, 2 inches high, no marks. *Carson Collection.*

Heights of these wood shakers left to right are 1-⅞, 1-¼ and 1-¾ inches. None are marked. *Carson Collection.*

These have a "Japan" paper label. They are 2-¾ inches high. *Courtesy of Cloud's Antique Mall, Castalia, Ohio.*

Metal bodies, plastic bottoms, leather-like wrappings. Height is 2 inches. The shakers are unmarked. *Carson Collection.*

Height of these is 3-⅜ inches. They are not marked. *Courtesy of Cloud's Antique Mall, Castalia, Ohio.*

Lots of souvenirs. The set at rear center is 4-¼ inches high. Only marked pair is at rear right, which has a "Japan" paper label. *Courtesy of Joyce Hollen.*

These boxed glass shakers are 2-¾ inches high. They are not marked. *Courtesy of Joyce Hollen.*

Wood, obviously bought in Mexico, and probably made there, too. Height is 3-¼ inches. *Carson Collection.*

Purchased in Greece in the early 1970s, these shakers are 3 inches high, unmarked. *Courtesy of Stan and Karen Zera.*

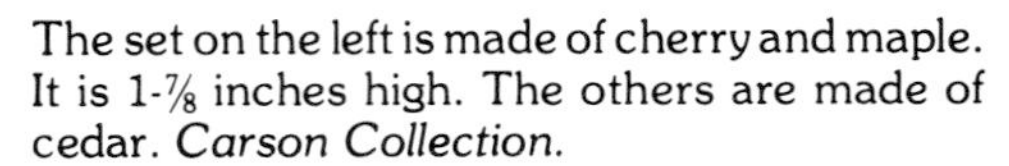

The set on the left is made of cherry and maple. It is 1-⅞ inches high. The others are made of cedar. *Carson Collection.*

Glass shakers, plastic tops, leather holder that hangs on the wall. The shakers, which are not marked, are 2-½ inches high. *Courtesy of Joyce Hollen.*

A pair of porcelain salt and pepper shakers, 2-⅛ inches high. Their mark indicates Bavarian origin. *Carson Collection.*

The set on the left is 2-¾ inches high, has a "Japan" inkstamp. The set on the right stands 2-3/16 inches, is not marked. *Carson Collection.*

This set is reminiscent of Pickard China, but it is not marked as such. The shakers are 5-⅞ inches high. The only mark is an "1196" written in the same gold with which they are decorated. *Courtesy of Cloud's Antique Mall, Castalia, Ohio.*

As I recall, this Marion Kay advertising set is made of milk glass. It is unmarked. Height is 4-½ inches. *Courtesy of Joyce Hollen.*

This set is only 1-⅛ inches high. It is made of sterling silver, and so marked. *Courtesy of Betty Laney.*

Chapter 14
Glass

It is not the purpose of this book to delve deeply into glass salt and pepper shakers. But an overview seems appropriate for one reason. As word spreads that you are a salt and pepper shaker collector, it is likely that eventually someone will present you with some glass shakers they want to dispose of. Because some art glass shakers are quite pricey, a little familiarity might lead to a profitable purchase, the proceeds of which could be put into your figural shaker collection.

Art Glass

Art glass salt shakers of the late Victorian Era are highly prized and eagerly collected. Prices have skyrocketed over the past few decades so that many now sell for hundreds or even thousands of dollars. As mentioned above, sooner or later you will probably run into some at a bargain, and you will want to be prepared when you do. In the event you do pick up a collection of them, I recommend *The World of Salt Shakers—Antique & Art Glass Value Guide* (second edition), by Mildred and Ralph Lechner, for identification and pricing. Information on how to obtain the book is listed in Appendix I Sources.

Another beautiful pair of Mt. Washington egg shape shakers with flat bottoms. *Courtesy of Cloud's Antique Mall, Castalia, Ohio.*

Mt. Washington fig salt shakers. The peppers that should be with them have rounded caps. These shakers are 3 inches high and about 100 years old. *Courtesy of Cloud's Antique Mall, Castalia, Ohio.*

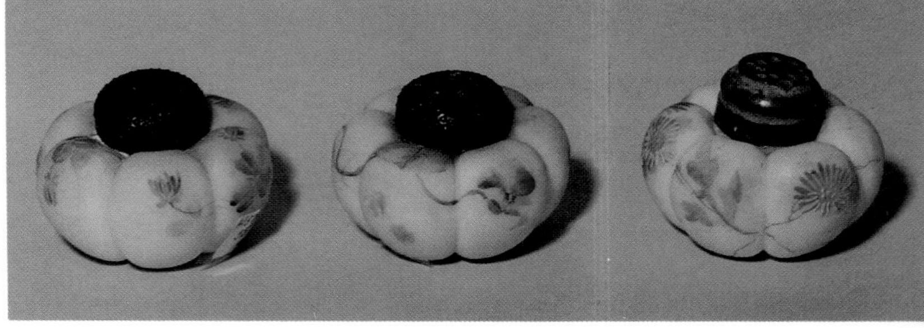

Collectors refer to this shape by Mt. Washington as Tomato. The shakers measure 1-½ x 2-⅝ inches. *Courtesy of Henrietta Hillabrand.*

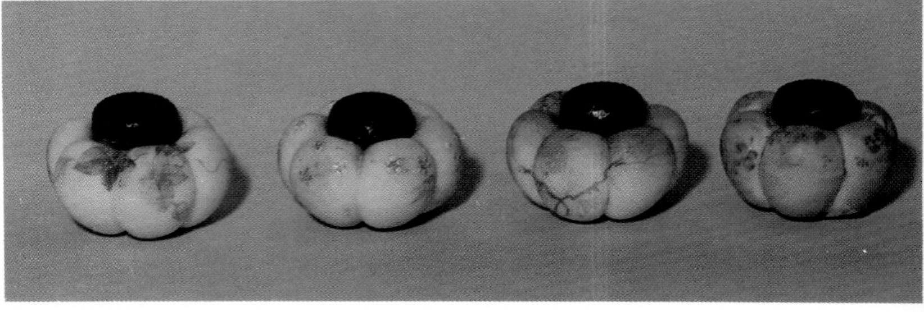

Same as above but with different decoration. *Courtesy of Henrietta Hillabrand.*

These are called Mt. Washington egg shape shakers with flat ends. The company made a similarly shaped shaker with a round bottom and a flat side. Height is 2-½ inches. *Courtesy of Henrietta Hillabrand.*

Challinor's (Challinor, Taylor and Company) Forget-me-not. The shakers are from the 1880s but the cap on the one on the right is not original. *Courtesy of Cloud's Antique Mall, Castalia, Ohio.*

Same set as above in white minus the tray. *Courtesy of Cloud's Antique Mall, Castalia, Ohio.*

Sunset pattern by Dithridge & Company. Height is 2-7/8 inches. *Courtesy of Betty Laney.*

This pattern is called Iowa, produced by the U.S. Glass Company around 1900. The shakers are 3-1/8 inches high. *Courtesy of Cloud's Antique Mall, Castalia, Ohio.*

Eagle Glass Company's Forget-me-not Pee Wee set, circa 1900. The shakers are 3 inches and 2 inches high. *Courtesy of Cloud's Antique Mall, Castalia, Ohio.*

On the left is Bulging 3 Petal, made by the Consolidated Lamp and Glass Company of Coraopolis, Pennsylvania around the turn of the century. The shakers are 2-1/4 inches high. Which of the tops, if either, is original is not known. On the right is Cotton Bale by the same company and dating from about 1895. Height is 3 inches. *Courtesy of Henrietta Hillabrand.*

Cone, also by Consolidated Lamp and Glass. Height is 3-¼ inches. It is unlikely that either cap is original. *Courtesy of Cloud's Antique Mall, Castalia, Ohio.*

These shakers are 3 inches high. Name is not known. *Courtesy of Cloud's Antique Mall, Castalia, Ohio.*

Pattern name and company are unknown, height is 3-⅝ inches. *Courtesy of Cloud's Antique Mall, Castalia, Ohio.*

Unknown, unmarked, 3-½ inches high. *Courtesy of Betty Laney.*

Another unknown entry, this pair of painted milk glass shakers is 3 inches high. *Courtesy of Henrietta Hillabrand.*

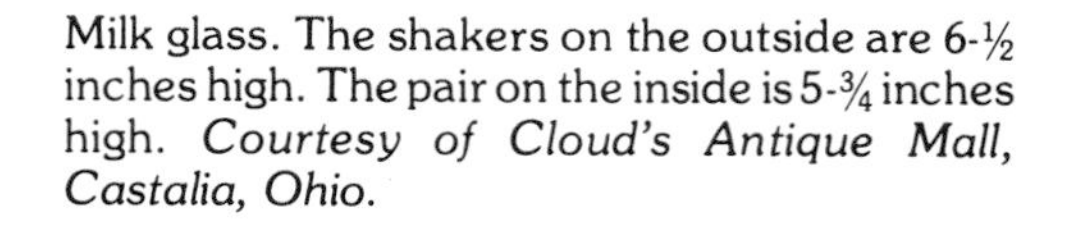

Milk glass. The shakers on the outside are 6-½ inches high. The pair on the inside is 5-¾ inches high. *Courtesy of Cloud's Antique Mall, Castalia, Ohio.*

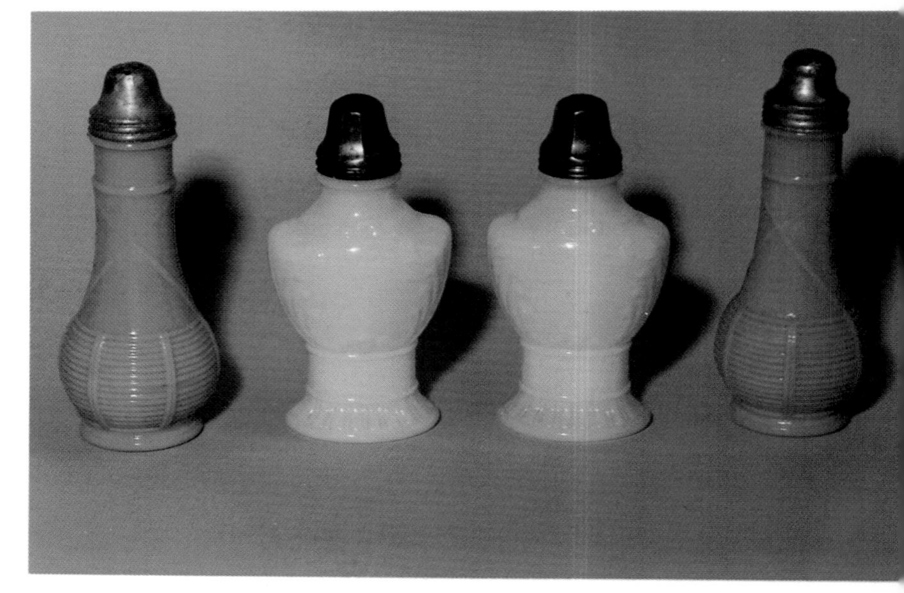

Miscellaneous Glass

Common glass shakers such as those used in restaurants are not very attractive and do not have much value. On the other hand, we saw them in just about every collection we photographed so a few are being shown here.

The tray holding these amber glass shakers is 6-½ inches long. *Courtesy of Cloud's Antique Mall, Castalia, Ohio.*

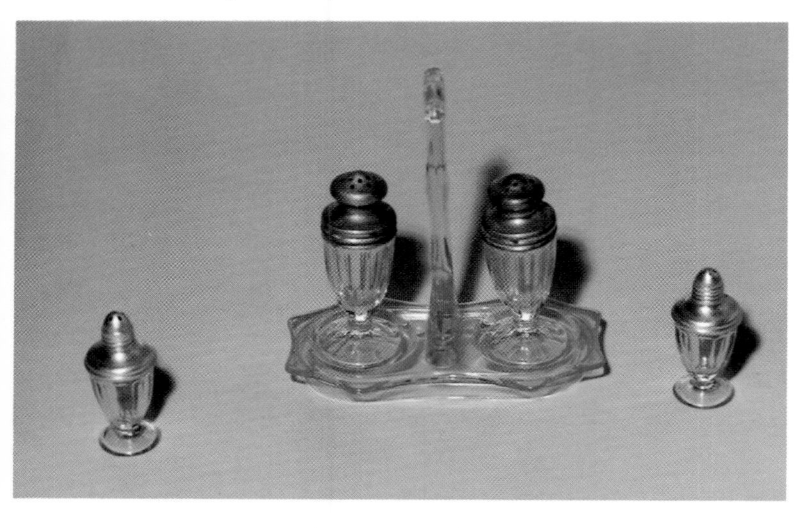

Depression glass shakers, those on the outside standing 2-¾ inches high, those on the inside 3-¾ inches. *Lilly Collection.*

The shakers are 2-⅞ inches high. The tray is 8 inches long, marked "Krome-Kraft / Farber Bros. / New York, NY" On three lines. *Courtesy of Cloud's Antique Mall, Castilia, Ohio.*

Milk glasss, 4-¼ inches high. *Courtesy of Cloud's Antique Mall, Castalia, Ohio.*

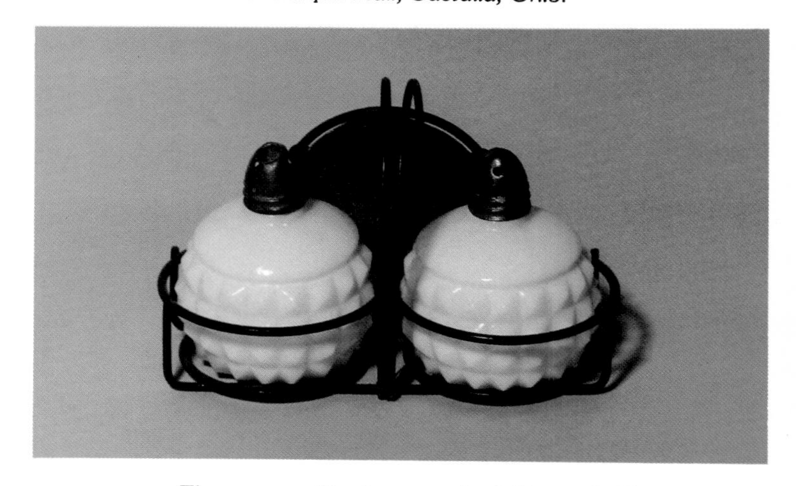

These are milk glass, marked "Japan." The shakers are 2-½ inches high. *Courtesy of Cloud's Antique Mall, Castalia, Ohio.*

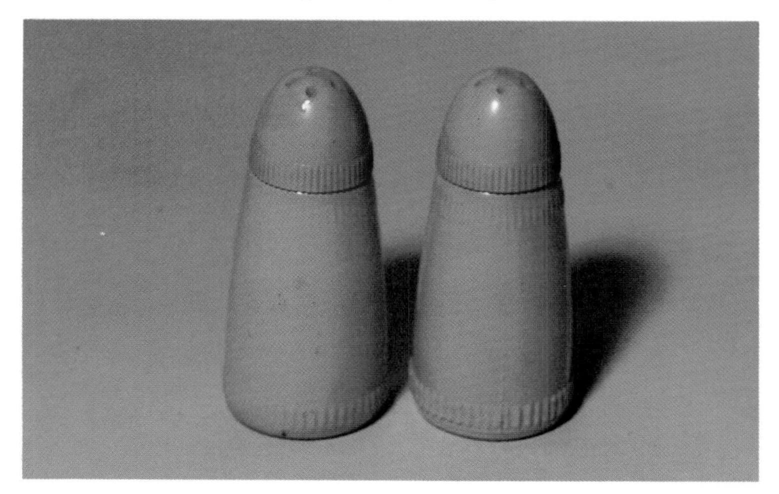

Painted clear glass bottoms, plastic tops, 3-¼ inches high. *Courtesy of Cloud's Antique Mall, Castalia, Ohio.*

Glass bottoms, metal collar, plastic tops. Height is 1-⅛ inches. *Carson Collection.*

These are 3-½ inches high, and very heavy for their size. Note the different tops. *Courtesy of Cloud's Antique Mall, Castalia, Ohio.*

Blown novelties of the Pilgrim Glass Company. Height is 4-½ inches. *Carson Collection.*

Cut glass shakers, 4-⅞ inches high, unmarked. *Courtesy of Joyce Hollen.*

This is Tiffin Teardrop. The salt and pepper are 3-⅝ inches high, the tray is 6 inches long. *Courtesy of Cloud's Antique Mall, Castalia, Ohio.*

The pair on the left and the pair on the right are etched. They are 2-¾ and 3-½ inches high. The pair in the center is cut, 3-½ inches high. One of the cut shakers has a plastic cap. *Courtesy of Cloud's Antique Mall, Castalia, Ohio.*

Pink Depression glass, 3 inches high. *Courtesy of Cloud's Antique Mall, Castalia, Ohio.*

Here is something novel—some salt and pepper shakers with salt and pepper in them. Heights left to right are 3, 3-¾ and 3-¼ inches. *Courtesy of Dan Schneider.*

Section IV: Origins

One of the biggest problems faced by figural and novelty salt and pepper shaker collectors is that of obtaining knowledge concerning the manufacture and marketing of pieces in their collections. Information is scant on many companies. Some seemed to have gone into and out of business so quickly that no record of them apparently remains. This is especially true of companies that manufactured plastic sets. Either they were operating for a very limited time, or were quickly bought out by larger firms wishing to avoid competition.

Chapter 15

Manufacturers, Importers, Distributors, Sellers and Artists

Listed below are the names of companies and individuals who have made or sold salt and pepper shakers in the past, or are doing so today. In some cases the information accompanying the name is fairly extensive. In other cases there is just the name with a note that no information was found.

Hopefully, the blanks will eventually be filled in by knowledgeable collectors who contribute articles to The Novelty Salt and Pepper Shakers Club newsletter, or by authors of future books on the subject. Or maybe someday I will write another book on shakers and have the opportunity to fill in some of them myself. Based on the people we have met, the friendships we have made, and the many outstanding collections we have seen while writing this book, that doesn't sound like a bad idea at all.

Ace Ace was apparently a company that imported and distributed products from Japan, as its paper labels on salt and pepper shakers show Japanese origin, but that is all that is known about it at the present time.

AGIFTCORP Agiftcorp is an acronym for American Gift Corporation, which is headquartered in Miami, Florida.

American Gift Corporation is one of the country's oldest importers of Pacific-Asian ceramics. It began in 1935, predating other well known shaker importers such as Lefton, Inarco and Napco by several years. According to a company spokesman, it was also one of the first companies to resume importing products from Japan after World War II late in 1945 or early in 1946. Most Agiftcorp shakers imported during that era bore only the non-identifying Occupied Japan mark, the spokesman said.

Today American Gift Corporation imports shakers from Taiwan and several other Pacific-Asian countries. The company designs all of its own shakers and each design is copyrighted. The foreign potteries then make the shakers to American Gift's specifications. The shakers are sold exclusively in tourist areas, the spokesman said, each shaker carrying the name of the city, area, state, etc. where it is purchased as a souvenir. The most commonly used mark, he said, is a copyright symbol with Agiftcorp written next to it.

Being located in Miami, the facilities of American Gift Corporation sustained some damage from Hurricane Andrew in August, 1992, but not enough to seriously affect daily transactions on a long term basis.

Aluminum Housewares Company, Inc. This firm specializes in importing and marketing kitchen gadgets and utensils, including salt and pepper shakers. It began operations in 1958. Aluminum Housewares Company is located in Maryland Heights, Missouri, a suburb of St. Louis.

According to James Seidel, executive vice president, the company still imports and sells salt and pepper shakers. Some are Aluminum Housewares' exclusive designs, others are purchased from available stock.

The old and new refrigerators shown in the color section are examples of the company's shakers. Seidel, who was with the Aluminum Housewares when the refrigerator shakers were copyrighted in 1976, said the story that circulates about them being a Westinghouse promotion is a myth. While the design may have been based on Westinghouse appliances, the shakers were marketed to mail order wholesale customers in bulk boxes, and to retailers in plastic blister packs, Seidel said.

American Bisque Company (ABCO) American Bisque Company, one of our country's most unsung potteries in this writer's opinion, was organized in Williamstown, West Virginia, in 1919. Its original product was bisque Kewpie dolls, which had been in short supply ever since European manufacturers shut down production in 1914 at the outbreak of World War I. Around 1921 the company introduced a line of planters for bulbs, and also expanded into vases and lamps.

In 1922, B.E. Allen, founder of the Sterling China Company, in Wellsville, Ohio, bought American Bisque. The firm remained in the Allen family until 1982.

Exactly when American Bisque began making salt and pepper shakers is not known, but it may have been just prior to World War II. In 1937 the pottery was flooded to a depth of 10 feet, and just about everything that had been part of American Bisque, except the name, was lost. According to Jenny B. Derwich and Dr. Mary Latos, in *Dictionary Guide to United States Pottery and Porcelain (19th and 20th Century)*, (Jenstan, 1984), the pottery is believed to have started making cookie jars upon recovering from the flood. Since most American Bisque salt and pepper shakers are simply miniature models of its cookie jars, it seems likely that shaker production would have been started about the same time.

It is often difficult, if not impossible, to pin down whether shakers normally credited to American Bisque were actually made by them, or by American Pottery Company, of Marietta, Ohio. The two firms were directly across the Ohio River from each other, and for several years American Pottery was co-owned by A.N. Allen, then owner of American Bisque, and J.B. Lenhart, American Bisque's sales manager. (See the entry on American Pottery Company for further details.)

American Bisque was sold to Bipin Mizra, in 1982. Mizra changed the name to American China Company. In 1983 the pottery was closed and dismantled.

American Gift Corporation *See AGIFTCORP*

American Pottery Company (Marietta, Ohio) The American Pottery Company was started in Byesville, Ohio, as the Stoin-Lee Pottery in 1942. Two years later, J.B. Lenhart, then employed as sales manager of American Bisque, bought the pottery and moved it to Marietta, Ohio, across the Ohio River from the American Bisque plant. Lenhart sold an interest in American Pottery to A.N. Allen, owner of American Bisque.

In many cases this co-ownership has made it nearly impossible to distinguish American Pottery Company products from those of American Bisque. Apparently both companies marketed some of their wares through the Leeds China Company, of Chicago, which was a Walt Disney Productions licensee from 1944 to 1954. Shakers made by American Pottery or American Bisque for Leeds would include the Disney characters such as Mickey Mouse, Pluto, Dumbo, etc. that were finished in clear (white) glaze with painted red and black decoration. Some are regular shakers, some are turnabouts. Most were made as accessories for matching cookie jars.

Lenhardt sold American Pottery in 1961. John F. Bonistall, formerly of the Shawnee Pottery, became general manager in 1962. The firm went out of business in 1965. Later the plant was razed.

American Pottery Company (Los Angeles) Not much is known about the American Pottery Company that was based in Los Angeles, but it is quite possible they made some salt and pepper shakers, and it is also possible that they could be identified. According to Tom Tumbusch in *Tomart's Illustrated Disneyana Catalog and Price Guide, Volume Four*, (Tomart Publications, 1987), American Pottery of Los Angeles was licensed by Walt Disney from 1943 to 1955 to make ceramic figurines and dinnerware sets. Tumbusch also states that, when Vernon Kilns gave up its Disney license for *Fantasia* figurines on July 22, 1942, all rights, molds, and inventory were assigned to American Pottery. Consequently, it seems probable that some Disney character shakers may have been made, and some non-Disney shakers, too, since we can reasonably assume dinnerware and figurines were the company's forte.

Evan K. Shaw was the owner of American Pottery. Later he bought Metlox Pottery, of Manhattan Beach, California.

ANCO This obvious acronym appears on some nodders, but efforts to identify and locate the company were unsuccessful.

An Enterprise Exclusive Located in Toronto, Ontario, this import company was started in the mid-1930s by a man named Harry Pearl. Pearl sold the business and retired in the early 1980s. The company was also known as Enterprise Sales and Distributors. Although neither An Enterprise Exclusive nor Enterprise Sales and Distributors was listed in the 1992 Toronto telephone directory, it may still be in business, the company name possibly undergoing a change with the change in ownership.

Besides the An Enterprise Exclusive paper label shown in the color section, the company also used an inkstamp: "ESD / Japan" on two lines, encircled by a wreath with a lion across the top of the wreath.

Annette's Nothing was found about the origin of this mark. It appears on a pair of brass shakers.

Arcadia Ceramics, Inc. Not surprisingly, Arcadia was located in Arcadia, California. It first appeared in the city directory in 1950 as Arcadia Ceramics Company, with Alfred and Esther Johnson shown as owners. Sometime in the next couple years, Alfred Johnson is said to have been killed in a plane crash, after which Esther Johnson apparently sold the business and moved from the area.

The 1952 directory shows the name changed to Arcadia Ceramics, Inc., John Bennett president, John Renaker secretary-treasurer. John Renaker was co-owner of Hagen-Renaker, a pottery in Monrovia, California.

By 1958, the year Arcadia Ceramics, Inc. closed, it too was located in Monrovia.

Arcadia is mainly known for miniature shakers, but it also made some table size shakers. It is likely the miniatures were influenced by Renaker. Hagen-Renaker is famous for its miniature animal figurines, a pursuit the pottery began in 1948.

Most Arcadia shakers were marked only with paper labels, many of which have long since disappeared. One key to identification is that many Arcadia shakers had crescent-shaped eyes cut into them which served as pore holes. (Note the eyes in the Arcadia miniature pickles on page 182) Also, a lot of Arcadia's shakers were sold as three-piece sets, quite often the tray having a figure in the center that served as a handle.

(Many thanks to California collector Marcia Smith for providing information about this company.)

Arcadia Export-Import This firm is known to have been in business during the 1950s. Very little additional information has been found. Lois Lehner, in *Lehner's Encyclopedia of U.S. Marks on Pottery, Porcelain & Clay* (Collector Books, 1988), shows a mark for Arcadia Export-Import that includes the words Newark and New York in that order. The company apparently either had offices in those two well known cities, or was based in the small town of Newark in upstate New York.

Arcadia was also shown as a trademark of Royal China & Giftware, a company located at 224 5th Avenue, New York, according to the 1980 *Trade Names Directory.*

Armita This company's name is associated with a pair of cherub shakers in Chapter 5, but no other information about it was found.

Artmark Chicago, Ltd. According to Patricia M. Quinn, director of marketing, this Chicago-based importer has been in business more than 40 years, during which time salt and pepper shakers have always been an important part of its housewares line. The company also markets giftware.

In the beginning the firm went by the name Lohzin and Born, but even then all its products were sold under the name Artmark, Quinn said.

Artmark depends on input from a variety of sources when choosing its shaker designs—buyers, sales representatives, customers and, of course, designers. In November, 1992 the company had more than 50 different sets of shakers on the market. From time to time it drops some models and adds others to keep its offerings fresh.

Quinn said fruit and vegetable sets, such as the boxed garden vegetables shown in Chapter 7, have always been in the Artmark line in one form or another. In the past they were made of plastic as well as ceramic. Years ago they were made with faces on them, Quinn said.

The good news for collectors about Artmark is that, according to the company, it intends to continue offering a wide range of shakers that run from traditional to whimsical.

Atlas At present, little is known about this company. A study of its products would seem to indicate it was active during the late 1940s and 1950s, that era when most salt and pepper shakers, and other ceramics imported from the Orient, came almost exclusively from Japan. It had apparently gone out of business, or had been absorbed by another company, by 1979 as it was not listed in that year's *Trade Names Directory*.

Aura Ceramics The company has been in business since June, 1981. It is owned by Larry Keiper and Jerry Policelli, neither of whom had any formal art education when they set up shop. Aura is located in Easton, Pennsylvania.

Operating out of a 5000 square foot building, Aura caters to the hobby ceramics market but does have some commercial accounts. For instance, since 1984 it has made a different Christmas ornament each year for the city of Easton. The city sells the ornaments in conjunction with its "World's Largest Candle"—94 feet high made of steel and plywood with a fiberglass flame—which annually decorates the town square throughout the holiday season.

Aura is included here, of course, because it produced the limited edition shakers that commemorated the 1988 and 1989 conventions of the Novelty Salt & Pepper Shakers Club. The shakers, "Raisin Nodder" and "Come Travel to Beautiful Ohio," shown in Chapter 4, were designed by Hubert and Clara McHugh, authors of *Goebel Salt & Pepper Shakers—Identification and Value Guide.*

Avon Products, Inc. The firm was founded in New York as the California Perfume Company by David Hall McConnell around 1886. The name change to Avon Products, Inc. took place on October 6, 1939.

At present all Avon ceramic products, including salt and pepper shakers, are made in Brazil by Ceramarte.

AVSCO Thus far, no information has been discovered about this plastics company other than it was located in Kansas City, Missouri.

Bavarian Bavaria is a state in southern Germany that has long been a center of ceramic production. The words "Bavaria" and "Bavarian" have been used by numerous manufacturers and importers throughout the years.

Bernard Studio Another pottery about which little is known. One of its paper labels indicates it was located in Fullerton, California. Based on that information, and the style of its ceramics, it would seem safe to assume Bernard Studio was active sometime between the 1930s and 1960s, during the heyday of the California pottery movement.

Beswick James Wright Beswick founded the pottery that still bears his name in Longton, Staffordshire, England in 1894. Primary ownership and control of the company was handed down through the Beswick family until 1969, when it was sold to Royal Doulton.

Beswick is probably best known for its naturalistic animal figurines, the first of which was made during the 1930s. It also makes Beatrix Potter animal characters.

Shaker production at Beswick has been quite limited when compared to many other companies, but those it has made have generally been outstanding. Beginning about 1936, most Beswick pottery was made with an impressed model number. By comparing that number with the chart in Appendix II, you can determine the year a particular set of shakers was introduced.

Blossom Valley A wonderful pair of fox shakers with this mark, copyrighted by Dave Nissen and bearing a strong resemblance to Regal China, are shown in the color section. Unfortunately, no additional information has been found.

Blue Ridge (Southern Potteries, Inc.) Blue Ridge is the name collectors commonly refer to when speaking of Southern Potteries, Inc., a firm that operated in Erwin, Tennessee from about 1917 to 1957. The company was started with capital supplied by the Carolina, Clinchfield and Ohio Railroad. The railroad's generosity stemmed from its need to develop industry along its line so its trains would have goods to transport. Early on the name was Clinchfield Pottery. This was changed to Southern Potteries about 1918.

The chief money-maker at Blue Ridge was dinnerware, which was made in hundreds, if not thousands, of different patterns. The bulk of the dinnerware was hand-decorated in bright, often gaudy, colors applied in simple designs, often flower patterns. For several decades prior to its demise, Blue Ridge sold huge amounts of dinnerware to large retailers such as Sears, Montgomery & Ward, Macy's and others. Most large retailers handled exclusive designs made especially for them.

While many Blue Ridge plates and other pieces of dinnerware are marked, the majority of salt and pepper shakers the company made are not. But that's not much of a problem. If you study the Blue Ridge pictures in this and other books, and familiarize yourself with the pottery at antique shows, the shakers will soon scream at you from several booths away.

Bona / Delfts / Bloup / Holland The mark appears on a pair of Dutch windmill shakers in Delft blue and white. That, unfortunately, is the extent of this author's knowledge of it.

Brayton Laguna Brayton Laguna Pottery operated in southern California from 1927 to 1968. It was started by Durlin E. Brayton at his home on the Pacific Coast Highway in Laguna Beach. In the beginning he displayed the pottery in his front yard, the ware catching the eye of both locals and tourists motoring along the heavily travelled highway. At the time Brayton started the business, he was working as a carpenter. Prior to that he had attended the Chicago Art Institute.

Pottery historians credit much of the success of the Brayton Laguna Pottery to Brayton's second wife, Ellen Webster Grieve Brayton, whom he married in 1936. A talented modeler with a good head for business, Webb Brayton, as she was known, designed and added an extensive line of mass produced handpainted figurines to the rather mundane dinnerware upon which Durlin Brayton had founded the business. As a result, Brayton Laguna evolved from a cottage business to a large commercial enterprise.

In 1938, a five-acre site was developed for the operation. Also that year, Walt Disney licensed Brayton to produce ceramic Disney characters, the first such license the entertainment giant granted. Business boomed, and during World War II as many as 150 people were employed. Webb Brayton died in 1948. Durlin Brayton passed away three years later. The company continued to operate until 1968 when it was permanently closed. Its buildings now house the Laguna Art Center.

How many different salt and pepper shakers Brayton Laguna made is a mystery. The only known sets in this book are the unmarked gingham dogs and calico cats. Davern, in Book II, pictures a pair of black people. It is possible there are others, since much Brayton Laguna pottery was sold unmarked. And quite often when it was marked, only a very light inkstamp was used. Another possibility is that pour holes may have been added to some of the numerous figurines Webb Brayton designed to turn them into salt and pepper shakers.

Dan Brechner & Company This mark appears on one of the married couples in Chapter 5. On the bottom of one of the shakers is a penciled date of 10-2-62. That is all that is known about it.

Brillium Metals Corporaiton According to the box shown on page 231, this firm was located in Jamaica, New York, and was in business prior to July 1, 1963.

Brinchell *See Colorflyte.*

Brinco *See Brinn's.*

Brinn's Located in Pittsburgh, Pennsylvania, the company has changed addresses within that city several times over the years, the moves apparently dictated by growth and expansion. Brinn's has been in business since at least 1934, as Lehner cites the company filing for a trademark registration in 1966 claiming use since 1934.

Brinn's has marketed a vast number of shakers, enough certainly to put it in contention for the status of America's largest shaker importer if there was a way to determine that type of thing.

Most Brinn's shakers are marked by paper label. Some of its paper labels, presumably more recent ones, use the acronym Brinco.

Burrough's Manufacturing Corporation This company, the maker of Burrite plastic, was located in Los Angeles. It was most likely active sometime between the end of World War II and the mid-1960s.

Cardinal China Company *See Cardinal, Inc.*

Cardinal, Inc. This is the company that markets salt and pepper shakers bearing the label Our Own Import. It was founded in 1946 in Carteret, New Jersey as the Cardinal China Company. Today it operates out of Port Reading, New Jersey not far from its original location. The move was made during the 1980s.

In the beginning Cardinal made some of its own pottery, but that didn't last long. Then it became a distributor, marketing products made by other American potteries, but often the Cardinal name was impressed. In recent years it has functioned primarily as an importer.

Carvanite This was apparently a trade name used by an early United States plastics manufacturer of the 1945 to 1965 era.

Ceramarte Ltd. Ceramarte Ltd. is a pottery in southern Brazil that employs approximately 1800 people, according to Tony Kreimer, vice president of marketing for Primex International Trading Corporation, New York. Primex is the exclusive representative of Ceramarte in the United States and Canada. Ceramarte has been in business approximately 25 years.

Its name, incidentally, is pronounced sar-a-mar-tay, not ce-ram-art as so many collectors and dealers pronounce it.

According to Kreimer, Ceramarte is currently the world's largest producer of beer steins. It also makes all of Avon's ceramic pieces, he said.

The pottery uses Brazilian clay to produce what Kreimer describes as a fine grain stoneware. He notes that is why Ceramarte pieces are so heavy when compared to most other modern ceramic products.

Salt and pepper shakers, such as the Hamm's Beer bears, were an outgrowth of other advertising pieces, Kreimer said. The pottery makes them on a somewhat limited basis, turns out just enough at one time to fill in the spaces between larger pieces during firing. Ceramarte is currently making Bud Man advertising shakers for Budweiser Beer, Kreimer said.

Ceramic Arts Studio When it comes to salt and pepper shakers manufactured in America, or anywhere else in the world for that matter, Ceramic Arts Studio, of Madison, Wisconsin was one of the premiere companies as far as quality and style are concerned.

The firm was started in 1941 by Ruben Sand and Lawrence Rabbit, who were both students at the University of Wisconsin at the time. To say they began on a shoestring would be an understatement. The pair relied on an old washing machine to strain clays and mix glazes, built their own kiln for which they also mixed the insulation, and jury-rigged a manual potter's wheel to give it electrical power. Ceramic Arts Studio used Wisconsin clays exclusively. It is the only major pottery known to have done so.

Many Ceramic Arts shakers have a "BH" stamped on the bottom as part of the mark. The initials stand for Betty Harrington, the modeler who provided the creativity that raised Ceramic Arts Studio above the also-rans. Her story is even more remarkable than that of Sands and Rabbit, the poor but determined entrepreneurs.

According to Sabra Olson Laumbach, author of *Harrington Figures* (Ferguson Communications, 1985), Betty Harrington discovered her talent by accident. In the fascinating account Laumbach relates, Harrington was working as a secretary for the state of Wisconsin in 1942 when she and her husband, Albert, built a new home in Madison. While digging a well for the home, some workmen ran into a layer of blue clay which the neighborhood children quickly used to form little animals and dishes. One of Harrington's two daughters presented a glob of the clay to Harrington, who, without ever having worked with clay before, shaped it into a correctly proportioned artistic figure of a kneeling nude girl holding a bowl. Pleased with her work, she decided to have the figure glazed and fired. After several turndowns, she ended up at the fledgling Ceramic Arts Studio where Ruben Sand agreed to fire it. Apparently recognizing Harrington's natural talent, Sand, instead of charging her for the work, hired her on the spot and the rest is history. Harrington worked for the firm for 13 years until Sand closed the business in 1955.

Many Ceramic Arts Studio salt and pepper shakers lack marks because there simply wasn't room for them. Keys to identification can be the rosy cheeks on people shakers (some collectors refer to Ceramic Arts shakers and figurines as "rosy cheeks"), the limited number of glaze colors the company used, and the high quality of the work. At the same time, you have to be aware of foreign imitations. Best examples of this are the elephants shown on page 104 and the Dutch children on page 45.

A footnote to the Harrington legacy can be seen in Chapter 4. At 77, Betty Harrington designed the Abraham Lincoln shaker set, "A Nod to Abe," that was used to commemorate the Novelty Salt and Pepper Shakers Club 5th Annual Convention in 1991.

Challenor, Taylor and Company The Challenor's Forget-Me-Not shakers shown in the glass section were made by Challenor, Taylor and Company in Tarentum, Pennsylvania around 1890. The company was formed in Pittsburgh in 1866, moved to Tarentum in 1884. A fire totally destroyed the plant in 1891, not long after it had merged with U.S. Glass. The factory was never rebuilt.

Charls Products, Inc. This import company was founded more than 50 years ago by Charles Finegush. It is still in business, located in Rockville, Maryland.

Chase No information has been found on this importer of handpainted Japanese ceramic products.

Chikaramachi No information was found on this Japanese porcelain firm. Its condiment sets would seem to indicate Chikaramachi was active sometime during the first half of the 20th century.

Clay Art Clay Art was established about 1978 by Jennifer McLain and Michael Zansagna, in San Francisco. Both founders had backgrounds in art.

The company designs its own products and has them made to its specifications overseas. As you have no doubt noticed above, or in your own collection, Clay Art markets only high quality shakers. Top sellers at Clay Art besides salt and pepper shakers are cookie jars, masks, banks, teapots, cups and saucers, and mugs.

In November, 1992, according to a recorded message on the company's telephone system, Clay Art was then marketing 70 different sets of shakers.

According to Clay Art designer Tom Biela, the company considers salt and pepper shaker collectors each time it develops a new design. "We try to make each design unique and humorous so it appeals to collectors. We find that shakers that appeal to collectors also appeal to the general public," said Biela.

In July, 1992 Clay Art sent Biela to the Novelty Salt and Pepper Shakers Club convention in Burlington, Vermont to mix with and speak to collectors about shakers. Wouldn't it be great if all companies were as tuned into collectors as Clay Art is!

Clover This mark appears on the one-piece plastic Tappan range shown above. That's as much as I have been able to find out about it.

CNC This paper label shows up on a pair of shakers in Chapter 5 that bear a strong resemblance to Hummel, but no other information has been discovered.

Colorflyte This is apparently a trademark for an early plastic, attributed on shakers to Brinchell, of St. Louis, Missouri. In a 1979 trade journal a "Color-Flyte" melamine dinnerware was listed under Lenox Plastics, Lexington United Corporation, St. Louis, but attempts to reach the company were unsuccessful.

A Commodore Product The 1975 edition of *Mass Retailing Merchandiser* listed a Commodore Import Corporation at 507 Flushing Avenue, Brooklyn, New York. Directory assistance did not show a company by that name in Brooklyn in 1992, so it cannot be said with authority that the A Commodore Product mark is related to the Commodore Import Corporation.

Consolidated Lamp & Glass Company When this glass company was founded in 1893, its plant was located in Fostoria, Ohio, and its main offices were in Pittsburgh, Pennsylvania. Within a couple years a fire reduced the Fostoria plant to ashes, after which the owners opted to rebuild in Coraopolis, Pennsylvania, a suburb of Pittsburgh. Financial problems and changes in ownership resulted in the factory undergoing several periods of inactivity over the years. Nevertheless, it stayed in business until 1964.

Couts No information has been found on this German mark.

Coventry Ware, Inc. This was a Barberton, Ohio firm that was in business from 1932 until sometime during the 1960s. Any Coventry shakers you have, however, were probably made between 1940 and about 1950, or so.

Coventry Ware, Inc., began as Dior Studios. It was started by Carrie Daum with the original focus being on art objects made of plaster. Initially, employees numbered about 10. Later the work force climbed to about 30. In 1936 the name was changed to Coventry Ware, Inc., in an effort to keep it from being perceived by the public as a photography studio. Production of fired ceramics began in 1940. Elaine Carlock (later of Cranberry Hill Studio, Pontiac, Michigan) was the chief modeler. According to Lehner, she worked for Coventry from 1940 to 1950. When cheap Pacific-Asian imported ceramics began to flow into the country following World War II, Coventry, like numerous other American potteries, found itself unable to successfully compete with them. At that point, the firm turned its attention to vinyl plaques, and is assumed to have lasted into the 1960s.

At the Springfield (Ohio) Antique Show and Flea Market Extravaganza in May, 1992, I came across a table of Coventry's well-known boy and girl with dog figurines, selling for $1 each. But they were different from those I had seen before. They were decorated quite nicely, but not as well as the Coventry figurines with which I was familiar. The lady who had them explained that when Coventry went out of business, someone bought the molds and used them to make the former Coventry figures, then applied their own decoration. There were no Coventry marks on the figures, so you have to assume all this was done above board. The incident is mentioned only because the same thing may have occurred with Coventry salt and pepper shaker molds, which might help explain future mysteries surrounding Coventry shakers you may discover that don't quite look like Coventry.

Craftsman China Not finding any information on this apparent importer was a disappointment, since the quality of its one-piece zebras and rhinos is outstanding.

C.R.H. Int'l, Inc. According to the company's salt and pepper shaker boxes, the firm is located at 33 High Street, Columbus, Ohio. All phone calls to it, however, have gone unanswered.

Dan-Dee Imports Dan-Dee Imports, which was located in New York in 1968, is still in business but has moved to Jersey City, New Jersey. According to a company spokeswoman, its emphasis is now on plush toys.

Dapol Plastics, Inc. Tumbusch shows this plastics company as a division of Plascor, Inc., of Worcester, Massachusetts, and licensed to produce Disney lawn decorations from 1957 to 1960, and 1971 to 1972.

A 1979 trade journal shows the same company, still in Worcester, but as a Division of Instruments Systems Corporation. Attempts to reach Dapol in 1992 were unsuccessful.

Davis Products No information available on this plastics company other than that it was located at 29 14th Street, Brooklyn, New York, and was in business in the early 1950s.

Decoware This mark is shown on the metal New Era potato chip can shakers, but all attempts to track it down were unsuccessful.

Dee Bee Imports No information known, other than the company imported handpainted Japanese shakers, and was apparently a Disney licensee for a time.

Delli, Inc. The only thing known about this company is that it was located in Menlo Park, California.

D Exclusive No reliable information at this time, other than it imported shakers made in Japan.

Dithridge & Company Edward D. Dithridge began working as a glass blower for the Fort Pitt Glass Works, Pittsburgh, Pennsylvania about 1850. By 1860 he owned most of the company's stock, by 1863 he owned all of it. He then changed the name to reflect his ownership.

Dithridge's sunset pattern shakers, shown above, were made a couple years either side of 1895.

In 1903 Dithridge & Company became a part of Pittsburgh Lamp, Brass and Glass Company, and the Dithridge name was no longer used.

Duncan Duncan is a ceramic mold company. While much of its line is geared toward the hobby ceramic market, commercial potteries often purchase molds from commercial mold companies to use in lieu of, or in addition to, their own creations. But whenever you see any commercial mold company shakers marked with a person's name or initials (often accompanied by a date), either incised, painted or glazed, it is a strong indication they were made by a hobby ceramist. That's not necessarily bad. A talented home ceramist with a passion for excellence and no clock to beat or quota to meet, will often turn out work that is superior to that of commercial potteries. Let your eye for quality be the judge.

Eagle Glass & Manufacturing Company According to the Lechners, James, Joseph, H.W. and S.O. Paull established this company in 1894. It was located in Wellsburg, West Virginia. Opal glass became one of its specialties.

Most Eagle opal glass was made between 1897 and 1906. Around 1925 the company stopped making glass but continued to market it, obtaining blanks from a plant nearby. In 1937 the company shifted its emphasis away from glass, choosing to work in metals.

Efco Ceramics Nothing was found on this company. Perhaps it was related to the company directly below.

Efcco Imports According to its label, this company was based in San Francisco. It is known to have imported Japanese black-glazed redware.

Elbee Art Products Company According to Derwich and Latos, the address of the Elbee Art Products Company was 717 Lakeside Avenue N.W., Cleveland, Ohio. The researchers found them listed in Cleveland City Directories in 1947, 1951 and 1954, but not listed in 1955. When included in the directory, the president of the company was Leslie Berk. Lillian Berk served as vice president, Abraham E. Gordon as secretary.

El Ranchito No information was found about this mark. Considering its name, the company may have been located in California or the southwest.

Emmel This is another mark that defied all efforts to identify it.

Empress Because you see this label quite often, you would think it would be easy to find out the story on it. Unfortunately, that didn't prove to be the case.

Enesco Enesco is an acronym for the N. Shure Company. Enesco was formed in 1959 as an import division of that firm. Today it is a wholly-owned subsidiary of Stanhome, Inc., of Westfield, Massachusetts, the parent company of Stanley Home Products.

Enesco is based in Elk Grove, Illinois, a suburb of Chicago. Its complex there consists of 455,000 square feet of warehouse space—that's more than 11 acres!—plus 33,000 square feet of office space. It has showrooms across the country in twelve major cities.

While Enesco has always stressed quality in its imports, many of which have been salt and pepper shakers, the workmanship has obviously gotten better over time. Some of its shakers from the early 1960s appear somewhat rough when compared to later efforts. Much of Enesco's quality can be attributed to it designing all its own products, rather than buying whatever happens to be available on foreign markets.

Essex China This paper label has been the cause of more than a little confusion among collectors, mainly because some take it to be the mark used by the Weller Pottery toward the end of its existence. In 1947, the majority of Weller stock was purchased by the Essex Wire Company, of Detroit, Michigan, which had rented space at Weller since 1945. Essex closed the pottery in 1948.

Now, forget all that because the Essex China paper label appears only on shakers made by the Shawnee Pottery Company, of Zanesville, Ohio. According to Lehner, registration for the mark was filed on January 4, 1944, Shawnee claiming use since September 16, 1943. How long after the filing the label was used has not been ascertained. Shawnee closed in 1961.

For more information see Shawnee Pottery Company.

Etc Seattle No information found; obviously located in Seattle, Washington.

Exco *See Inco.*

Fairway Nothing is known about this company except that it was an importer of Japanese ceramics.

Fern This paper label appears on several sets of shakers but no clue to its identity has been uncovered.

F & F *See Fiedler & Fiedler Mold & Die Works.*

Fiedler & Fiedler Mold and Die Works. This plastics company, located in Dayton, Ohio, made several sets of shakers such as Willie and Nillie, Fifi and Fido, and the very rare Luzianne Mammy. But it is probably best known for its Aunt Jemima-Uncle Mose line that was made as premiums for the Quaker Oats Company's Aunt Jemima Pancake Mix during the 1950s.

Actually, it was an entire line of Aunt Jemima premiums. Besides the salt and pepper shakers in table and range-size, a cookie jar, syrup, creamer and sugar, and a seven-piece (including the metal rack) spice set were made.

According to Arthur F. Marquette, in *Brands, Trademarks and Good Will—The Story of The Quaker Oats Company* (McGraw-Hill, 1967), four million of the Aunt Jemima-Uncle Mose shakers were distributed. He does not say how many were table size and how many were range size, but it's obvious many more smaller ones were made than large ones. More than one million of the syrups were given away, and 150,000 cookie jars, Marquette states.

Fitz & Floyd Fitz & Floyd is an importing company that was established by Pat Fitzpatrick and Robert Floyd in 1960. It is headquartered in Dallas, Texas. It has offices in Japan and Taiwan.

The company has always been known for importing fine ceramics. In fact, the next time someone starts talking down to you about recently imported shakers in your collection, you might want to familiarize them with Fitz & Floyd. According to

Mike Posgay and Ian Warner in *The World of Head Vase Planters* (Antique Publications, 1992), both presidents Reagan and Bush called upon the company to be the official supplier of White House china.

Fitz & Floyd's mark is the often seen F F, usually impressed, sometimes on a paper label.

Adrian Forrette No information was found about this mark, which appears on the plastic one-piece Mt. Rushmore.

Fostoria Glass Company Fostoria was based in Fostoria, Ohio when it was founded in 1887. But production costs forced a move to Moundsville, West Virginia just four years later. It remained there until shutting down in 1986.

At its peak, around 1950, Fostoria employed more than 900 people, and turned out 7.5 million pieces of glass annually, according to the Lechners.

Frankoma Potteries Frankoma Potteries, of Sapulpa, Oklahoma, was an outgrowth of the Frank Potteries, a small studio shop founded by John Frank in Norman, Oklahoma, in 1933.

Frank graduated from the Chicago Art Institute in 1927, moved to Norman in August of that year to organize and teach in the Ceramic Arts Department of the University of Oklahoma. He resigned from the university in 1936 to devote full-time to pottery making. According to Phyllis and Tom Bess, in *Frankoma Treasures* (self-published, Tulsa, Oklahoma, 1983), Frank incorporated Frankoma Potteries in February, 1934, which is assumed to be the date the name was changed. He remained in Norman about two years after resigning his teaching position, then moved his family and business to Sapulpa, a small town 15 miles southwest of Tulsa, in February, 1938. By June 7, 1938 Frank had his pottery up and running.

Approximately five months later, on November 10, 1938, Frankoma Potteries was dealt what came very close to being a death blow. According to the Bess's, the building caught fire that night. Because it was located outside the city limits, the Sapulpa Fire Department refused to answer the call. Frank and his family were forced to stand by helplessly as they watched their future burn to the ground. But Frank was able to convince a local banker to take a chance on him, and started rebuilding almost immediately.

Frank's daughter, Jonice Frank (later Jonice Frank Nelson), became president and chairman of the board of Frankoma Pottery in 1970. Holding a Bachelor of Fine Arts degree in Sculpture from the University of Oklahoma, she had begun working at Frankoma 10 years earlier. John Frank died in 1973 at age 68. On September 26, 1983, the pottery was again destroyed by fire. It was partially rebuilt and operated on a part-time basis until 1991 when it was sold.

You can roughly date Frankoma shakers a couple of ways. Through 1952, all Frankoma pottery was made of clay mined in Oklahoma's Arbuckle Mountains. When fired, it turned a sandy-looking, yellowish-tan. Sometime in 1953, Frank began securing his clay from Sugarloaf Hill, which was nearer to Sapulpa. The Sugarloaf Hill clay fired to a brick red color. The changeover was completed near the end of 1954. Consequently, Frankoma shakers with the yellowish clay were made from the pottery's inception through 1954, which those with the red body were made from 1953 on.

The glaze is another clue. You are probably aware of the two-color glazes for which Frankoma was known—tan and brown, green and brown, etc. But it may surprise you to find out they were actually one-color. According to the Bess's, this is because Frank and his associate, J.W. Daughtery, put their heads together and developed what is called a rutile glaze. Rutile is a mineral that contains titanium dioxide. A rutile glaze allows the color of the body (brown) to show through, creating the illusion of a two-color glaze. Frank obtained his rutile from mines in the United States, until the mines were closed around 1969 or 1970. Then he had to get it from mines in Australia, which have a little different type of rutile. When used as a glaze ingredient, the Australian rutile does not result in as bold a color as the American rutile does. Therefore, green, blue, gold and brown glazes applied before 1970 have bolder, brighter colors than those applied after 1970. You will probably need to make a side-by-side comparison to tell the difference.

Gaby Nothing was found that would help solve the mystery of this mark.

George-Goode Corporation *See Josef Originals.*

Gerold Porzellan, Bavaria Ralph and Terry Kovel, in *Kovel's New Dictionary of Marks* (Crown, 1986), show Gerold as a porcelain factory in Tettau, Bavaria, in former Western Germany, that has been operating since 1937.

Gibson California Arts No information found.

Gibson's There was a Gibson & Son's pottery that operated in Burslem, Staffordshire, England from 1885 to sometime past 1972, according to the Kovels. Also, a 1979 trade journal showed "Gibson's" English teapots among the items handled by Fisher, Bruce & Company of Mt. Laurel, New Jersey. But since no company by that name is listed in the city's 1992 telephone directory, this mark remains somewhat of a mystery.

Gifco A Gifco, Inc. was listed in a 1981 trade journal as being in Northbrook, Illinois. No other information, other than it imported plastic onion salt and peppers.

Giftco Some references to a company by this name were found, but none that would seem to relate to salt and pepper shakers.

Giftcraft No information found.

Gladding, McBean & Company Founded in 1875 as a sewer tile manufacturer, this company became a major player in the California pottery game through its numerous lines of dinnerware, the first of which appeared in 1934. Some Gladding, McBean dinnerware lines included figural shakers as accessories.

One of its most popular patterns was Franciscan Apple. The shakers of this pattern are shown in the color section. A little known fact about Franciscan Apple is that it may be the only dinnerware pattern that was ever won in a divorce. According to Ann Gilbert McDonald in *All about Weller* (Antique Publications, 1989), Franciscan Apple started out as Weller Zona in 1911. In 1933 when then-Weller president Frederic J. Grant and Ethel Weller, daughter of the founder, were divorced, Grant's right to reproduce Zona was included in the settlement. Grant went on to become Gladding, McBean's first art director and promptly restyled Zona into the ever popular Franciscan Apple.

In 1962, Gladding McBean became a division of Lock Joint Pipe Company, which changed the name of the parent corporation to Interpace in 1963. Josiah Wedgwood and Son, Ltd. bought Interpace's Franciscan Division in 1979, closed it in 1984. Throughout most of Gladding, McBean's history, its major operations were in Los Angeles and Glendale, California.

Gnco No information found.

Goebel The W. Goebel Porzellanfabrik, or porcelain factory, was opened in Oeslau, Bavaria, Germany in 1871. Oeslau later became Rodental, West Germany.

Probably most associated with Hummel figurines by the general public, Goebel has produced literally hundreds of figural salt and pepper shakers over the years beginning in the 1920s. Goebel shakers exhibit top quality in design, material and workmanship. They often have a hefty price tag attached to them when you see them at antique shows.

The company has employed a variety of marks including impressions, inkstamps, paper-foil labels and clear plastic labels. Several of the marks and many of the shakers are shown in this book.

However, for a more complete review of Goebel shakers than would be possible in any generalized volume, I strongly recommend that you also purchase *Goebel Salt and Pepper Shakers—Identification and Value Guide,* by Hubert and Clara McHugh. Published in 1992, the book is the most thorough study of Goebel shakers ever undertaken, and was written by the world's foremost authorities on the subject. Information on how to obtain *Goebel Salt & Pepper Shakers* is listed in Appendix I Sources.

Goldammer Ceramics Currently there is no information available about this San Francisco importer.

Goldman Morgan *See Lego*

Gourmand Stoneware This is a mark used by ENESCO.

Grant, S. No information was found on this designer.

Grantcrest This is a mark that was used by the W.T. Grant Company on both foreign and domestic china.

The company was started by William Thomas Grant in 1906 as a 25-cent store, which meant nothing was sold at a price higher than a quarter. The limit was raised to $1 about twelve years later and abandoned in 1940. The company was purchased by Ames Department Stores in 1985.

Most of the items upon which this mark is seen appear to be from the 1950s.

Grindley Artware Manufacturing Company Sebring, Ohio was the home of this pottery. Founded by Arthur Grindley Sr. in 1933, it operated continuously until 1947 when a fire destroyed it. The plant was rebuilt afterward, but then went out of business about 1952. Trade journals listed it as a maker of ceramic novelties.

The few pairs of Grindley shakers shown may be only the proverbial tip of the iceberg as far as the pottery's shaker designs are concerned. Ohioans Richard and Susan Oravitz, who have been collecting and researching Grindley for several years, say every time they think they have a lock on the pottery's style, something completely different turns up with a Grindley paper label. Grindley seemed to rely heavily on the production of figurines, which, as in the case of Ceramic Arts Studio, may have been made as both figurines and shakers.

Hamilton No information found.

Harper Pottery Nothing was found to shed any light on this firm, other than that appearing on its mark, which shows it was located in California. That's unfortunate because it appears the Harper Pottery did some very good work. The lack of information suggests it was short-lived.

Hazel Atlas Known mainly for containers and Depression glass, this company was formed in 1902 by the merger of the Hazel Glass Company and the Atlas Glass and Metal Company, both of Washington, Pennsylvania.

HH (one on top of the other) *See Holt Howard.*

H. Kato Apparently a Japanese pottery, possibly a Japanese exporter. Because the mark is found on Occupied Japan shakers, it can be assumed the company was active sometime between the end of World War II and April 28, 1952 when the American occupation ended. How long it was active during that time, or whether it was in business on either side of that window, is unknown.

Hillbottom Pottery Located in Alfred Station, New York, the Hillbottom Pottery has been in business since about 1967, according to owner Bruce Green, who operates it on a part time basis. Alfred Station is 60 miles south of Rochester.

Hillbottom specializes in stoneware. Green made the limited edition of 47 stoneware shakers (see Chapter 4) that were used to commemorate the first convention of the Salt Shakers Collectors Club, held in Hornell, New York in 1986. Like all Hillbottom pottery, each of the shakers was hand thrown.

Green's full-time job is teaching art at Alfred-Almond High School, in Almond, New York. He holds a Master of Arts Degree from the University of Illinois at Urbana-Champaign, Urbana, Illinois, and a Master of Fine Arts Degree in Ceramics from Alfred University College of Ceramics, Alfred, New York.

In case you haven't heard of Alfred University, you might be interested in knowing that its considerable influence, while probably not recognizable, can undoubtedly be seen in your salt and pepper shaker collection. Alfred is the largest ceramic engineering school in the world. Way back in 1900, it became the first school to offer a master of fine arts degree in ceramic design. And about 40 years ago it led the charge to promote the rejuvenation of stoneware as an art medium. The financing of the school is rather unique; the university is a private institution, but its College of Ceramics is funded by the state of New York.

Holt Howard This import company was established in 1949 in Stamford, Connecticut by Grant Holt and John W. Howard, with the latter serving in the role of president. By 1969, when it was purchased by General Housewares Corporation, of Hyannis, Massachusetts, it had sprawled to include Holt-Howard Associates, Inc., Holt-Howard International, and Holt-Howard, Canada Ltd. In 1990 General Housewares sold its Holt-Howard line to Charles Donell and Richard Rakaukas, who operate it as Kay Dee Designs, out of Hope Valley, Rhode Island.

Holt Howard shakers from the early and middle years of the company often exhibit cleverness in design that was sadly lacking in the products of many other importers.

The focus of Kay Dee Designs is coordinated kitchen accessories, going from mugs and salt and pepper shakers right up through towels and curtains. In 1991 the company offered eight solid color sets of shakers in hunter green, French blue, navy, peach, mulberry, teal, black and white. They were accessories for a complete line that included a 3-piece canister set, cookie jar, 4-piece mug set, spoon rest, napkin holder and a utility jar. The shakers were not figurals, but square in shape with round tops. It also offered some barrel-shape shakers, basically white with flower designs.

Homer Laughlin This company was started by its namesake in East Liverpool, Ohio in 1869. Originally it was a stoneware and yellow ware manufacturer.

In the mid-1930s Homer Laughlin commissioned famed

British ceramics designer Fredrick Rhead to design what became its Fiesta line, the best known of the numerous dinnerware lines Laughlin manufactured. The company introduced Fiesta at the Pittsburgh Pottery and Glass Show in January, 1936. It remained in production until 1969 when it was restyled. The restyled version was discontinued on New Years Day, 1973. Fiesta was revived in 1986 in different colors than it was made in previously, and is currently still in production.

Harlequin and Riviera are but two of the numerous other lines of dinnerware that Homer Laughlin produced that included salt and pepper shakers.

A.E. Hull Pottery Company Addis E. Hull formed the Hull Pottery in Crooksville, Ohio in 1905. It operated until 1986.

Its best known and most coveted salt and pepper shakers are the Little Red Riding Hoods, which came in three sizes—3¼, 4½ and 5½ inches—and were made to compliment an entire line of Little Red Riding Hood pottery that included cookie jars, canisters, banks, lamps, and many other items. The 4½ inch shakers are the hardest to find and most expensive.

In his book, *Collecting Hull Pottery's "Little Red Riding Hood"* (L-W Books, 1989), noted pottery author Mark Supnick shows a fourth pair he attributes to the series, an undecorated kneeling Little Red Riding Hood.

An air of mystery currently surrounds the Little Red Riding Hood pieces. It has long been known that the patent for its design, Design Patent 135,889, was issued on July 29, 1943 to Louise Elizabeth Bauer, with the Hull Pottery Company listed as the assignor. It has also long been known that the Regal China Company, of Antioch, Illinois did some decorating work for Hull, including the decoration on the Little Red Riding Hoods.

In the above mentioned book, however, Supnick declares his belief that only an early cookie jar marked "#967 Hull Ware Little Red Riding Hood Patent Applied for USA," and a dresser jar marked "Hull Ware #932 USA," were actually made at the Hull factory. He says all the rest, including the salt and peppers, were in reality made by Regal for Hull. And his argument is compelling. He notes that only the specific cookie jar mentioned, along with the dresser jar, is made of the off-white bisque for which Hull was known, while everything else in the Little Red Riding Hood line is made of the bright white bisque for which Regal was known.

Perhaps someday a contract between Regal and Hull will surface proving Supnick right. Or perhaps an invoice from Regal to Hull for slip (liquified clay used for making molded pottery) will turn up proving him wrong. Until that day comes, Supnick's theory will likely remain a subject of lively debate whenever and wherever shaker collectors congregate.

Iaac Ceramics No information was found on this concern. Its work appears to be quite recent.

Imperial Porcelain Imperial Porcelain had a rather short run when compared to other Ohio potteries such as Weller and McCoy. It was in business only from 1946 to 1960. But it left a mark on the ceramics industry in the form of several lines of pottery—the Paul Webb Mountain Boys, Al Capp Dogpatch series, and the Americana Folklore Miniatures—that won't be erased anytime soon.

According to Derwich and Latos, a trio of doctors founded the pottery. They were Dwight Smith, Fred Philips and J. Diehl Sulberger.

Paul Genter, a ceramic artist of note, was hired, and with Paul Webb created the Mountain Boys series which became an immediate success. Webb was the artist who is credited with developing Lil' Abner for Al Capp. He was trained at the Pennsylvania Art Academy. Webb laid down the Mountain Boys series on paper, Genter transformed Webb's drawings into clay.

Inarco Inarco is an acronym for International Artware Corporation, an importing firm established in Cleveland, Ohio about 1960 or 1961 by Irwin Garber. Garber was familiar with acronyms—he had previously worked for Napco (National Potteries Corporation) in nearby Bedford.

Inarco specialized in importing Japanese pottery of its own design, much of it coming from the creative talents of Garber himself.

According to Posgay and Warner, a 1970 trade journal showed Inarco as the parent company of Moskatel of Cleveland, Inc., Moskatel of Los Angeles, Inc., L.J. Baker Specialty Company, Inc. of Des Moines, Iowa, and Lewis Ribbon Corporation, New York. While an educated guess can be made about the activities of the ribbon company, I have no idea about the business of the others and mention them only on the slim chance that some day a pair of shakers may show up bearing one of their names.

By 1985 Inarco, the company started by a renegade Napco employee, had come full circle by being absorbed by Napco, which moved it to Jacksonville, Florida one year later.

Inco N.Y.C. This mark appears on the metal Corn Palace shakers. The company was undoubtedly based in New York, and may have been related to Exco, another mark that appears on metal Corn Palace shakers.

Iowa City Glass Located, not surprisingly, in Iowa City, Iowa, The Iowa City Flint Glass Manufacturing Company was incorporated on April 30, 1880. By the spring of 1881 a plant had been built and was in production, staffed mostly by glass workers brought in from New York, Pennsylvania, and Ohio. A brochure dated November 1, 1881, put out by the Iowa City Board of Trade, boasted of a labor force of 150 men, and the capacity to turn out "a [railroad] carload of glass per day." That sounds great except for one thing; the company went belly up the following summer, which allowed it a production run of only 15 months.

As its name implies, Iowa City specialized in flint glass.

Jasco You see this name on all kinds of imported pottery, but trying to run it down is another matter. In a 1980 trade journal there was a listing for Jasco Products, Inc., 50 Webster Ave., New Rochelle, NY 10801. However, there was no listing in the 1992 telephone directory.

Josef Originals The company was started by Muriel Josef George in Arcadia, California in 1946. The main focus was ceramic birthday girls, a product the firm still makes today.

When inferior Japanese copies of the birthday girls flooded the market in the early 1950s, George turned the tables on her underhanded competitors by going to Japan herself and setting up shop there. In Japan she could put out basically the same product as she could in California, but at lower cost. Most importantly, she could market her first class birthday girls at a price competitive with the inferior Japanese copies.

George's Japan adventure was in conjunction with a California jobber named George Goode, and the company eventually became known as the George-Goode Corporation. It is still in business, but Muriel George has retired.

Josef Originals' California plant remained in operation from 1946 to 1955. Pieces made there have an incised Josef Originals mark. Pieces made overseas carry an oval black and gold paper label that says "Josef Originals." Quite often it is accompanied by a very small but similarly shaped and colored label saying "Japan," "Hong Kong," "Taiwan" or "Korea."

Exactly how many sets of shakers can be attributed to Josef Originals is unknown. While none that could be positively identified appear in the book, I did run across a set too late to include it. It is a pair of tan horses, about 2½ inches high. One is sitting, the other is recumbent (all four legs tucked under it). The sitting horse holds an orange flower in its mouth, the recumbent one has a yellow bee on its tail. Each has a Josef Original paper label. Each is nicely executed. Since shaker-size animal figurines have been an important part of the Josef line, it's probable there are more shakers out there waiting to be discovered.

Jsny This is another of those names you see all over the place, but which I had no success in locating.

Juanita Ware Juanita Ware was made by Ohio potter Bert Crawford. In partnership with Greta Corey, Crawford operated the Triangle Novelty Company in Carrollton from 1933 until 1939 when they moved the pottery to Malvern. There they remained in business until Corey retired in 1949. At that time Crawford moved to Dalton, Ohio, where he formed Crawford's Pottery and made Juanita Ware until around 1969. He died December 4, 1972 at 81.

Few pieces of Juanita Ware are marked, but once seen it is instantly recognizable by its swirled appearance. (The swirl is in the bisque, not in the glaze.) The shakers shown in this book are brown and white, the only colors I've ever personally encountered. But an article by Grace Allison in the February 26, 1977 edition of *Tri-State Trader,* (now *AntiqueWeek)* states that it also came in shades of red, pink, yellow and green. They must be beautiful.

Kass John B. Kass was born in Bavaria, Germany in 1892. He worked for a time at the Royal Copenhagen Pottery, Copenhagen, Denmark, prior to arriving in New York in 1912. He found his way to East Liverpool, Ohio, where he worked at various local potteries before starting his own as a backyard operation in 1932, or 1935, depending upon which of several references you choose to believe.

John B. Kass died of a heart attack in 1946, the morning after a significant but non-devastating fire had occurred at the pottery. The Kass Pottery was at its peak at the time, employing about 50 people.

John Kass Jr. took over management of the pottery upon his father's death, changing the business to a decorating concern in the mid-1950s, and running it until suffering a debilitating stroke in 1972. In 1973 the younger Kass rented the plant to a partnership which called itself Arcadia Pottery (not to be confused with the two Arcadias listed at the beginning of this chapter), but it lasted only a couple years. John Kass Jr. died in 1983; the pottery was demolished in 1986.

Most of the Kass pieces I have seen have been finished in gloss glazes very close to what would be called pastels, and have always had the same color strength regardless what color they have been. Yellow and blue appear to be the most common. Kass pottery is often seen with gold decoration. Little of it was marked.

If Kass shakers are not present in your part of the country, don't be surprised. An interesting article, "Kass China," by well known American pottery expert Elizabeth Boyce, appeared in *The New Glaze* in December, 1987. The article was based upon an interview Boyce conducted with June Kass Jackson, John B. Kass Sr's only surviving child. It seemed to say that Kass Pottery was a somewhat localized phenomenon. The article states that for a short time Kass items were sent all over the country. But it blames the decline of the pottery during the 1960s on the opening of Interstate 70, which spelled an end to the numerous souvenir stores along U.S. Routes 30 and 40.

That would seem to indicate two things. First, that it was in those stores where the majority of Kass shakers were sold. Second, if the main route between the east and west coasts from your neck of the woods prior to the mid-1960s was not USR 30 or USR 40, it is unlikely that very many Kass shakers were brought to your area by their original purchasers.

Kato *See H. Kato.*

Kelvins Treasures A 1979 trade journal listed this or a similar trademark as being associated with Circle Import-Export Company, of Los Angeles, a firm specializing in bone china. Attempts to reach the company in 1992 were not successful.

Kenmar No information was found pertaining to this mark.

Kenmore Kenmore is a trademark used by Sears for a line of appliances.

Alfred E. Knobler & Company, Inc. Alfred E. Knobler graduated from Virginia Tech during the Great Depression with a degree in Ceramic Engineering. Unable to find a position in his chosen field, he took a job in the sales department of the Trenton Potteries Company, Trenton, New Jersey, where he soon discovered he had a great talent for selling ceramics.

He left Trenton when World War II broke out, employed his degree serving the War Department. Shortly after the war, in 1946, he started Alfred E. Knobler & Company which wholesaled ceramics and glass to retail markets. Later the company became known as Knobler Imports. Today the firm is called Knobler International. Its sales offices and warehouses are located in Moonachie, New Jersey. Its showroom is in New York. Knobler is still active in the company, serving as CEO.

According to a company spokeswoman, salt and pepper shakers have always been an important part of the Knobler line, and still continue to be.

For more on Alfred E. Knobler, see Pilgrim Glass Corporation.

Kobid Corporation No information found.

Kreiss This is a fairly common import mark, but no information was found. The paper label on the devil and angel shakers shown in the color section indicates the company was active in 1955.

Krome-Kraft According to Kevin McConnell in *Collecting Art Deco* (Schiffer, 1990), Krome-Kraft was a trademark used on chromium-plated accessories made by Farber Brothers, a New York firm that was in business from 1915 to 1965. McConnell states the Krome-Kraft line enjoyed greatest success during the 1920s and 1930s. Many Krome-Kraft items had glass or ceramic inserts made by Fenton, Cambridge, Lenox and other companies.

Leader Nothing was found relating to this mark.

Lefton An importing firm, the George Zoltan Lefton China Company is anchored in Chicago. It was started in 1940, one year after its founder, George Z. Lefton, came to the United States from his native Hungary. In Hungary, Lefton manufactured clothing but collected porcelain. In the United States he decided to make ceramics his living.

In the beginning, and for a time after World War II, Lefton imported goods exclusively from Japan. This base was later broadened to include Taiwan, Mexico, Italy and other countries.

Lefton salt and pepper shakers are nearly always of good quality. This is especially true of those made in Japan during the late 1940s and early 1950s when many other importers were settling for mediocrity. Most Lefton shakers are well marked with inkstamps or paper labels that include the company name.

Lego Imports Lego is the trademark of Goldman & Morgan, a New York importer. The only information a call to the company produced was that the name has now been changed to LaVie International.

The 1980 *Trade Names Directory* showed Lego Imports dealing in giftware, favors and jewelry boxes.

Lipper International, Inc. *See Lipper & Mann.*

Lipper & Mann According to Posgay and Warner, this company was in business as early as 1949, at which time its main offices were located in New York. The apparent namesakes were Seymour Mann, who was listed as president in 1949 and 1953 trade journals, and Hal Lipper, shown as sales manager in 1953.

The authors note a name change to Lipper International, Inc. by 1970 with H.A. Lipper as president. The change probably took place in the mid-1960s as Mann formed Seymour Mann, Inc. in 1966. A 1979 trade journal showed the company's address as Wallingford, Connecticut, where it is still located today.

According to a company spokeswoman, the only shakers currently being sold by Lipper International are accessories for its Blue Danube dinnerware line.

E.W. Loew Company This Erie, Pennsylvania plastics firm was founded by Ernest W. Loew in 1962. Loew said the Filter Queen vacuum sweeper shakers shown in Chapter 10 were made about 1967. Between 1962 and 1972 the firm made several sets of advertising shakers for various companies, Loew said. E.W. Loew is still in business but has not made shakers since 1972.

Ludowici Celadon Ludowici Celadon was probably one of the unlikeliest potteries to ever produce salt and pepper shakers. Specializing in roofing tile, the New Lexington, Ohio firm switched to domestic wares during World War II when continued dependence upon the construction industry would have probably driven it out of business.

Cookie jars were the company's specialty during that time. According to Lehner, as many as 70 employees were kept busy turning them out. Cookie jars patented by the company included the Belmont lion, Fluffy cat, and a Dutch boy and girl. Also several Disney jars such as the Dumbo turnabout and Donald Duck-Joe Carioca turnabout. Consequently, any salt and pepper shakers that are identical to the Ludowici Celadon cookie jars can be attributed to Ludowici Celadon—with reservations.

When Ludowici Celadon ceased domestic ware production at the end of the war, it apparently sold its molds to American Bisque, of Williamstown, West Virginia, which at the time had strong ties to American Pottery Company, just across the Ohio River in Marietta, Ohio. Or perhaps it sold them to American Pottery. Either way, salt and pepper shakers normally attributed to Ludowici Celadon may have been made by that company, or by American Pottery or American Bisque. At present, I am not aware of any reliable means to determine which of the three made them.

Lugenes No information has been found on this mark of Japanese origin.

Seymour Mann, Inc. The company was started by Seymour Mann in 1966. It is located in New York. Mann, who has been in the housewares and tableware business for more than 40 years, was formerly associated with Lipper & Mann.

The business is a family affair. Mann's wife, Eva, serves as creative director of the company. If you are thinking nepotism, that's not the case. Eva Mann is probably one of the most accomplished art directors of any import concern in the country and does, in fact, have paintings hanging in the Metropolitan Museum of Art.

Seymour Mann still markets salt and pepper shakers, along with a variety of table—and giftware, collectible dolls and high styled (high quality) Christmas accessories.

MAPCO No information was found on this company.

Mark L Exclusive No information was found on this mark.

Mar Mar This is another company that proved elusive when searching for information about it.

Maruri According to Posgay and Warner, Maruri was a Japanese pottery that filled orders for Lefton, and, presumably, other American importers. Today, according to the same authors, Maruri has abandoned Japan in favor of Taiwan, Mexico and Malaysia, and specializes in porcelain.

McCoy Sorting out the history of the McCoy potteries is almost as challenging as diagraming the family trees of European royalty, but here it is in a nutshell.

First on the block was W. Nelson McCoy who started a stoneware pottery in Putnam, Ohio, now part of Zanesville, in 1848. W. Nelson McCoy taught the business to his son, J.W. McCoy, who founded the J.W. McCoy Pottery in Roseville, Ohio in 1899. In 1910 J.W. McCoy, in partnership with his son, Nelson McCoy, formed a separate company, the Nelson McCoy Sanitary Stoneware Company, also in Roseville. The following year the J.W. McCoy Pottery accepted capital from its general manager, George Brush, and became known as the Brush-McCoy Pottery. This association lasted until 1925 when the McCoy family withdrew from the Brush-McCoy Pottery to focus its attention on the Nelson McCoy Sanitary Stoneware Company, the name of which was changed at some point to the Nelson McCoy Pottery Company. The former Brush-McCoy Pottery continued on as the Brush Pottery. In 1967 Chase Enterprises bought the Nelson McCoy Pottery Company. Chase sold it to Lancaster Colony Corporation in 1974. Later it was acquired by the Sebring, Ohio firm Designer Accents, which operated it as Nelson McCoy Ceramics. At present, the Roseville plant is still standing, but has been idle for several years.

If the above paragraph seems hard to understand, reading it over 25 or 50 times might help.

Shakers were apparently never a priority at McCoy, which seems a little strange because kitchen artware was one of its specialties. It made scads of cookie jars, lots of pitchers, and a considerable number of creamers and sugars.

The cabbage shakers shown in Chapter 7 were introduced in 1954 to match a grease jar of the same design. A similarly sized pair of shakers depicting a cucumber and pepper were made in the same colors.

McKee Glass Company The McKee Glass Company began life in 1853, in Pittsburgh, Pennsylvania as McKee & Brothers Glass Works. In 1889 it was moved approximately 27 miles east to what later developed into Jeanette, Pennsylvania, named for Jeanette McKee, wife of H. Sellers McKee, the owner of the company. It was probably around that time that the name was changed to McKee-Jeanette Glass Works, sometimes called McKee-Jeanette Glass Company.

In 1899 the McKee factory became part of National Glass, a loosely knit combine similar to U.S. Glass that included the Indiana Tumbler & Goblet Works, Northwood Glass Works, and about a dozen other companies. National Glass lasted only a few years. In 1910, by which time National Glass had sailed off into oblivion, McKee was reorganized as McKee Glass Company, a name it apparently had already been using for several years. It became a division of the Thatcher Glass Company in 1951, and was purchased by the Jeanette (glass) Corporation 10 years later.

Now that we have gone through all of that, you can forget it, as the main point to remember about collectible McKee Glass shakers is that most of the better examples were made from roughly 1889 to 1899.

Meridan No information found.

Milford Pottery This is the company that made the Klay Kraft states series, but I was not able to find any information on it.

Miyad No information was found about this mark.

C.F. Monroe Company Not a glass manufacturing company but a glass decorating firm, the C.F. Monroe Company was located in Meriden, Connecticut. It was established by C.F. Monroe as a glass importing business in 1880. Decorating began in 1882, and incorporation papers were filed in 1892. The company went out of business in 1916.

C.F. Monroe bought much of its glass from Pairpoint (Mt. Washington), and Dithridge & Co. European suppliers were also used.

The best known Monroe product is Wavecrest, a trademark registration for which was issued in 1898. Kelva was registered in 1904, while the third type of glass associated with the company, Nakara, was never registered. All three are highly desirable, eagerly sought, and very expensive.

Mt. Washington Glass Company The Mt. Washington Glass Company was started in New Bedford, Massachusetts in 1869 as the Mt. Washington Glass Works. The original owner was William L. Libbey, who left in 1872.

The name was changed to Mt. Washington Glass Company in 1876. The firm was acquired by the Pairpoint Manufacturing Company in 1894. Pairpoint went out of business in 1938.

Mt. Washington/Pairpoint was one of the most prolific makers of art glass shakers.

Muth & Sons This Buffalo, New York firm, maker of nearly all of the older glass beer bottle shakers, appears to have vanished without a trace.

NapcoThis importer, also known as National Potteries Corporation, was formed in Bedford, Ohio in 1938. Bedford is a suburb of Cleveland, the city to which the company moved sometime during the 1940s or very early 1950s. In 1984, it made a much farther move to Jacksonville, Florida.

Napco bought out its hometown competitor, Inarco, in 1985, and moved Inarco to Jacksonville in 1986. In recent years Napco has largely abandoned its giftware operation, choosing instead to put its emphasis on the florist trade.

But that doesn't diminish the company's star in relation to novelty salt and pepper shakers. Throughout its history Napco imported some of the best foreign shakers sold on the American market. The ones from Japan are especially well done. Consequently, whenever you see Napco's familiar artist's palette paper label, or one of its several inkstamps, rest assured you will be getting good quality for your buck.

Nasco An acronym for National Silver Company, a New York marketing concern. Exactly when the company was founded is not known. However, Gladding, McBean & Company was making pottery with a Nasco mark as early as 1934.

Many other potteries—Canonsburg, Fredericksburg, French Saxon China Company, W.S. George, Santa Anita and Southern to name a few—also turned out orders for National Silver. Some of these pieces were marked, others weren't.

National Silver has also used the mark, "N.S. Co." Black and silver, and black and gold paper labels have been used, too.

The status of the company today has not been determined.

National Silver Company *See Nasco.*

Nec Nothing was found on this mark.

New——land Ceramics Corp. Your guess is as good as mine here. The mark was partially obliterated on the only example that was photographed. Newfoundland? New England? New mystery!

New Line Industries, Inc. According to the mark on the Magic Chef plastic chefs, this concern was located in New York at one time, but it's not there today under that name according to Directory Assistance.

Niko China Obviously a Japanese pottery, but no information was found.

Nippon This is the mark used on Japanese porcelain from the time the McKinley Tariff Act went into effect in 1891, until an amendment in 1921 required the use of the words "Made in Japan."

The price of Nippon porcelain is generally much higher than the price of Made in Japan porcelain. Many people collect Nippon porcelain under the false assumption that Nippon is also of higher quality than Made in Japan porcelain. While that is no doubt true if you compare Nippon to items the Japanese exported in the 1950s and later, it is not true of items exported within a few years of the change when the only thing different was the inkstamp on the bottom. The result of this mentality is that you can often purchase Made in Japan shakers identical to Nippon shakers at a fraction of the cost of the Nippon pieces. And once they are on the shelf, who is going to look at the bottom anyway?

Dave Nissen (Blossom Valley) I would love to know more about this mark as the shakers it is on are of top quality, but no clues were found.

Norcrest An import business that was started in Oregon in 1924 under the name Northwest Trading Company. Hide Naito was the founder; original facility was a gift shop. The year 1955 saw the name changed to Norcrest China Company in a effort to keep the firm from being perceived as regional, since by that time it was selling wares throughout the country.

Today the company is based in Portland, and imports a variety of products from a variety of countries.

Noritake The Noritake Company dates from 1904 when its registered name was Nippon Gomei Kaisha, which was changed in 1917 to Nippon Toki Kabushiki Toki. Morimura Brothers, of New York, was the main American distributor. The M within a wreath you often see marked on Noritake signifies Morimura.

The Azalea shakers and condiment sets shown above were originally premiums given away free by the Larkin Company as an incentive to buy its Sweet Home Soap. The campaign began in 1916, ended sometime in the 1930s. Seventy different pieces of azalea were made. While azalea shakers are still reasonably priced, some rarer azalea pieces sell for hundreds of dollars.

As a footnote, the Larkin Company also gave Buffalo Pottery's famous Deldare Ware as a premium during the winter of 1922-23, during Deldare's second incarnation. (Deldare was first made from 1908 to 1910, revived from 1923 to 1925.) I haven't been able to ascertain whether any regular Deldare Ware shakers were ever made. But at least one pair of emerald Deldare shakers were put out as they are shown in *The Book of Buffalo Pottery* (Schiffer, 1987), by Seymour and Violet Altman.

Nov. Co. This mark was found on both metal and ceramic shakers, but no additional information about it has been found.

Oci Omnibus Collection International is a division of Fitz and Floyd. It was created in 1983 as a vehicle to market medium priced gift and gourmet ware.

Occupied Japan This mark was used on pottery and other products made in Japan for export to the United States during the U.S. occupation following World War II, which lasted until April 28, 1952.

However, the Occupied Japan mark wasn't a hard and fast rule. According to Gene Florence, in *The Collectors Encyclopedia of Occupied Japan Collectibles* (Collector Books, 1976), products coming out of Japan during that period had to be marked in one of four ways: "Occupied Japan," "Made in Occupied Japan," "Made in Japan," or "Japan." Consequently, it is possible that many shakers made during the American occupation were not marked as such. Indeed, as shown above, sometimes only the boxes they were shipped in carried the occupied mark, while the shakers themselves were marked with an unidentifying "Japan."

In my opinion, Occupied Japan is largely a contrived collectible, collectible only because it was made for a relatively short time, and because several books have been written about it. While some Occupied Japan shakers may be truly exquisite, most of the time the quality just isn't there. The best advice, which is no different for Occupied Japan than for any other classification of shakers, is to carefully discriminate when choosing the pieces you incorporate into your collection.

Oil City Glass Company There seems little doubt this company must have been located in or near Oil City, Pennsylvania, but my search for information about it was not successful.

Olinco Nothing was found on this apparent importer of Japanese goods.

Old King Cole The mark on the plastic chefs shown in Chapter 5 indicates this firm was located in Canton, Ohio. That's all that is known about it.

Omc Another obvious importer of Japanese shakers on which no information has surfaced.

Omnibus A Fitz and Floyd mark for its Omnibus Collection International Division, which was formed in 1983.

Otagiri Mercantile Company, Inc. Otagiri is a San Francisco importer that wholesales housewares and giftware items to upscale department stores.

The company was started in 1946 by Japanese immigrant James Otagiri, who spent most of World War II as a prisoner in an American interment camp. While imprisoned, Otagiri taught Japanese at the University of Colorado Language School. His own education was gained through attending the Berkeley Public Schools and the Berkeley School of Commerce. Otagiri died about 1976.

Like Fitz and Floyd, and a couple other current importers, Otagiri enjoys a reputation for excellence. All of their shakers and other ceramic products are top quality.

Our Own Import *See Cardinal, Inc.*

Pac An importer of Japanese ceramics; no other information found.

Palomar You see this mark quite a bit on shakers, but I haven't been successful in finding any additional information. Palomar, California seems like it might be a good place to start searching.

Parkcraft Best known for its 48- and 50-states series, Parkcraft also made series to commemorate the days of the week (seven sets), months of the year (12 sets), famous people (seven sets), famous cities (18 sets), and nursery rhymes (12 miniature sets).

The company was established in Burlington, Iowa in 1955 by Bob and Marianne Arhold who run Heather House, a mail order concern through which Parkcraft shakers were marketed. While Bob Arhold designed the Parkcraft shakers, most of them were made by Taneycomo Ceramics, Hollister, Missouri. The days of the week and miniature nursery rhyme series were made in Japan.

The 48-state series was produced about 1957, the 50-state series about 1968. Some of the companion pieces were changed for the later series. An entire rundown of both series is shown in Appendix III.

Parksmith Corporation I had no luck in locating this company. Since its mark shows an address with a zone code (New York 3, New York), we know it was located in that city, and was active prior to July 1, 1963 when ZIP Code went into effect.

Par-Tee Toy Chest The mark on the plastic kittens with meowers shows this company located in Windsor, Ontario, Canada. Whether it signifies a store, importer, exporter, distributor or manufacturer is unknown.

Pearl China Another company for which there seems to be only a partial and confusing history. George Singer operated a distributing company in East Liverpool, Ohio named Pearl China during the 1930s. About the middle of that decade, apparently in partnership with his brother Dennis, Singer leased a rundown plant from the Hall China Company, also of East Liverpool, in order to have more control over his product. The factory apparently operated under the name Pioneer Pottery. In 1958, Pioneer Pottery was sold to Craftmaster of Toledo, Ohio. Craftmaster was acquired by General Mills 10 years later.

Pearl China is probably best known for its pastel glazes, especially yellow, the color often found on its mammy and chef salt and pepper shakers, and matching cookie jars.

Perma Ware This mark was found on the bottom of the plastic McCormick-Deering shakers, which, unfortunately, is the extent of my knowledge of it.

Pilgrim Glass Corporation Still in business today, Pilgrim's factory is located in Ceredo, West Virginia while its sales office is in Moonachie, New Jersey.

Pilgrim began in Huntington, West Virginia as the Tri-State Glass Manufacturing Company, basically a garage-size operation owned by Walter Bailey. Alfred Knobler, a name not unfamiliar to shaker collectors, bought the company in 1949, changed the name to Pilgrim. Knobler moved it to Huntington in 1956. Pilgrim, according to the company, is today's largest manufacturer of cranberry glass. Glass animals and art glass eggs are other popular lines.

Knobler, incidentally, must have been the world's greatest salesman. Prior to purchasing the plant, which had a problem with low natural gas pressure in winter, Knobler talked Columbia Gas into running a new line to the factory under two miles of city streets, at Columbia's expense!

Poinsettia Studios No information except that the company was located somewhere in California.

Price Imports The founder of this company, Albert E. Price, must have been a tremendous business man considering that he successfully got the firm off the ground in the catastrophic year of 1929. At that time the company was located in Philadelphia, and dealt exclusively in jewelry. Eventually it began dealing in giftware, too.

At some point problems within the Price family resulted in a split in the company. One part of the family retained the jewelry business, another part won control of the giftware keeping the name in the process.

In 1970 Price Imports moved to Bellmawr, New Jersey. It is still there today, being run by descendants of Albert E. Price.

Purinton This company fired its first kiln of pottery on December 7, 1941, probably the only high point of that historic day which President Roosevelt correctly predicted would live in infamy. The operation was located in Shippenville, Pennsylvania. It lasted until 1959. Bernard S. Purinton, formerly of Wellsville, Ohio was the principal owner of the company, along with a group of investors from Shippenville.

Purinton pottery was hand decorated under the glaze, and fired only once. Bernard Purinton developed a slip mixture that made the greenware more durable than most, which allowed the one-firing process to operate profitably with a minimum of breakage. At its peak, Purinton employed about 150 people.

PY (in ellipse) This is a fairly common mark about which no information was found. However, one intriguing clue was uncovered. A mouse in Chapter 6 carries the PY mark, and also has a remnant of an Enesco paper label. Whether PY signifies an Enesco line, a pottery or exporter from which Enesco bought ceramics, or something else is not known.

Quality Product This mark appears on some nodders, but no other information was found.

A Quality Product Possibly a more complete name for the entry just above, or perhaps a separate company. Either way, attempts to locate it were unsuccessful.

Ransburg The Harper J. Ransburg Company was a decorating and distributing firm, founded by Harper J. Ransburg in 1908. It was located in Toledo, Ohio at that time, but moved to Indianapolis, Indiana four years later.

Most Ransburg shakers display heavily painted decoration, and were obviously made to match the company's cookie jars and other kitchen utility pieces. Ransburg's first cookie jar, the 207, was introduced around 1930. The shape remained popular until at least the mid-1950s, as one was made with a Davy Crockett theme. The shakers were probably made during this same time period, too.

To a large extent, Ransburg was a regional company, the bulk of its products having been marketed in the midwest.

Red Wing Red Wing Potteries, Inc. was born in Red Wing, Minnesota, in 1878 as the Red Wing Stoneware Company. Following several mergers and changes in name, Red Wing Potteries, Inc. was decided upon in 1936. That name lasted until 1967 when the company went out of business due to labor difficulties.

Reebs, Cour, O. Nothing was uncovered that would help lead to the identity of this mark.

Regal China Company Some confusion exists concerning exactly when this company was founded. According to most sources it was 1938. But Al Cembura and Constance Avery, in *Jim Beam Regal China Go-Withs,* indicate it was at least as early as 1930, because founder Herman Kravitz hired Catherine Miller as a decorator that year. Miller's name, "© C. Miller," appears on some Regal products as the copyright holder. She is said to have been one of the best decorators of the period.

But regardless of when the company was founded, it was located in Chicago, and functioned as a decorating firm. With the advent of World War II, which was already ravaging Europe in the late 1930s, many potteries began turning out products for the war effort—ceramic parts for communications and electrical equipment, for instance—and Regal eventually found itself unable to obtain pottery to decorate. Thus, in 1940, it moved to Antioch, Illinois, a small town about 50 miles north of Chicago, where its new facilities would allow it to pour and fire its own bisque. Sometime during the 1940s Regal was purchased by the Royal China and Novelty Company, of Chicago. Royal China acted as Regal China's distributor.

Regal made salt and pepper shakers, cookie jars, coffee carafes, ash trays, lamps and other pieces. At one point, more than 60 percent of its production was devoted to lamps. Regal made its first Jim Beam bottle in 1955 beginning a line that was so successful that in 1968 the James B. Beam Distilling Company took over Regal, which functioned as a wholly-owned subsidiary until closing its doors during the summer of 1992.

The best known Regal salt and pepper shakers are the Van Telligen Snuggle-Hugs. Children's illustrator Ruth Van Telligen, later Ruth Van Telligen Bendel, designed the shakers for Royal China and Novelty, a patent being issued in her name on May 6, 1949. The Snuggle-Hugs were made until 1963 when a fire at the plant destroyed all of the molds. Some Snuggle-Hugs were made in both table size and range size, the range size being much rarer. A few, such as the love bugs, were made in three sizes. A very rare matching cookie jar was made to go with the peek-a-boo Snuggle-Hugs shown in the color section.

Snuggle-Hugs can be rough dated simply by looking at their bottoms. Those that say "Bendel" all have a patent date of 1958, and can be assumed to have been made between that year and 1963. Those marked "Van Telligen" all say "Patent Pending," and were made as early as 1949.

Shakers were made for the Old MacDonald canister set in the form of boy and girl faces, feed sacks with bovines on them, and in the shape of butter churns. As with the Snuggle-Hugs, Regal made other coordinated sets of salt and pepper shakers and cookie jars. Among these are Goldilocks, Alice in Wonderland, Humpty Dumpty, and Puss-n-Boots. You can distinguish the Regal Puss-n-Boots from the more common Shawnee Puss-n-Boots by the ornamentation on the hat. The ones made by Regal sport a fish, the ones made by Shawnee a flying bird.

Check the Hull Pottery entry for information on the Little Red Riding Hood series, which was, at the very least, decorated by Regal.

Regaline The name appears on a Dutch boy and girl napkin and shaker set. No other information found.

Relco No doubt an importer, but no other information was found.

Relco Creations No information was found; possibly a label used by the company listed just above.

Rite-Lite Corporation This mark on glass light bulb shakers shows the company located in Clarendon, Pennsylvania, but attempts to reach it there in 1992 were not successful.

Ron Gordon Designs, Inc. No information found.

Rosemeade The Wahpeton Pottery Company, maker of Rosemeade, was established in 1940 in Wahpeton, North Dakota by Laura A. Taylor and Robert J. Hughes.

Taylor was the artist. Born and reared in Rosemeade Township, near Wapheton, she had studied pottery making at State Teachers College in Valley City, North Dakota. Also at the University of North Dakota where she served as student assistant in the ceramics department in 1933. Prior to starting Wapheton, she served for approximately four years as state supervisor of the Federal Clay Project at Mandan, North Dakota under the auspices of the Works Progress Administration. The sandy color clay used to make Rosemeade came from Mandan.

Hughes, owner of Wahpeton's Globe-Gazette Printing Company and Gift Shop, supplied the capital and handled the business side of the venture. He and Taylor married in 1943.

Laura Taylor Hughes' forte in shakers was animals, probably a result of her rural upbringing. The only "people" shaker she is known to have made is Paul Bunyan.

Laura Taylor Hughes died of cancer in 1959. Production at the Wahpeton Pottery Company continued through the end of 1961, the showroom remained open until September 1964.

Royal This mark, shown on handpainted Japanese shakers, remains a mystery.

Royal Winton According to the Kovels, this pottery is in operation in Stoke, Staffordshire, England, where it has been located since 1886. Earthenware is the main focus.

Rubicon, Inc. This company's name appears on glass and plastic shakers. Included with one of its marks is a date of 1958.

St. Labre Indian School According to Tony Foote, controller at the St. Labre Indian School in Ashland, Montana, the familiar St. Labre shakers were assembled at the school by Guild Arts and Crafts, during the 1970s. They were sent to potential donors with appeals for money, similar to the way the Disabled American Veterans send key ring size license plates. Foote said the promotion lasted a couple years, during which time millions of shakers were mailed. Less common but also made were a sugar bowl and a napkin holder, Foote said.

The St. Labre Indian School was founded in 1884. It currently has approximately 650 students, mostly Cheyenne Indians, in grades kindergarten through 12. It is affiliated with two other schools, Pretty Eagle in Xavier, Montana, and St. Charles in Pryor, Montana. Students at Pretty Eagle and St. Charles are largely Crow Indians, Foote said.

Sarsaparilla According to Leslie Sackin, president, he and his wife, Joyce, who is an artist, founded Sarsaparilla in 1976. It is located in West New York, New Jersey across the Hudson River from Manhattan. Prior to 1976, the Sackins were in the antique business.

The concepts for all Sarsaparilla shakers originate with the Sackins. In some cases they turn the concept over to a designer, in others they make the original model themselves.

Sarsaparilla's catalog changes each season, Sackin said. The shakers offered to wholesalers in any particular catalog will be composed of a mixture of new designs and reissues. For instance, when I talked to Sackin in November 1992, Sarsaparilla was bringing out a line of railroad-related shakers. A lamp set and a crossing set in the line were new, an engineer set was a reissue.

Sears Roebuck & Company Richard Warren Sears started his first mail order firm, the R.W. Sears Watch Company, in 1886 in Minneapolis. After moving it to Chicago, he sold it for $100,000 in 1889. Then, when retirement didn't suit him—he was only 26 years old—he moved back to Minneapolis and went into a similar business with A.C. Roebuck, his former watch repairman. The original name was A.C. Roebuck & Company. Later it was changed to Sears, Roebuck & Company.

Over the years Sears has contracted with many foreign and domestic potteries to produce kitchen artware to be sold under the Sears name. It has also sold nondescript metal shakers from time to time.

Sdd *See Sarsaparilla.*

Shafford Most famous for its black cat line, the Shafford Company was established in 1952 by S.H. Hartman, to act as a sales agency for his U.S. Asiatic Company, Inc., which he had begun three years earlier. Like so many other importers, Hartman was based in New York.

According to Posgay and Warner, Hartman gave exclusive right to market his products in Canada to Arnold Fisher, owner of the Maso Company in Montreal, and the name Shafford was derived by combining the first letter of Hartman's first and middle names, along with the first letter of Fisher's first and last names, and adding the word ford.

Shawnee Pottery Company The former plant of the defunct American Encaustic Tiling Company, in Zanesville, Ohio became the home of the Shawnee Pottery Company when it was established in 1937. While many potteries started in backyards and garages, that wasn't the case with Shawnee. According to company advertising, the plant had the capacity to turn out 100,000 pieces of pottery per day right from day one.

The first president and general manager of Shawnee was Addis E. Hull, Jr., former president and general manager of the A.E. Hull Pottery Company, and son of its founder. Hull had a degree in Ceramic Engineering from Ohio State University. He had been president of Hull since his father's death in 1930. Shawnee signed Hull to a four-year contract at $6000 per year plus 5 percent of the net profits. Hull resigned in 1950 to go to Western Stoneware Company.

Albert P. Braid replaced Hull until 1954 when John Bonistall took over. Bonistall remained in the position until Shawnee closed in 1961. At that time Bonistall bought some of the Shawnee molds for his own company, Terrace Ceramics.

Many of Shawnee's shakers were made as accessories for its popular line of cookie jars.

SH This mark, along with the word "Japan" is found on some shakers, but nothing else is known about it.

Sigma No concrete information was found on this recent importer.

Silvestri This import company is based in Chicago. It was founded in 1940. During the 1950s it began specializing in

Christmas-oriented housewares and giftware, a focus it has maintained to this day.

Southern Potteries *See Blue Ridge.*

Spin Original The mark couldn't be any clearer, but no information about the company was found.

Srp, Sandy Ohioan Sandy Srp began making miniature salt and pepper shakers commercially in 1990 after being a ceramic hobbyist and certified ceramic instructor for many years.

According to the artist, she modifies commercially available figurine molds to make her shakers, also uses a commercial slip for pouring. Srp's miniatures are somewhat different than most in that every one is made so it may be used. (Most folks don't use them, of course.) While all of her creations are limited editions, she does not make the same number of sets for each edition. Srp estimates she has made no more than 35 of any particular set. Most editions average 20 to 25 sets, but some are less. The least number Srp has made of any set is three, the purpose being to have a very, very limited edition for the Novelty Salt and Pepper Shakers Club to sell at auction during its convention in Chicago in 1991.

Srp markets her miniature shakers through the Novelty Salt and Pepper Shakers Club, giving first crack to members of the Ohio Chapter at their quarterly meetings. She sells most sets for $12 to $25, the average being $15. For some sets, however, depending upon the amount of labor required to make them, she charges as much as $35.

If you would like to add some Sandy Srp miniatures to your collection, you can find her address in Appendix I Sources.

Stanford Pottery, Inc. A Sebring, Ohio firm that was started by George Stanford Sr., George Stanford Jr., W.J. Stanford and L.L. Root in 1945.

Stanford is best known for its corn line, which was very similar to Shawnee's. On Stanford corn shakers, the leaves form an S and P to differentiate between the salt and pepper. The company is also known to have made ceramic products for National Silver Company, which bought items from many different potteries.

Fire destroyed the Stanford Pottery on January 21, 1961. According to Lehner, however, Stanford continued in business as a sales organization until at least 1965.

Starke Design, Inc. No information about this mark that is found on early plastic shakers, other than it was located in Brooklyn, New York. As with several other companies included in this chapter, one of its marks shows a zone number in the address (Brooklyn 3, New York), so it was obviously active sometime before ZIP Code went into effect in 1963.

Sterling The word sterling on silver salt and pepper shakers signifies they are composed of an alloy that is .925 fine silver by weight. Other silver terms you might find handy are: coin silver, .900 fine; Britannia silver, .9584 fine; and pure silver, .999 fine.

Most silver shakers and other dinner utensils are made of sterling, but the others have been used on occasion. Another term you will sometimes see is "weighted sterling." This means that the article itself is made of sterling, but that its hollow base has been filled with a heavier substance to provide stability.

Be leery of any other named silver such as "German silver" or "Mexican silver." These terms have no universal meaning, and most objects so marked contain less than 5 percent fine silver by weight

The term "silver plate" designates a piece made of a base metal to which a coat of silver plating has been bonded, most often by the electroplating process. Silver plated articles normally contain less than 10 percent silver by weight, in most cases much less.

Suder-lon A Chicago-based firm about which no information was found.

Superior Plastics, Inc. A 1979 trade journal showed this company located in Chicago, but that is all that is known about it at this time.

Superlon Products, Inc. Little is known about this company whose name is found on plastic salt and pepper shakers. Different shakers show different addresses. One pair is marked New York 3, New York while another says Chicago 3, USA. Either way, it is safe to assume the company was active sometime prior to July 1, 1963, the date the U.S. Post Office Department abolished the old zone system and inaugurated ZIP Code.

Sutton Ware Nothing was found except the information, on one of the sets of shakers shown above, that the company was located in Canada.

Takahshi / S.F. To date, no information has been found on this presumably Japanese company.

Tasco Nothing was found on this company. One pair of shakers is marked 1981 Hong Kong.

Taylor, Smith and Taylor Commonly called TST by collectors, Taylor, Smith and Taylor was primarily a dinnerware manufacturer. It was founded at the turn of the century by Homer J., John N. and W.L. Taylor, W.L. Smith and Joseph G. Lee. It's plant was in Chester, West Virginia while East Liverpool, Ohio served as the site of its offices.

The Taylors bought out Lee in 1903, Smith bought out the Taylors in 1906, and Anchor Hocking Glass Company bought the whole package in 1976. At that time the name was changed to Ceramic Division of Anchor Hocking. The plant was closed in 1982.

Besides the LuRay shakers shown in the color section, TST made shakers to compliment many other lines of dinnerware. One line that is quite popular with collectors, and for which shakers were made, is Vistosa, a brightly colored dinnerware not unlike Homer Laughlin's Fiesta. Vistosa was manufactured in the 1930s and 1940s.

Thames An importer of Japanese ceramics; no other information available.

Thrifco Obviously an importer, but that is about all that is known. Thrifco was apparently a subsidiary of, or the same company as, the next listing, Thrift Ceramics, as both used the same palette-shape paper labels. It could be that the firm changed its name after a period of time. Another possibility is that one was an export company set up overseas to ship goods to America, while the other was an import firm set up here to receive and distribute them.

Thrift Ceramics *See Thrifco.*

Tiffin Glass The Tiffin Glass factory, in Tiffin, Ohio was built in 1892 by U.S. Glass. U.S. Glass was a cooperative of 18 glass companies that was formed one year earlier as a survival effort during a time when increased costs, intense competition and labor unrest where ravaging the industry.

The combine eroded over the years as factories dropped out or went broke. In 1935, when only the Tiffin and Pittsburgh plants remained, the U.S. Glass headquarters were moved to

Tiffin. Sixteen years later only the Tiffin plant was left, and it declared bankruptcy in 1963. Employees purchased and revived it, sold it to Continental Can in 1966. Three years after that it was sold to Interpace, which closed the doors of Tiffin Glass for the final time in 1980.

In the beginning all Tiffin glass was handmade; many hand operations continued to the end. Throughout its 88-year history Tiffin was known for producing top quality glassware.

Treasure Craft This pottery was started by Alfred A. Levin in 1945 in Gardena, California in a single car garage. Today it is located in Compton, California, and has grown from the one-car garage to a 250,000 square foot facility. In 1988 Treasure Craft was purchased by Pfaltzgraff, of York, Pennsylvania. Bruce Levin, the founder's son, serves as Treasure Craft's president.

Treasure Craft's forte has always been matte glazes in earth tone colors (mainly brown), which have been tremendously popular with retail consumers but have not generated much interest among collectors.

Be that as it may, Treasure Craft is one of the few remaining survivors in an industry that included such revered names as Shawnee, American Bisque and Regal China. And for that reason alone it would seem that any representative collection of salt and pepper shakers would include a few Treasure Craft sets.

Treasure Masters Corporation Treasure Masters is located in Derry, New Hampshire. Half a dozen calls to the company resulted in no information other than its creative director is a lady named Lynn White.

Treasure Masters was listed in the 1979 edition of *Trade Names Directory,* so we may safely assume it was founded no later than 1978. It may have started much earlier. Paper labels indicate that at one point the company imported bone china shakers from Taiwan.

Trevewood Pottery Trevewood was located in Roseville, Ohio, across the street from the Brush Pottery Company. It was owned by Ruth Prindle and Margaret Bolantz.

According to Steve Sanford, nephew of Ruth Prindle, the pottery was active from the early 1940s to the mid-1950s. It employed about six people. In 1954 Prindle and Bolantz sold the business to a man named Dorsey, and retired. Sanford said sometime later the building burned.

(If Sanford's name sounds familiar, it is because he is co-author, with his wife Martha, of the excellent book, *The Guide to Brush-McCoy Pottery,* self-published in 1992.)

Trevewood's specialty was go-with salt and pepper shakers, a modest number of which Sanford has acquired. His Trevewood collection presently consists of the following sets: Hammer and large tack; toothbrush and toothpaste; hatchet and stump; pipe and pipe stand; bowling ball and bowling pin; mouse and cheese; sow and piglet; chicken and basket; roasted pig and platter; roasted turkey and platter. All but one set is marked. "Quite often only the larger member of a pair will be marked," he said.

Trice Brothers Beyond the fact that it was located in England and made cottage ware, little is known about this company.

Trico China Apparently a Japanese pottery that, by the appearance of its work, may have been active during the 1920s or 1930s.

Trish Ceramics According to Melva Davern, founder and past president of the Novelty Salt & Pepper Shakers Club, Trish Ceramics was a home ceramics business in Pittsburgh, Pennsylvania. It made the limited edition shakers, "Reach Out to Other Collectors," for the club's 2nd Annual Convention in 1987. The people who ran the business have since moved, Davern said, and could not be located.

Tucker Products Corporation This company was founded in 1946. It is located in Leominster, Massachusetts. Today it is called Tucker Housewares, Inc. Mobil is the parent corporation.

According to Frank DiSesa, vice president of research and development, the plastic toaster shown above was marketed between 1958 and 1962 as part of a children's tea set. Between 15 and 20 different tea sets were sold, the number of pieces varying between 20 and 85. DiSesa said a plastic television salt and pepper was included in some of the sets. Presumably, this would be the one on which the shakers move up and down when you turn the volume control knob.

Tucker still makes a few salt and pepper shakers as accessories to its kitchen line, but now its main focus is on containers. According to DiSesa, Tucker is currently the world's largest manufacturers of plastic trash cans. Plastic laundry baskets are another well known product. So are plastic totes, or inventory boxes as antique dealers often call them. Which puts antique dealers who sell salt and peppers at shows in a potentially ironic situation—they sell items made by a company 30 years ago, and rely on the products that same company makes today to protect the vintage items while they are in transit.

Twin Winton Shaker collectors currently tend to view this California pottery as they do its neighbor and main competitor, Treasure Craft—dull, brown and boring.

The pottery was founded by twins Don and Ross Winton in 1946 in Pasadena, where the brothers had previously made pottery from the mid-1930s to 1942 when they enlisted in the military during the Second World War. Their older brother, Bruce, acted as business manager of the company, and in 1952 bought out Don and Ross who found freelance designing more to their favor. Much of their freelance work, however, was done for Twin Winton.

Bruce moved the plant to El Monte, California about 1953. In 1964, the company was moved to San Juan Capistrano, where it occupied the same plant Brad Keeler had been building for his pottery when he suffered a fatal heart attack in 1952. Twin Winton was sold during the mid-1970s; Treasure Craft acquired its molds.

Many Twin Winton salt and pepper shakers were made as accessories for cookie jars and other kitchen artware.

UCAGCO Based in New Orleans, the United China & Glass Company can trace its roots all the way back to 1850 when it was founded by a man named Abe Mayar as Abe Mayar & Company. The trademark UCAGCO, however, was never used until 1935. By then the firm had come under the control of Mayar's nephew, J.W. Moses, who had renamed it United China & Glass Company in 1908.

Early in his tenure, Moses's focus was limited to wholesaling in a few southern states, Mexico and South America. That scope eventually broadened. A few years prior to the outbreak of World War II, Moses began importing goods from Japan under the watchful eye of S.A. Stolaroff who handled the Japanese end of the business. The imports stopped during the war, but afterward Stolaroff obtained the very first contract with the Japanese when trade reopened. While working in Japan, Stolaroff served double duty by designing much of UCAGCO's product line. Stolaroff, and other investors, bought the company in 1946. They sold it to Sammons Enterprises in 1956. Stolaroff served as president until his retirement in 1962.

From my experience with salt and pepper shakers and other

ceramic products, a UCAGCO paper label on the secondary market is generally a sure sign that the piece to which it is attached will strike a nice balance between high quality and reasonable pricing. On the retail market, the same is true of the company's current paper label which is "U / CG / C" on three lines.

United China & Glass Company *See UCAGCO.*

UCGC *See UCAGCO.*

Vallona Starr According to Lehner, Vallona Starr was active in Los Angeles as early as 1945 under the name Triangle Studios. By 1949 the company had moved to El Monte, California. The name was changed to Vallona Starr in 1951. Lehner did not find a listing for it after 1953. Valeria De Marsa and Everett S. Frost were shown as the owners.

Besides salt and pepper shakers, the company is known to have manufactured cookie jars, sugars and creamers, vases, bowls and boxes.

Vandor This import company, located in Salt Lake City, Utah, was started during the late 1950s by Ted Van Doorn. The founder sold the company in the 1980s. Like most American importers of merit, Vandor obtains some of its designs from its own design department, others from freelancers. According to a company spokesman, as the economy worsened in the early 1990s, Vandor scaled back its staff to the point where only about ten employees remained.

This company, along with Fitz and Floyd and a couple others, has been the standard bearer for excellence in imported ceramics during the 1980s and 1990s.

Vernon Kilns Vernon Kilns was one of the top names in California pottery dinnerware from the 1930s through the 1950s. The company was started by Faye G. Bennison in 1930 when he bought the Poxon China Company in Vernon. For a time Bennison called it Vernon Potteries. When he changed the name to Vernon Kilns is disputed. While different sources cite dates ranging from the mid-1930s to the late-1940s, the best evidence points to the earlier dates being closer to the truth. Bennison retired in 1955, the company closed in 1958.

Many Vernon Kilns dinnerware shakers are available, but more interesting to readers of this book would be the Hop and Lo shakers shown on page 182. They were made under contract to Walt Disney, and are from Disney's two-hour animated classic, *Fantasia,* which debuted in 1940.

According to Tumbusch, Vernon Kilns held a Disney license from October 10, 1940 to July 22, 1942. During that time the company made 36 figurines, six bowls and two vases based on *Fantasia.* It made another six figurines based on *Dumbo* and *The Reluctant Dragon.* Hop and Lo were the only Disney shakers produced. Molds for the entire line were made from the same models the Disney studio's animators used when making the film. In July 1942 all rights, inventory and molds for the Disney products were assigned to the American Pottery Company, of Los Angeles, according to Tumbusch.

VG This mark appears on the large ceramic ladybugs and also on plastic canning jar shakers made in Hong Kong. Beyond that, I haven't succeeded in finding out anything about it.

Victoria Ceramics No information that would provide clues to this importer's origin were found.

Victory Glass Company Best known for candy containers (which the lantern shown above started out as), the Victory Glass Company was located in Jeanette, Pennsylvania. It was formed in 1919, bought out by the J.H. Millstein Company, also of Jeanette, in 1955.

Wade The Wade Group of Potteries was formed in 1958 through the merger of A.J. Wade, Wade Heath & Company, Ltd., George Wade and Son, Ltd., and Wade (Ulster) Ltd. A fifth company was formed and added to the group in 1969, Wade (PDM) Ltd. The Wade Group is based in Burslem, Stoke-on-Trent, England. Its roots go back to 1810.

Wade has not made a great many shakers, and most of those it has made would not fit into the figural category. Nevertheless, they go well with a figural collection, especially the condiment sets.

For a wealth of information on Wade, I recommend *The World of Wade—Collectible Pottery and Porcelain* (Antique Publications, 1988), by Ian Warner with Mike Posgay. Information on how to obtain the book is listed in Appendix I Sources.

Wahpeton Pottery Company *See Rosemeade.*

Wales A rather prolific importer of Japanese ceramics during the early postwar era, but I wasn't successful in finding out more about it.

Josiah Wedgwood & Sons Ltd. Josiah Wedgwood founded the English pottery that still bears his name in 1759. The company is located in Barlston, Stoke-on-Trent, England.

Best known for jasperware, Wedgwood also makes or has made black basalt, creamware (queen's ware) and bone china, among other types of ceramics.

Many people who own Wedgwood jasperware, whether shakers or otherwise, fret about cleaning it, feeling that immersing it in water will somehow ruin it. But that's not so. Jasperware is actually stoneware, which is impervious to water. However, according to literature packed with Wedgwood jasper, the company recommends that those who choose to clean it in an automatic dishwasher, use Calgonite, Dishwasher All or Cascade in water no hotter than 150 degrees F.

Whirley Industries This company is located in Warren, Pennsylvania. Its history is unusual when you consider it might have made one of the shakers in your collection. So we'll start at the beginning.

According to Robert Sokolski, president, Whirley was started in 1959 as a coin-operated laundry. Over the next six years it concentrated on building and operating laundry and dry cleaning stores, and gas stations with on-site car washes. The company still operates Whirley Wash Laundries and/or Sparkle Car Washes in Warren, Bradford and Kane, Pennsylvania, and Jamestown, New York.

In 1966, with larger facilities available, the focus changed to designing and manufacturing automatic pressure car wash equipment. Approximately 200 units were sold throughout the United States.

Then in 1968 Whirley made the one-piece plastic tiger shown in Chapter 6, and at the same time began to de-emphasize car wash manufacturing. Today the company's main line is plastic souvenirs, gifts and novelties that are sold at approximately 5000 retail stores across the country and in every state in the Union. Whirley maintains a 45,000 square foot manufacturing facility and a 15,000 square foot distribution center.

According to a Whirley spokeswoman, the plastic one-piece is the only shaker Whirley ever made, but it was finished in several different styles, not just as a tiger.

Wihoa's Ceramics *See Wisecarver, Rick.*

Wisecarver, Rick Rick Wisecarver is a self-employed ceramic artist in Roseville, Ohio. He has been in the ceramics business since 1970. He starts at the very beginning with a block of clay, follows through to the very end to the final glaze or gold firing.

Wisecarver calls his business Wihoa's Ceramics. The address is listed in Appendix I Sources.

W.N. Co Ceramics This paper label was found on a pair of Las Vegas souvenir shakers. The label indicated the shakers were made in Japan. There was also a pencil date of 1984.

K. Wolfe Studios K. Wolfe Studios is owned and operated by Kathy and Doug Wolfe of West Bloomfield, Michigan, a suburb of Detroit. They are both sculptors, but Doug also works a full-time job as a sergeant on the Royal Oak, Michigan Police Department. Kathy began working in ceramics in high school in the mid to late 1960s, and followed up with commercial art courses while attending community colleges in southern Michigan.

Some of the shakers the Wolfe's make are modeled by themselves, others are standard molds purchased from mold companies. While there is no exact number set for the shakers they model themselves, Kathy Wolfe says most of them are limited to editions of about 100. By then the mold has generally worn down to the point it can no longer be used. The number of shakers made from commercial molds is unlimited. They continue to make them as long as the public shows interest. Shakers are sold at craft shows, and also by mail order. They are also sold as commemorative sets, such as the Mackinac Bridge, shown on page 198, which was made for a convention of the Michigan Chapter of the Novelty Salt & Pepper Shakers Club.

One of K. Wolfe Studios latest pairs of shakers is the Big Boy and hamburger, shown on page 71.

The address of K. Wolfe Studios can be found in Appendix I Sources.

Appendix I: Sources

Books

An Introduction to Novelty Salt and Pepper Shakers
Frances Harris
Tomorrow's Antiques Today
Sylvia P. Tompkins
25C Center Drive
Lancaster, Pennsylvania 17601

Beautiful Rosemeade
Shirley L. Sampson and Irene J. Harms
Irene J. Harms
2316 West 18th Street
Sioux Falls, South Dakota 57104

Collecting Hull's Little Red Riding Hood
Mark Supnick
L-W Book Sales
P.O. Box 69
Gas City, Indiana 46933

Collector's Guide to Shawnee Pottery
Duane and Janice Vanderbilt
Collector Books
P.O. Box 3009
Paducah, Kentucky 42002-3009

Goebel Salt & Pepper Shakers, Identification and Value Guide
Hubert and Clara McHugh
Vintage Machinery Company
1850 West Main Street
Stroudsburg, Pennsylvania 18360

Great Shakes
Gideon Bosker
The Salt and Pepper Man
Larry Carey
P.O. Box 329
Mechanicsburg, Pennsylvania 17055

Harrington Figurines (Ceramic Arts Studio)
Sabra Olson Laumbach
Sabra Olson Laumbach
c/o Antiques Minnesota, Inc.
1516 Cake Street
Minneapolis, Minnesota 55407

Purinton Pottery I and II
Pat Dole
P.O. Box 4782
Birmingham, Alabama 35206

Salt and Pepper Shakers--Figural and Novelty, Series I and II
Melva Davern
Collector Books
P.O. Box 3009
Paducah, Kentucky 42002-3009

Salt & Pepper Shakers I, II, and III
by Helene Guarnaccia
Collector Books
P.O. Box 3009
Paducah, Kentucky 42002-3009

The World of Salt Shakers--Antique & Art Glass Value Guide
by Mildred and Ralph Lechner
Collector Books
P.O. Box 3009
Paducah, Kentucky 42002-3009

Ceramic Artists

K. Wolfe Studios
3551 Walnut Lake Road
West Bloomfield, Michigan 48323

Sandy Srp
Sandy's Small Salts
10878 Chardon Road
Chardon, Ohio 44024

Rick Wisecarver
43 Maple Street
Roseville, Ohio 43777

Clubs

Novelty Salt and Pepper Shakers Club
c/o Irene Thornburg, Membership Coordinator
581 Joy Road
Battle Creek, Michigan 49017

Shawnee Pottery Collector's Club
c/o Pam Curran
P.O. Box 713
New Smyrna Beach, Florida 32170

Salt and Pepper Shaker Sellers

Jazz'e Junque Shop
3831 N. Lincoln
Chicago, Illinois 60613

The Salt and Pepper Man
Larry Carey
P.O. Box 329
Mechanicsburg, Pennsylvania 17055

Tomorrow's Antiques Today
Sylvia P. Tompkins
25C Center Drive
Lancaster PA 17601

Judy Posner
R.D. #1 Box 273
Effort, PA 18330
(Mail order catalog available on request)

Appendix II: Beswick Model Numbers by Year of Introduction

The chart below gives the year of introduction for all Beswick designs from 1936 through 1985. The chart applies for all Beswick pottery, not just salt and pepper shakers and condiment sets. Note how new designs fell to almost nothing during World War II.

Model Numbers	Year	Model Numbers	Year	Model Numbers	Year
1 - 377	Undated	1227 - 1279	1952	2256 - 2202	1969
378 - 460	1936	1280 - 1322	1953	2203 - 2349	1970
461 - 567	1937	1323 - 1362	1954	2350 - 2396	1971
568 - 672	1938	1363 - 1391	1955	2397 - 2443	1972
673 - 794	1939	1392 - 1468	1956	2444 - 2500	1973
795 - 880	1940	1469 - 1516	1957	2501 - 2522	1974
881 - 968	1941	1517 - 1576	1958	2523 - 2554	1975
969 - 990	1942	1577 - 1667	1959	2555 - 2582	1976
991 - 1000	1943	1668 - 1732	1960	2583 - 2607	1977
1001 - 1013	1944	1733 - 1792	1961	2608 - 2636	1978
1014 - 1042	1945	1793 - 1861	1962	2637 - 2666	1979
1043 - 1082	1946	1862 - 1920	1963	2667 - 2700	1980
1083 - 1107	1947	1921 - 1996	1964	2701 - 2750	1981
1108 - 1141	1948	1997 - 2053	1965	2751 - 2804	1982
1142 - 1180	1949	2054 - 2086	1966	2805 - 2846	1983
1181 - 1209	1950	2087 - 2176	1967	2847 - 2899	1984
1210 - 1226	1951	2177 - 2255	1968	2900 - 2940	1985

Appendix III: Parkcraft States

As stated in the text, Parkcraft made two series of states, one around 1957, when there were only 48 states in the Union, and another around 1968, by which time Alaska and Hawaii had been admitted to make 50. Also, several of the mates for the states were changed for the 1968 set. Listed below is each state along with its mate for both sets. Wherever a change is involved, the entries are in capital letters.

State	**1957**	**1968**
Alaska	----	Igloo
ALABAMA	COTTON	WATERMELON
Arizona	Cactus	Cactus
Arkansas	Razorback	Razorback
CALIFORNIA	BATHING BEAUTY	ORANGE
Colorado	Pack mule	Pack mule
Connecticut	Graduation cap	Graduation cap
Delaware	Lighthouse	Lighthouse
FLORIDA	BATHING BEAUTY	FISH
Georgia	Rebel cap	Rebel cap
Hawaii	----	Native girl
Idaho	Potato	Potato
ILLINOIS	CORN	ABRAHAM LINCOLN
Indiana	Race car	Race car
Iowa	Corn	Corn
Kansas	Wheat	Wheat
Kentucky	Whiskey jug	Whiskey jug
LOUISIANA	COTTON	SUGAR SHACK
MAINE	LIGHTHOUSE	PINE TREE
Maryland	Oyster	Oyster
Massachusetts	Bean pot	Bean pot
Michigan	Car	Car
Minnesota	Canoe	Canoe
MISSISSIPPI	COTTON	STEAMBOAT
Missouri	Mule	Mule
Montana	Revolver	Revolver
NEBRASKA	CORN	COWBOY BOOTS
Nevada	Ace of Spades	Ace of Spades
New Hampshire	Snowman	Snowman
New Jersey	Miss America	Miss America
New Mexico	Pueblo	Pueblo
New York	Statue of Liberty	Statue of Liberty
North Carolina	Cigarettes	Cigarettes
NORTH DAKOTA	WHEAT	OIL WELL
Ohio	Tire	Tire
OKLAHOMA	OIL WELL	AMERICAN INDIAN
Oregon	Duck	Duck
PENNSYLVANIA	COAL	LIBERTY BELL
Rhode Island	Rooster	Rooster
SOUTH CAROLINA	LIGHTHOUSE	COTTON
South Dakota	Pheasant	Pheasant
TENNESSEE	COTTON	HORSE HEAD
TEXAS	OIL WELL	COWBOY HAT
Utah	Covered wagon	Covered wagon
Vermont	Maple syrup bucket	Maple syrup bucket
Virginia	Ham	Ham
Washington	Apple	Apple
West Virginia	Coal	Coal
Wisconsin	Cheese	Cheese
Wyoming	Bronco	Bronco

Appendix IV, United States Invention Patent Numbers

The chart below shows the first patent number issued by the United States government each year for a 100 year period from 1860 through 1959. By comparing the patent number on a pair of shakers to the number on the chart, you can easily determine when the shakers were patented, but not necessarily when they were made since patent dates and dates of manufacture may vary by several years in some cases.

Year	Number	Year	Number	Year	Number	Year	Number	Year	Number
1860	26,642	1880	223,211	1900	640,167	1920	1,326,899	1940	2,185,170
1861	31,005	1881	236,137	1901	664,827	1921	1,364,063	1941	2,227,418
1862	34,045	1882	251,685	1902	690,385	1922	1,401,948	1942	2,268,540
1863	37,266	1883	269,820	1903	717,521	1923	1,440,362	1943	2,307,007
1864	41,047	1884	291,016	1904	748,567	1924	1,478,966	1944	2,338,081
1865	45,685	1885	310,163	1905	778,834	1925	1,521,590	1945	2,366,154
1866	51,784	1886	333,494	1906	808,618	1926	1,568,040	1946	2,391,856
1867	60,658	1887	355,291	1907	839,799	1927	1,612,700	1947	2,413,675
1868	72,959	1888	375,720	1908	875,679	1928	1,654,521	1948	2,433,824
1869	85,503	1889	395,305	1909	908,436	1929	1,696,897	1949	2,457,797
1870	98,460	1890	418,665	1910	945,010	1930	1,742,181	1950	2,492,944
1871	110,617	1891	443,987	1911	980,178	1931	1,787,424	1951	2,536,016
1872	122,304	1892	466,315	1912	1,013,095	1932	1,839,190	1952	2,580,379
1873	134,504	1893	488,976	1913	1,049,326	1933	1,892,663	1953	2,624,046
1874	146,120	1894	511,744	1914	1,083,267	1934	1,941,449	1954	2,664,562
1875	158,350	1895	531,619	1915	1,123,212	1935	1,985,878	1955	2,698,434
1876	171,641	1896	552,502	1916	1,166,419	1936	2,026,516	1956	2,728,913
1877	185,813	1897	574,369	1917	1,210,389	1937	2,066,309	1957	2,775,762
1878	198,733	1898	596,467	1918	1,251,458	1938	2,104,004	1958	2,818,567
1879	211,078	1899	616,871	1919	1,290,027	1939	2,142,080	1959	2,866,973

Appendix V: United States Design Patent Numbers

This chart shows the first U.S. design patent number issued each year from 1860 through 1959. Design patent numbers are often marked "Des. Pat...." on salt and pepper shakers and other kitchen ware.

Year	Number	Year	Number	Year	Number	Year	Number	Year	Number
1860	1,183	1882	12,647	1904	36,723	1926	69,170	1948	148,267
1861	1,366	1883	13,508	1905	37,280	1927	71,722	1949	152,235
1862	1,508	1884	14,538	1906	37,766	1928	74,159	1950	156,686
1863	1,703	1885	15,678	1907	38,391	1929	77,347	1951	161,404
1864	1,879	1886	16,451	1908	38,980	1930	80,254	1952	165,568
1865	2,018	1887	17,046	1909	39,737	1931	82,966	1953	168,527
1866	2,239	1888	17,995	1910	40,424	1932	85,903	1954	171,241
1867	2,533	1889	18,830	1911	41,063	1933	88,847	1955	173,777
1868	2,858	1890	19,533	1912	43,073	1934	91,258	1956	176,490
1869	3,304	1891	20,439	1913	45,098	1935	94,179	1957	179,467
1870	3,810	1892	21,275	1914	46,813	1936	98,045	1958	181,829
1871	4,547	1893	22,092	1915	48,358	1937	102,601	1959	184,204
1872	5,452	1894	22,994	1916	50,117	1938	107,738		
1873	6,336	1895	23,922	1917	51,629	1939	112,765		
1874	7,083	1896	25,037	1918	52,836	1940	118,358		
1875	7,969	1897	26,482	1919	54,359	1941	124,503		
1876	8,884	1898	28,113	1920	56,844	1942	130,989		
1877	9,686	1899	29,916	1921	60,121	1943	134,717		
1878	10,385	1900	32,055	1922	56,844	1944	136,946		
1879	10,975	1901	33,813	1923	61,748	1945	139,862		
1880	11,567	1902	35,547	1924	63,675	1946	143,386		
1881	12,082	1903	36,198	1925	66,346	1947	146,165		

Bibliography

Allison, Grace, "Juanita Ware," *Tri-State Trader News Supplement,* April, 1977.

Altman, Seymour and Violet, *Book of Buffalo Pottery, The,* Crown, New York, 1969.

American Export Register: 1990, Thomas International Publishing, Inc., New York, 1990.

Boyce, Elizabeth, "Kass China," *The New Glaze,* December, 1987.

Brands and Their Companies—9th Edition, Donna Wood, Editor, Gale Research, Inc., New York, 1991.

Campbell, Carol, "More Arcadia Ceramics Identified," *Novelty Salt & Pepper Shakers Club Newsletter,* May, 1992.

Cembura, Al, and Avery, Constance, *Jim Beam Regal China Go-Withs,* self-published, Berkeley, California, 1979.

Chipman, Jack, *Collector's Encyclopedia of California Pottery,* Collector Books, Paducah, Kentucky, 1992.

Companies and Their Brands 9th Edition, Donna Wood, Editor, Gale Research, Inc., New York, 1991.

Congdon-Martin, Douglas, *Images in Black—150 Years of Black Collectibles,* Schiffer Publishing Ltd., West Chester, Pennsylvania, 1990.

Cunningham, Jo, *Collector's Encyclopedia of American Dinnerware, The,* Collector Books, Paducah, Kentucky, 1982.

Davern, Melva, *Salt & Pepper Shakers—Figural and Novelty,* Collector Books, Paducah, Kentucky, 1985.

______ *Salt & Pepper Shakers—Figural and Novelty, Second Series,* Collector Books, Paducah, Kentucky, 1990.

Derwich, Jenny B., and Latos, Dr. Mary, *Dictionary Guide to United States Pottery & Porcelain (19th and 20th Century),* Jenstan, Franklin, Michigan, 1984.

Directory of Corporate Affiliations—1991, National Register Publishing Company, Wilamette, Illinois, 1990.

Ehrmann, Eric W., and Miller, Robert L., *M.I. Hummel, The Golden Anniversary Album,* Galahad Books, New York, 1984.

Florence, Gene, *Collector's Encyclopedia of Depression Glass, The,* Collector Books, Paducah, Kentucky, 1982.

______ *Collectors Encyclopedia of Occupied Japan Collectibles, The,* Collector Books, Paducah, Kentucky, 1976.

______ *Collector's Encyclopedia of Occupied Japan Collectibles—Third Series, The,* Collector Books, Paducah, Kentucky, 1987.

______ *Kitchen Glassware of the Depression Years,* Collector Books, Paducah, Kentucky, 1981.

Guarnaccia, Helene, *Salt & Pepper Shakers,* Collector Books, Paducah, Kentucky, 1990.

______ *Salt & Pepper Shakers II,* Collector Books, Paducah, Kentucky, 1990.

______ *Salt & Pepper Shakers III,* Collector Books, Paducah, Kentucky, 1991

Harris, Frances, *Novelty Salt and Pepper Shakers,* Tabby Publications, England, 1991.

Heide, Robert, and Gilman, John, *Cartoon Collectibles,* Doubleday, Garden City, New York, 1983.

______ *Popular Art Deco,* Abbeville Press, New York, 1991.

Huxford, Sharon and Bob, *Collector's Encyclopedia of Fiesta, The* Collector Books, Paducah, Kentucky, 1984.

______ *Collectors Encyclopedia of McCoy Pottery, The,* Collector Books, Paducah, Kentucky, 1980.

______ *Schroeder's Antique Price Guide--11th Edition,* Collector Books, Paducah, Kentucky, 1993.

Klamkin, Marian, *Collector's Guide to Depression Glass, The,* Hawthorn Books, Inc., New York, 1973.

Kovel, Ralph and Terry, *Advertising Collectibles Price List,* Crown, New York, 1986.

______ *Kovel's Antiques & Collectibles Price List—24th Edition,* Crown, New York, 1991.

Laumbach, Sabra Olson, *Harrington Figurines,* Ferguson Communications, Hillsdale, Michigan, 1985.

Lechner, Mildred and Ralph, *The World of Salt Shakers—Antique & Art Glass Value Guide,* Collector Books, Paducah, Kentucky, 1992.

Lehner, Lois, *Lehner's Encyclopedia of U.S. Marks on Pottery, Porcelain, and Clay,* Collector Books, Paducah, Kentucky, 1988.

______ *Ohio Pottery and Glass Marks and Manufacturers,* Wallace-Homestead Book Company, Des Moines, Iowa, 1978.

Lindenberger, Jan, *Collecting Plastics, A Handbook and Price Guide,* Schiffer Publishing Ltd., West Chester, Pennsylvania, 1992.

______ *Black Memorabilia for the Kitchen—A Handbook and Price Guide,* Schiffer Publishing Ltd., West Chester, Pennsylvania, 1992.

Longest, David, *Character Toys and Collectibles,* Collector Books, Paducah, Kentucky, 1987.

______ *Character Toys and Collectibles, Second Series,* Collector Books, Paducah, Kentucky, 1987.

Marquette, Arthur F., *Brands, Trademarks and Goodwill—The Story of the Quaker Oats Company,* McGraw-Hill, New York, 1967.
Matzke, Edwin P., "Pepper," *Encyclopedia Americana,* Americana Corporation, Danbury, Connecticut, 1979.
May, Harvey, *Beswick Collector's Handbook, The,* Kevin Francis Publishing, Ltd., London, 1986.
McConnell, Kevin, *Collecting Art Deco,* Schiffer Publishing Ltd., West Chester, Pennsylvania, 1990.
McHugh, Hubert and Clara, *Goebel Salt & Pepper Shakers—Identification and Value Guide,* Vintage Machinery Company, Stroudsburg, Pennsylvania, 1992.
McNulty, Lyndi Stewart, *Price Guide to Plastic Collectibles,* Wallace-Homestead, Radnor, Pennsylvania, 1987.
O'Connor, Muriel, "The Campbell Kids," Novelty Salt and Pepper Shakers Club Newsletter, November 1991.
Peterson, Arthur G., *Glass Salt Shakers,* Wallace-Homestead Book Co., Des Moines, Iowa, 1970.
Posgay, Mike, and Warner, Ian, *World of Head Vase Planters, The,* antique Publications, Marietta, Ohio, 1992.
Read, William T., "Salt," *Encyclopedia Americana,* Americana Corporation, Danbury, Connecticut, 1979.
Rinker, Harry L., *Warman's Americana & Collectibles,* 5th edition, Wallace-Homestead, Radnor, Pennsylvania, 1991.
Roerig, Fred and Joyce Herndon, *Collector's Encyclopedia of Cookie Jars, The,* Collector Books, Paducah, Kentucky, 1991.
Samson, Shirley L., and Harms, Irene J., *Beautiful Rosemeade,* private printing, Sioux Falls, South Dakota, 1986.
Schneider, Mike, *Animal Figures,* Schiffer Publishing Ltd., West Chester, Pennsylvania, 1990.
______ *Complete Cookie Jar Book, The,* Schiffer Publishing Ltd., West Chester, Pennsylvania, 1991.
Smith, Marcia, "Arcadia Ceramics, Inc.," *Novelty Salt & Pepper Shakers Newsletter,* February, 1991.
Stern, Michael, *Stern's Guide to Disney Collectibles,* Collector Books, Paducah, Kentucky, 1989.
______ *Stern's Guide to Disney Collectibles, Second Series,* Collector Books, Paducah, Kentucky, 1991.
Supnick, Mark E., *Collecting Hull's Little Red Riding Hood,* L-W Book Sales, Gas City, Indiana, 1989.
______ *Collecting Shawnee Pottery,* L-W Book Sales, Gas City, Indiana, 1989.
Tumbusch, Tom, *Tomart's Illustrated Disneyana Catalog and Price Guide, Condensed Edition,,* Wallace-Homestead, Radnor, Pennsylvania, 1989.
______ *Tomart's Illustrated Disneyana Catalog and Price Guide, Volume III,* Tomart Publications, Dayton, Ohio, 1985.
Vanderbilt, Duane and Janice, *Collector's Guide to Shawnee Pottery,* Collector Books, Paducah, Kentucky, 1992.
Weatherman, Hazel Marie, *Colored Glassware of the Depression Era 2,* Weatherman Glassbooks, Ozark, Missouri, 1982.
Young, Jackie, *Black Collectables—Mammy and Her Friends,* Schiffer Publishing Ltd., West Chester, Pennsylvania, 1988.

Index

accordions, Japan, 228
ace of spades and Nevada,
 Parkcraft, 236
A Commodore Product
 Oriental couple, 43
Santa & Mrs. Claus, 73
acorns
 unknown, 175
 wood, 175
 wood composition, 175
Adam and Eve
 Enesco, 11
 Japan, 64
Advertising
 Ace Lunch Box, 248
 Atlantic, 193
 Aunt Jemina and
 Uncle Mose 31
 beer bottles, 13, 211-212
 Big Boy, Japan, 71
 Big Boy and hamburger, 71
 Blue Bonnet Sue, 70
 Borden's, 93
 Campbell Kids, 27
 Campbell's soup, 27, 212
 Captain Bob-Lo, 33
 Chicken of the Sea, 157
 Cities Service, 193
 Coca-Cola, 210
 Conoco, 193
 Coors, 211
 DiMaggio's, 217
 Elmer and Elsie, 93
 Elsie and Elmer, 93
 Electrolux, 240
 Ernie the Keebler elf, 80
 Esso, 18, 193
 Eveready, 234
 F & F dog and cat, 91
 Farm Bureau, 87
 Fido and Fifi, 91
 Fifi and Fido, 91
 Filter Queen vacuum, 203
 Ford rotunda, 13
 gas pumps, 192-193
 General Electric, 205
 Golden Age, 210
 Gulf, 193
 Hamm's beer, 8
 Happy Homer, 71
 Harvestore silos, 18, 197
 Hershey's chocolate kisses, 224
 Higgins ink, 212
 Hot N Kold, 27
 Howard Johnson's, 110
 Italian Swiss Colony, 210
 Kellogg's, 71
 Kelvinator, 203
 Kenmore, 246
 Lennox Furnance Company, 33
 Lenny Lennox, 33
 Luzianne coffee, 31
 Magic Chef, 27
 Marion Kay, 251
 Mason, 210
 McCormick-Deering, 206
 Milemaster, 193
 Morton's salt, 14
 Mr. Peanut, 183-184
 Nabisco, 70
 New Era potato chips, 212
 Nipper, 18, 101
 Northwest Airlines, 14
 Oil City Glass Company, 195
 Peerless beer, 80
 Pepsi, 210
 Philgas, 195
 Phillips 66, 193
 Pillsbury, 27
 Planter's, 183-184
 RCA Victor, 18, 101
 Red Brand fence, 18
 Red Lobster, 246
 Richfield, 193
 Rite-Lite Corporation, 13
 Saf-T-Cup, 223
 Sandeman brandy, 71
 Schlitz, 211
 Seagram's, 234
 Sealtest, 211
 Seven-Up, 210, 211
 sheet metal, 235
 Shell, 193
 Sinclair, 193
 Sleep Guard blanket
 controls, 206
 Snap and Pop, Kellogg's, 71
 Sohio, 193
 Staggs-Bilt Homes, 71
 Standard Oil, 192
 Stokley's, 172
 Strasburg Railroad, 33
 Styron, 248
 Sunshine biscuits, 26
 Tappan, 27, 201
 Tastee Freeze, 70
 Tee and Eff, 70
 Texaco, 193
 Uncle Mose and
 Aunt Jemima, 31
 Wolfschmidt vodka, 179
 Zephyr, 192
Afloat on the Mississippi, Wisecarver, 24
African heads, unknown, 47
Agiftcorp
 owls, 141
Agnew and Nixon, K. Wolfe, 69
aircraft carrier and airplane, unknown, 190
airplane and aircraft carrier, unknown, 190
alabaster pepper mill and salt shaker, unknown, 216
alarm clocks
 Japan, 230
 with faces, 230
 wood, 230
alligator and
 black boy, Ceramic
 Arts Studio, 59
 suitcase, Vandor, 149
 umbrella, Vandor, 149
 woman Empress, 15
alligators
 Japan, 148, 149
 miniatures, 148
 one-piece, 148
 Srp 148
 unknown, 148, 149
 Vandor, 149
Aluminum Housewares
 refrigerators, 202
American Bisque
 bears, 84, 95
 cats, 88
 cows, 95
 dogs, 98
Amish
 boy and girl, 57
 couple, 49
 girl and boy, 57 in chair, 49
 with wheelbarrow, 49
An Enterprise Exclusive
 pixies, 79
angel and devil,
 Kreiss, 75
 unknown, 75
angels,
 Japan, 75
 miniatures, 75
 Srp, 75
 unknown, 75
A Nod to Abe, Regal, 25
anvil and hammer, unknown, 239
anvils, metal, 240
Applause
 Autumn Seranade, 233
 mark, 233
apple, Otagiri, 183
apple and, school books
 Srp, 171
 Washington state, Parkcraft, 19
 worm, unknown, 119
apples
 Arcadia, 171
 Franciscan, 171
 Gladding McBean, 171
 in baskets, 171
 Japan, 170, 174, 176
 Occupied Japan, 171
 on tray, 172
 Otagiri, 170
 plastic, 171
 unknown, 171
 with faces, 172
 wood, 170
A Quality Product
 birds on log, 142
 cars, 189
 Dutch couple on wagon, 44
 Eisenhower Lock, 199
 fish, 154
 Indian nodders, 40
 Indians in moccasins, 39
 1og and birds, 142
 nodders, 17, 40, 83
 skull nodders, 17
Arcadia Ceramics
 apples on tray, 171
 cucumbers/pickles, 182
 strawberries, 173
ark, Enesco, 190
armadillos, unknown, 82
Armita
 cherubs, 75
 mark, 76
arrow and hearts, unknown, 208
artichoke condiment, Japan, 177
artichoke and green onions, Japan, 177
Artmark
 bear and fish, 83
 beets, 180
 chocolate bunnies, 223
 corn, 180
 Cow Jumped Over the Moon, 94
 dog and hydrant condiment, 99
 dogs, 100
 Easter eggs, 223
 mark, 94
 mugs, 215
 oil lamps, 225
 peppers, 180
asparagus, unknown, 178
Atlas
 monkeys, 82
atomizer and dresser jar, unknown, 232
Aura Ceramics
 Come Travel
 To Beautiful Ohio, 24
 Raisin Nodders, 24
automobiles (see cars)
Autumn Seranade, Applause, 233
AVSCO
 mark, 195
 Philgas tanks, 195
Babe and Paul Bunyan
 Japan, 67
 Rosemeade, 67
 unknown, 67
babies
 candy cane, 54
 Clay Art, 54
 Enesco, 54
 in buggy, 54
 in cradle, 54
 Inarco, 53
 nodder, 53
 peek-a-boo, 53, 54
 Regal China, 53
 Sarsaparilla, 53
 snow, 54
 Srp, 54
 Van Tellingen, 53
baby, New Years and Father Time,
 Napco, 71
bag of groceries and Daisy Duck,
 Good Company, 134
baked potatoes
 Japan, 186
 unknown, 186
Bakelite
 non-figural, 247
 Washington Monument, 199
Bambi and Thumper, Enesco, 119
banana as person, Napco, 168, 176
bananas
 chalkware, 168
 Holt Howard, 167
 Napco, 168
 unknown, 168, 173
 with faces, Holt Howard, 167
bananas and grapes on tray, unknown, 175
banjos
 metal, 228
 Sterling silver, 228
 unknown, 229
barbeque grill, unknown, 29, 202
bartender, drinker and horse, Japan, 36
baseball and glove, unknown, 217
baseballs, unknown, 217
Bashful and Valentine Boy Lover,
 Ceramic Arts Studio, 55
basket and chicks, Taiwan, 131
bathing beauty and Florida, Parkcraft, 236
batteries,
 Eveready, 234
 JSNY, 234
 metal, 234
bean pot and Massachusetts,
 Parkcraft, 238
beans, Victor Goldman, 178
beans (can) and cookbook, unknown, 223
bear and
 cub, Ceramic Arts Studio, 16
 fish, Artmark, 83
 garbage can, Japan, 83
 hunter, Enesco, 37
bears
 American Bisque, 84, 95
 A Quality Product, 83
 Ceramarte, 8
 Ceramic Arts Studio, 16
 chefs, Taiwan, 85
 China, 84
 Christmas, Otagiri, 84
 Hamm's, Ceramarte, 8
 Japan, 16, 82-85
 JTC, 84
 nodders, 82
 on bicycle, unknown, 83
 on tricycle, unknown, 84
 one-piece, 85
 Otagiri, 84
 PAC, 86
 redware, 82
 Regal, 85
 Rosemeade, 83
 Smokey, 85
 Srp, 85
 Thanksgiving, Otagiri, 84
 unknown, 83-85
 Van Tellingen, 85
 Victoria Ceramics, 82
 Winnie-the-Pooh, unknown, 84
 with bibs, American Bisque, 84

beavers, Victoria, 86
beehives, Brinn's, 150
bees
Japan, 150
unknown, 150
beets,
Artmark, 180
Germany, 186
Holt Howard, 178
Japan, 182
one-piece, 177
unknown, 177
beets and cauliflower, Japan, 185
before and after couple, Niko, 51
begging dogs, Rosemeade, 96
bellhop and maid, Germany, 35
bellows, unknown, 239
bells,
Disneyland, 207
unknown, 229, 230
bench and piano
Five and Dime, 228
Nec, 228
Bernard Studios
bread and butter, 239
fish, 154
football and megaphone, 218
hotdogs, 222
mark. 240
Beswick
Laurel and Hardy, 70
mark, 14
Sairey Gamp
and Mr. Micawber, 70
Shakespeare's House
condiment, 14
bicycle, metal with glass shakers, 190
Big Boy and hamburger,
Japan, 71
K. Wolfe, 71
Billy and Butch, Ceramic Arts Studio, 97
binoculars, unknown, 229
binoculars and case, unknown, 230
bird on nest,
China, 143
condiment, 143
unknown, 131
birds on log, A Quality Product, 142
birdhouse and bird, unknown, 144
birdhouses, Japan, 198
birthday cake and,
gift, unknown, 223
slice, unknown, 223
birthday card and gift, unknown, 216
bison (see: buffalo)
blacks
African heads, 47
A Float on the Mississippi, 24
A Nod To Abe, 25
Aunt Jemima and
Uncle Mose, 31
babies, 55
butler, 16
butler and maid, 31
chefs, 28
child on alligator, 59
child on toilet, 58
child on elephant, 59
children, 29, 59
children eating watermelon, 60
children on raft, 24
cook with coffee pots, 30
head and watermelon, 59
Luzianne mammy, 31
maid and butler, 31
mammy and chef, 21,
mammies, 31, 29, 30
man and woman, 51,52
porter and bags, 34
Sambo and tiger, 68
with turbines, 43
wood heads, 30
black walnuts, 224
blenders, plastic, 203
Blossom Valley
foxes, 106
mark, 106
bluebirds
Japan, 142
plastic, 143
unknown, 143
Bluebonnet Sue, Taiwan, 70
Blue Ridge
chickens, 127, 128
flowers, 167
boat with
Dutch Kids, unknown, 45
Owl and Pussycat, Fitz and
Floyd, 88
eskimos, unknown, 49
bobbies, unknown, 32
bobbing heads, Oriental couple, A
Commodore Product, 43
Bobby Shaftoe, Japan, 69
bobwhite quail, Red Wing, 138
Boldman-Morgan iceboxes, 203
bolt and wrench, unknown, 218, 239
bolts and nuts, unknown, 240
book and telephone, unknown, 16, 205
books, Japan, 229
books and apple, Srp, 171
boot and Nebraska, Parkcraft, 237
boots, metal, 219
Bo Peep and sailor, Shawnee, 67
bottles
beer, 13, 212
ink, 8
milk, 211
soda, 210, 211
snuff, 213
whiskey, 210, 211
wood, 213
bowling ball and pin,
condiment, 218
plastic, 218
unknown, 218, 239
wood, 218
boxers, Japan, 36
boxing gloves, unknown, 217
boy and
dog, Japan, 59
dog, Regal, 59
dogs with cart, Japan, 59
chair, Ceramic Arts Studio, 55
girl, Shafford, 57
girl in basket, Japan, 56
rocket, Enesco, 59
thunder mug, unknown, 62
boy urinating, unknown, 62
Brayton Laguna gingham dog and
calico cat, 97
bread and butter, Bernard Studios, 239
bread on napkin holder, Brinn's, 222
bride and groom
before and after, 50
Japan, 50-51
Occupied Japan, 50
unknown, 51
briefcase and hat, unknown, 220
Brillium Metal Corporation
frying pans, 232
watches, 231
Brinn's
beehives, 150
bread napkin set, 222
deer, 95
dog and doghouse, 100
frogs, 109
goats, 108
mice in corn, 113
owls, 141
Santa and Mrs. Claus, 73
broom and dustpan, unknown, 240
brush and comb, unknown, 232
buffalo
and Buffalo Bill Cody, unknown, 70
Indian, Srp, 38
Buffalo Bill and buffalo, unknown, 70
buffalos,
Rosemeade, 86
unknown, 86
bucket and Vermont, Parkcraft, 238
bug band, Japan, 151, 152
bugs, Japan, 150, 151
bugs with instruments,
Japan, 151, 152
Buildings
barn and silo,
Japan, 197
unknown, 197
Corn Palace, Exco, 196
Empire State Building, metal, 196
Empire State Building and King
Kong,
Five and Dime, 81
Sarsaparilla, 81
Ford rotunda, unknown, 13
gingerbread house, unknown, 197
Harvestore silos, unknown, 18,
197
Hearst Castle, unknown, 197
house, unknown, 197
igloo and eskimo
Japan, 48
Srp, 10, 48
unknown, 197
1ighthouse and cabin, Japan, 198
1ighthouses, Japan, 197
log cabin, metal, 12
Mormon Temple and Tabernacle
Japan, 196
metal, 196
Outhouses, Japan, 198
pagodas, unknown, 43
Ponderosa Ranch, unknown, 196
Seattle Space Needle
Century Souvenir Company,
199
Etc/Seatle, 199
Japan, 199
metal, 199
Shakespeare's house condiment,
Beswick, 14
silo and barn
Japan, 197
unknown, 197
silos, 18, 197
Texas State Capitol, unknown, 196
Washington Monument, Bakelite,
199
Windmills
Five and Dime, 200
Holland, 200
Occupied Japan, 200
plastic, 200
Sarsaparilla, 200
unknown, 200
Bulging 3-Petal, Consolidated Lamp
and Glass, 253
bullfighter and bull
nodder, 36
PAC, 36
unknown, 36
bullet and gun, unknown, 234
bullets, wood, 232
bums
Japan, 60
unknown, 60
bun and hotdog, Knobler, 222
bunnies, *(see: rabbits)*
Bunyan, Paul
Ceramic Arts Studio, 67
Japan, 67
Rosemeade, 67
unknown, 67
Burrite mugs, 214
Burroughs Manufacturing Corporation
mugs, 214
buses
Japan, 188
metal, 188
Butch and Billy, Ceramic Arts Studio, 97
butler, Trico China, 16
butter and bread, Bernard Studios, 239
buzzard and cactus, Vandor, 142
cabbage
Enesco, 178
Japan, 177
McCoy, 178
one-piece, 177
Rosemeade, 181
unknown, l78
cabbage and turnip, unknown, 185
caboose and locomotive, unknown, 187
cacti
Goebel, 164
Japan, 164
metal, 164
one-piece, 164
unknown, 164
cactus and
Arizona, Parkcraft, 236
buzzard, Vandor, 142
coyote, Vandor, 125
donkey, Srp, 103
skull, Vandor, 125
vulture, Vandor, 142
cactus in flower pot, unknown, 165
cake and slice, unknown, 223
cake with Marilyn Monroe, Clay Art, 69
calico cat and gingham dog
Brayton Laguna, 97
Ceramic Arts Studio, 93
camel
nodder, 86
plastic, 87
camels
Ceramic Arts Studio, 86
Norcrest, 86
Victoria, 86
camera, unknown, 36
camera and photographer, unknown, 36
Campbell Kids
F & F, 27
plastic, 27
cannibal and dinner
Japan, 65
unknown, 65
cannon and
cannonballs, unknown, 218
Civil War soldier, Japan, 32
cannons, unknown, 234
canoe and Minnesota, Parkcraft, 236
canoe (base)
with Indians,
Japan, 37
unmarked, 37
with eskimos,
unknown, 49
candle and
candleholder, unknown, 224
1antern, Japan, 225
candles in holder, unknown, 224
candy canes as people, Japan, 223
candy container (lantern), 226
canning jars
plastic, 210
unknown, 210
V.G., 210
cans, beer, 211
cap (rebel) and Georgia, Parkcraft, 237
Cap'n Pepper and Old Salty
busts, 33
Our Own Import, 32
plastic, 33
wood, 33
captain and tugboat, unknown, 33
Captain Bob-Lo, Japan, 33
car and
cats condiment, Japan, 87
Indiana, Parkcraft, 237
Michigan, Parkcraft, 236
Mickey Mouse, Good Company,
114
trailer, unknown, 187
trailer and Ohio, Aura, 24

with pig driving, Japan, 189
with rabbit driving, unmarked, 189
card and Nevada, Parkcraft, 236
carpet sweeper, metal, 204
carrot and rabbit, Japan, 119, 179
carrots
Giftcraft, 185
Japan, 177, 179, 181
unknown, 179
Victoria, 179
cars
A Quality Product, 189
Maserati, 189
metal, 188, 189
Parksmith, 189
plastic, 189
cart with boy and dog, Japan, 59
case and binoculars, unknown, 230
cat-a-lac condiment, Japan, 87
cat and
creamer, Ceramic Arts Studio, 91
dog, Srp, 91
yarn
Japan, 90, 91
unknown, 90
cats
Agiftcorp, 90
American Bisque, 88
black, 11, 20, 89, 91
Brinn's, 87
calico (and gingham dog)
Brayton Laguna, 97
Ceramic Arts Studio, 98
Ceramic Arts Studio, 88, 91, 98
chefs, 90
Cleo (and Pinocchio)
condiment, Japan, 87
Enesco, 20
F & F, 91
Fifi (and Fido), F & F, 91
Goebel, 77, 89
Holt Howard, 90
Hong Kong, 92
Japan, 87, 88, 90-92
Lego, 90
long, 20
nodder, 88
one-piece, Enesco, 20
plastic, 92
Rosemeade, 88
Shafford, 89
Shawnee, 88
Srp, 91
Taiwan, 91
Thai and Thai-Thai, Ceramic Arts Studio, 88
unknown, 11, 87, 90
Wales, 89
wood, 90
cats in
bags, Japan, 87
baskets,
American Bisque, 88
Goebel, 89
Japan, 92
car, Japan, 87
shoe, Japan, 11
cats on
shoes, Lego, 90
cats with
newspapers, unknown, 87
umbrellas, unknown, 90
water bowl, Japan, 87
yarn, unknown, 90
cattle
Artmark, 94
Elsie and Elmer, unknown, 93
Lefton, 93
Lego, 93
one-piece
Empress, 94
Japan, 94
Rosemeade, 92
cattle on feed bags, Regal, 94
cattle with teapots, unknown, 94
culiflower and beets, Japan, 185
celery, Japan, 178, 181
cello and music book, unknown, 223
cellos, Japan, 228
Century Souvenir Company
Seattle Space Needle, 199
Ceramic Arts Studio
alligator and black boy, 59
Bashful and Valentine Boy Lover, 55
bear and cub, 16
Billy and Butch, 97
black boy and alligator, 59
black boy and elephant, 59
boy on chair, 55
Butch and Billy, 97
calico cat and gingham dog, 93
camels, 86
cat and dog, 93
cats, 38, 91
chair and boy, 55
chair and girl, 55
chickens, 127
Chinese couple, 41, 42
clown and dog, 35
coral and seahorse, 162
covered wagon and ox, 191
deer, 95
Dem and Rep, 104
doe and fawn, 95
dog and cat, 93
dog and clown, 35
dogs, 97, 98
donkey and elephant, 104
donkeys, 103
Dutch couple, 45
Eithiopian palace guards, 43
elephant and black boy, 59
elephant and donkey, 104
eskimos, 48
fawn and doe, 95
fish, 156
fox and goose, 107
French couple, 46
frog and mushroom, 146
gingham dog and calico cat, 93
giraffes, 107
girl on chair, 55
goose and fox, 107
gorilla and baby, 16
harem girl and sultan, 43
Hiawatha and Minehaha, 38
horseheads, 109
Indians, 38
kangaroo and joey, 16
1ambs, 121
1eaves and oak sprites, 79
1eopards, 112
1ions, 112
marks, 45, 46, 47, 98, 104
Minehaha and Hiawatha, 38
mushroom and frog, 146
mushroom and pixie, 79
Oak sprites on leaves, 79
Ox and covered wagon, 191
parakeets, 142
Paul Bunyan and tree, 67
penguins, 145
pigs, 116
pillow and poodle, 98
pixie and mushroom, 79
pomeranians, 98
poodle and pillow, 98
poodles, 98
rabbits, 118
Rep and Dem, 104
Sambo and tiger, 68
Santa Claus and tree, 74
scotties, 98
Scotch couple, 46
seahorse and coral, 162
sheep, 121
skunks, 121
spaniels, 97
Spanish couple, 1
sultan and harem girl, 43
Suzette and pillow, 98
Swedish couple, 46
Thai and Thai-Thai, 87
tiger and Sambo, 68
tree and Paul Bunyan, 67
tree and Santa Claus, 74
Valentine Boy Lover and Bashful, 55
Valentine Boy Lover and Willing, 55
Willing and Valentine Boy Lover, 55
zebra,, 125
cereal box and raisins, Aura, 24
Cereal Sisters, Japan, 222
Ceramarte
Hamm's bears, 8
mark, 8
chair and
boy, Ceramic Arts Studio, 55
couch, unknown, 231
cow, Vandor, 94
dog, unknown, 103
Earth, Clay Art, 233
girl, Ceramic Arts Studio, 55
President Kennedy, 69
woman
Norcrest, 52
unknown, 52
chairs, Japan, 231
chalkware,
bananas, unknown, 168
farmer and wife, unknown, 34
fish, unknown, 158
Indian kissers, unknown, 40
1emon and orange, unknown, 170
1emons, unknown, 170
Oranges, unknown, 169
pears, unknown, 173
strawberries, unknown, 168
Charls Products
1ocomotive and tender, 187
mark, 138
Chase
mark, 139
turkeys, 139, 140
Chase Imports, giraffes, 107
cheese and,
mouse, unknown, 113
Wisconsin, Parkcraft, 236
chef and
cook, one-piece, 20
refrigerator, unknown, 27
hat and head, unknown, 28
chefs
cats, wood, 90
bears, Taiwan, 85
black, 29, 30
Campbell Kids, F & F, 27
condiment, Shafford, 27
corn and pineapple, Grantcrest, Hong Kong, 28
Hot N Kold, unknown, 27
in barbeque grill, Rubicon, 29
Japan, 28
Magic Chef,
milk glass, unknown, 27
plastic, Old King Cole, 27
mama and papa pizza, unknown, 28
Pillsbury Doughboy, unknown, 27
pineapple and corn, Grantcrest, 176
Poinsettia Studios, 29
Sunshine Biscuits, unknown, 26
Tappan, plastic, 27
wood, 29
cherubs, Armita, 75
chick and egg
Japan, 131
unknown, 132
chicken and
Rhode Island, Parkcraft, 238
soup bowl, Vandor, 130
Chicken of the Sea, 157
chicken on nest, unknown, 103, 131
chickens
Blue Ridge, 127
Ceramic Arts Studio, 127
condiment, 128
C.R.H. International, 131
Enesco, 130
Germany, 130
Goebel, 128
Holt Howard, 130, 132
Japan, 128-132
nodders, 128
one-piece, 128
Rosemeade, 130
Shawnee, 127
soft sculpture, 131
Southern Potteries, 127
Taiwan, 130
unknown, 130
Vandor, 130
Wisecarver, 129
chicks
hatching, Japan, 132
in basket, Taiwan, 131
Napco, 131
with eggcups, Holt Howard, 132
Chik flowers, 166
Chikaramachi
condiment-non-figural, 242
mark, 242
children
flower children, Napco, 56
flower girls, Goebel, 57
Hummel-like, CNC, 56
mother's little helper, Enesco, 58
New York World's Fair 1964, unknown, 57
praying,
Hong Kong, 55
Japan, 55
Lego, 55
plastic, 55
with patched clothing, Japan, 64
chimney sweeps, Goebel, 34
chocolate bunnies, Artmark, 223
chocolate kisses, unknown, 224
chopping block and cleaver, unknown, 239
chopping blocks, wood, 231
Christmas
dogs, Japan, 100
girl, Holt Howard, 64
trees, unknown, 6
cigarette pack and North Carolina, Parkcraft, 238
Cinderella on coffee pot, Japan, 203
Claus, Santa and Mrs.
Brinn's, 72
Ceramic Arts Studio, 74
Commodore, 73
Holt Howard, 74
Kelvin, 72
Lefton, 72-73
Lipper & Mann, 73
Midwestern Home Products, 74
Napco, 73
plastic, 72
Srp, 73
Clay Art
babies, 54
Earth/Easy Chair, 233
Marilyn Monroe, 69
Marilyn Monroe in cake, 69
mark, 233
refrigerator and woman, 65
woman and refrigerator, 65
cleaver and chopping block, unknown, 239
Cleo and Pinocchio, Goebel, 77
clipper ships, unknown, 191

clock and
- mouse, unknown, 114
- Victrola, unknown, 101

clocks
- Japan, 230
- wood, 230

clothes pins with faces, plastic, 235
cloverleafs, unknown, 164
Clover Company, Tappan stove, 201
clown and dog, Ceramic Arts Studio, 35
clowns
- Germany, 35
- Japan, 208
- unmarked, 35

CNC
- Hummel-like children, 56
- mark, 56

coal and West Virginia, Parkcraft, 237
coconut and palm tree, unknown, 163
coffee and donut, unknown, 221, 223
coffee grinders, metal, 202
coffee pot and stove, unknown, 202
coffee pots,
- as people, unknown, 209
- Disney, 208
- glass, 209
- Japan, 208
- metal, 209
- plastic, 207, 209
- unknown, 208, 251
- with Cinderella, Japan, 207, 208
- with cook, unknown, 30
- wood, 13, 214

Colonial Oil Lamps, plastic, 225
comb and brush, unknown, 232
Come Travel to Beautiful Ohio, Aura, 24
Commodore
- dogs, 102
- mark, 73
- Santa and Mrs. Claus, 73

concertinas, Japan, 228
condiment sets
- artichokes, Japan, 177
- Azalea, Noritake, 241
- Shakespeare's house, Beswick, 14
- bird and babies, Japan, 143
- bird nodders, unknown, 144
- bowling pins and ball, plastic, 218
- bullfighter and bull nodders. unknown, 36
- car and cats, Japan, 87
- chefs, Shafford, 27
- chickens, unknown, 128
- covered wagons, Japan, 191
- dog and hydrant, Artmark, 99
- English cottage, Trice Brothers, 197
- flowers, Japan, 165
- fish, Goebel, 159
- 1obster, Japan, 161
- Munich condiment, Goebel, 58
- horse and rider nodders, unknown, 144
- non-figural
  - azalea, Noritake, 241
  - Bramble, Wade, 167
  - Chikaramachi, 242
  - Goebel, 241
  - Japan, 242
  - Nippon, 242
  - Noritake, 15, 241
  - U.S. Army, 242
- 0riental people, Japan, 42
- pixie, Japan, 78
- pixies, Japan, 14
- rub a dub dub, unknown, 68
- toilet and thundermugs, Japan, 231
- tomatoes, Japan, 18$

conductor, Japan, 33
Cone, Consolidated Lamp and Glass Company, 254
cookbook and beans, unknown, 223
cooking pots, wood, 213
cool claret, Enesco, 175
cop and prostitute, metal, 32
coral and seahorse, Ceramic Arts Studio, 162
corks, France, 210
corn
- as chef, Grantcrest, 176
- as pitchers, unknown, 181
- Artmark, 180
- Duncan Mold, 181
- Goebel, 180
- Japan, 177, 180, 181
- Napco, 176
- one-piece, 177
- Stanfordware, 180
- Ucagco, 178
- unknown, 179, 181
- with faces, Japan, 176, 180
- with skirt, Napco, 176

cornucopia, unknown, 175
corset, unknown, 219
cotton bale, Consolidated Lamp and Glass Company, 253
cotton boll and South Carolina, Parkcraft, 238
couch and chair, unknown, 231
Coventry
- fruit stacks, 174
- Father Time and New Years Baby, 71

covered wagon and
- horse, Japan, 191
- Ox, Ceramic Arts Studio, 191
- Utah, Parkcraft, 237

covered wagons,
- condiment, 191
- Japan, 192
- unknown, 191, 192

cow, one-piece, Empress, 94
cow and
- Amish woman, metal, 34
- calf, Japan, 94
- one-piece, 34
- chair, Yandor, 94
- pig, Srp, 94
- milkcan, unknown, 99
- moon, Artmark, 94

Cow Jumped Over the Moon, Artmark, 94
cowboy, Japan, 35
cowboy and girl, unmarked, 18
cowboy and Indian mice, unknown, 113
cowboy boot and Nebraska, Parkcraft, 237
cowboy hat and Texas, Parkcraft, 236
cowgirl and boy, unmarked, 18
cows,
- American Bisque, 95
- Artmark, 94

cows on feedbags, Regal, 94
coyote and cactus, Vandor, 125
crabs, unknown, 160
cradles, metal, 240
Craftsman
- rhinoceros, one-piece, 120
- zebra, one-piece, 126

crate with oranges, plastic, 169
creel and fish, unknown, 160
creamer with bowl of strawberries, Japan, 173
C.R.H. International
- chickens, 131

crocodile, *(see: alligator,)*
croquet set, plastic, 219
cucumbers,
- Arcadia, 182
- Japan, 177, 181
- Occupied Japan, 181
- one-piece, 175, 177
- Rosemeade, 181
- with faces, unknown, 181

cup and saucer, Japan, 16
cups, wood, 214
cut glass, 256
Daisy Duck and bag of groceries, Good Company, 134
dalmations, unknown, 100
Davis Products
- piano, plastic, 228
- water lilies, plastic, 166

Deary is weary, Enesco, 35
decks of cards, wood, 219
Decoware
- mark, 212
- New Era potato chip cans, 112

Dee Bee Company, kangaroos, 112
deer,
- Brinn's, 95
- Ceramic Arts Studio, 95
- Japan, 95, 96
- Kass, 95
- Rosemeade, 95

derby hats, unknown, 220
devil and angel
- Kreiss, 75
- unknown, 75

Delft, non-figural, 245
dice and pixies, Enesco, 77
dinosaurs, Japan, 149
Disney
- Bambi and Thumper, Enesco, 119
- cereal sisters, Japan, 222
- Cinderella on coffee pot, unknown, 208
- Daisy Duck and bag of groceries, Good Company, 134
- Donald Duck,
  - Good Company, 134
  - Leeds, 134
  - unknown, 134
- Donald Duck and Ludwig Von Drake, Inarco, 134
- Dumbo, unknown, 105
- elephant and Mowagli, Enesco, 68
- Flower, Goebel, 119
- Hop and Lo, Vernon Kilns, 182
- Kanga and Roo, unknown, 111
- Ludwig Von Drake and Donald Duck, Inarco, 134
- mark, 207
- Mickey Mouse in car, Good Company, 114
- Mickey and Minnie Mouse,
  - Applause, 114
  - Good Company, 114
  - unknown, 114
- Mowagli and elephant, Enesco, 68
- mushrooms (Hop and Lo,), 182
- Pluto, American Bisque, 98
- Rabbit and Winnie-the-Pooh, unknown, 84
- Snow White and Dopey, unknown, 68
- Thumper eyes closed, Goebel, 119
- Thumper eyes open, Goebel, 119
- Tinkerbell on bell, 207
- Tweedle Dum and Tweedle Dee, Regal, 230
- Winnie-the-Pooh and Rabbit, unknown, 84

dog, Nipper, 18
dog and
- boy,
  - Japan, 59
  - Regal, 59
- cat,
  - Srp, 91
- chair, unknown, 103
- clown, Ceramic Arts Studio, 35

doghouse
- Brinn's, 100
- Ceramic Arts Studio, 98
- Disney, 98
- Japan, 99
- unknown, 105

eskimo, K. Wolfe, 48
hydrant
- Artmark, 99
- condiment, 99
- Japan, 99
- unknown, 99

dog heads, Rosemeade, 96
dogs
- American Bisque, 98
- Artmark, 100
- Brayton, 97
- Ceramic Arts Studio, 97, 98
- Commodore, 102
- Enesco, 102
- GNCO, 102
- Goebel, 11, 100
- Japan, 11, 59, 96, 100, 102
- JSNY, 100, 102
- Lego, 102
- Napco, 101, 105
- Nipper, plastic, 101
- Shawnee, 97
- Spuds McKenzie, Napco, 101
- unknown, 99-102
- Vandor, 100

dolphin
- chalkware, 159
- unknown, 158

donkey and elephant
- Ceramic Arts Studio, 104
- chalkware, 104

donkeys
- Ceramic Arts Studio, 103
- Palomar, 103
- Relco, 103
- Srp, 103
- Thames, 103

donkey with
- cream and sugar, Palomar, 103
- vinegar and oil, Thames, 103
- unknown, 103

donut and coffee, unknown 221, 223
Dopey and Snow White, unknown, 68
dragon, unknown, 149
Dreamy and doghouse, unknown, 99
dresser jar and atomizer, unknown, 232
drum and
- frog, unmarked, 19
- monkey, Japan, 19

drummer (base) and drums (shakers), unknown, 36
drums (Indian), Victoria, 38
drums and
- Caribbean drummer, unknown, 47
- Indian drummer, Japan, 37
- tigers, Japan, 124

duck and Oregon, Parkcraft, 237
ducks
- Daisy, 133
- Donald, 134
- Goebel, 132
- huggers Japan, 135
- Leeds, 134
- Ludwig Von Drake, 134
- nodders, 133
- Occupied Japan, 133, 135
- one-piece, 20
- Regal, 135
- Rosemeade, 140
- Royal, 133
- Shawnee, 141
- Srp, 133
- unknown, 133, 134
- Van Tellingen, 135
- with instruments, 134

Dumbo
- American Bisque, 105
- redware, 105

Duncan Mold Company
- corn, 181

dustpan and broom, unknown, 240
Easter eggs, Artmark, 223

Eff and Tee, unknown, 70
EFFCO
whales, 124
egg and
chick,
Japan, 131
unmarked, 132
frying pan, unknown, 221
eggplant, unknown, 182
Eisenhower Lock, A Quality Product, 199
Elbee Art
legs, 66
mammies, 31
mark, 49
pilgrims, 49
seals, 120
squirrels, 122
electric blanket controls, plastic, 205
elephant and
black boy, Ceramic Arts Studio, 59
donkey, Ceramic Arts Studio, 104
chalkware, 104
Mowagli, Enesco, 68
elephants
American Bisque, 105
Ceramic Arts Studio, 104
Ceramic Arts Studio Impostors, 104
Dumbo, 105
Meridan, 105
Palomar, 105
Rosemeade, 104
Tambo and Temblno, 104
elf, Keebler, Taiwan, 80
El Ranchito
mark, 190
typewriter and ink bottle, 190
Elsie and Elmer, unknown, 93
Elvis Presley, unknown, 70
Empress
cow one-piece, 94
lemons, 170
Oranges, 170
man grabbing woman's breast, 62
mark, 170
roses, 166
Smokey-the-Bear, S5
woman and alligator, 15
Enesco
Adam and Eve, 11
baby faces, 11
Bambi and Thumper, 119
boy and rocket, 59
cabbage, 178
cats, black, 20
chickens, 130
cool claret, 175
Deary is weary, 35
dogs, 102
fruit stacks, 175
farm boy and girl,
flowers, 165
tulips, 165
golden girls, 58
grinder and spinning wheel, 206
hippos, 108
mark, 20, 168, 202
mice, 113
mother's little helper, 58
Noah's ark, 190
Pinocchio, 77
pixies, 79
pixies with dice, 77
peaches on tray, 172
pears on tray, 172
pigs, 117
pineapples with faces, 168
potatoes, 186
rocket and boy, 59
snails, 133
stove, 202
Thumper and Bambi, 119
turkeys, 140
eskimo and
dog, K. Wolfe, 48
igloo, Srp, 48
eskimos
Ceramic Arts Studio, 48
IAAC, 49
Japan, 18, 49
Norcrest, 49
Srp, 48
unknown, 48
Victoria, 48
Etc Seattle
Seattle Space Needle, 199
Exclusive
mark, 136
pelicans, 136
EXCO
Corn Palace, 196
Fairway
hillbilly, 61
Farber Brothers
glass shakers with metal tray, 255
farmer and wife,
chalkware, 34
Goldammer, 34
Father Time, Coventry, 71
Father Time and New Years Baby, Napco, 71
feed bags, Regal, 94
feet
Japan, 65, 66
unknown, 65, 66
with faces, 65
Fern
guitars, 226
hammer and anvil, 240
smoking pipes, 240
Fido and Fifi, F & F, 91
F & F
Aunt Jemima and Uncle Mose, 31
Campbell Kids, 27
dog and cat, 91
Fido and Fifi, 91
Luzianne Mammy, 31
mark, 31
soup cups, 65
Fiesta, 243
fighting leopards, Ceramic Arts Studio, 112
Filter Queen, plastic, 203
film containers, plastic, 216
fire hydrant and dog condiment, Artmark, 99
fireplaces, Japan, 231
fish
A Quality Product, 154
Bernard Studios, 154
Ceramic Arts Studio, 156
chalkware, 158
Chicken of the Sea, 157
China, 158
Goebel, 156, 159
Grindley, 155
Japan, 154, 156-160
metal, 158
Napco, 154, 156
nodders, 160
Occupied Japan, 157
one-piece, 160
Poinsettia Studios, 157
Rosemeade, 3, 155, 156, 159
Treasure Craft, 160
unknown, 101, 154, 156, 157, 158
fish and
bear, Artmark, 80
creel, 160
fisherman, Enesco, 37
fisherman and fish, Enesco, 37
Fitz & Floyd
mark, 88, 153
kangaroos, 112
Owl and pussycat, 88
snails, 153
Five and Dime
giraffe, 107
Hollywood sign,
juke box, 229 .
King Kong and Empire State Building, 81
mark, 82, 200
piano and bench, 223
Spuds McKenzie., 101
flags, Parkcraft, 249
flamingoes, Rosemeade, 136
flickertails, Rosemeade, 108
Flower (skunk), Goebel, 119
flower children, Napco, 56
flower girls, Goebel, 57
flower pot and
cactus, unknown, 165
sprinkling can, plastic, 240
flowers,
Blue Ridge, 167
condiment, 165, 167
Davis Products, 166
Elbee Art, 166
Enesco, 165
Japan, 165, 166
Occupied Japan, 165
PY, 166
Relco, 165
Rosemeade, 167
Shawnee, 10, 166, 167
Southern Potteries, 167
Wade, 167
flying saucer and Jetsons, Vandor, 66
food grinder, unknown, 204
football and
helmet, unknown, 217, 218
megaphone, unknown, 218
football players, one headless, Japan, 37
footballs, Japan, 217
foreign costumes
Amish, metal, 49
Caribbean drummer, unknown, 47
Chinese, Ceramic Arts Studio, 42
cowboy and cowgirl, Napco, 39
Dutch,
Ceramic Arts Studio, 43, 45
Germany, 44
Goebel, 44, 45
One-piece, unknown, 20
Napco, 55
plastic, 44
Regaline, 44
Shawnee, 44
Ethiopian palace guards, Ceramic Arts Studio, 43
eskimo
K. Wolfe, 10
unknown, 48, 197
French, Ceramic Arts Studio, 47
harem girl and sultan, Ceramic Arts Studio, 43
hula girl, Victoria, 49
Indians
A Quality Product, 39, 41
Ceramic Arts Studio, 38
chalkware, 40
Japan, 37-41
Mark L Exclusive, 40
metal, 41
Napco, 39
nodders, 40
plastic, 41
Royal, 40 Srp, 38
unknown, 37, 38, 39, 40
Victoria, 38
Mexicans, plastic, 48
Oriental,
Ceramic Arts Studio, 42
China, 42 Commodore, 43
condiments, 42
Japan, 42, 43
Pilgrims,
Elbee Art, 49
Japan, 49
Scotch,
Ceramic Arts Studio, 46
Hong Kong, 46
Miyad, 46
Napco, 47
Norcrest, 46
O.C., 46
plastic, 46
Spanish,
Ceramic Arts Studio, 1
Japan, 47
Napco, 47
snake charmer nodder, unknown, 43
Sultan and harem girl, Ceramic Arts Studio, 43
Swedish, Ceramic Arts Studio, 46
Swiss, Shawnee, 47
Forget-me-not
Challinor's, 253
Eagle Glass, 253
forks and spoons as people
Japan, 221
unknown, 221
Forette, Adrian, Mt. Rushmore, 199
four-eyed nodders, unknown, 51
fox and goose, Ceramic Arts Studio, 107
foxes
Blossom Valley, 106
Enesco, 105
Japan, 105
Nissen, Dave, 106
Palomar, 106
Relco, 105
snazy, 105
unknown, 106
Francie,
mark, 226
musical notes, 226
Franciscian apple, Gladdening, McBean, 171
Frankoma
wagon wheels, 192
frog and
drum, unmarked, 19
mushroom,
Ceramic Arts Studio, 146
Japan, 146
unknown, 146
frog hobos, Japan, 147
frog musicians, unknown, 147
frogs
Brinn's, 109
Japan, 21, 146-148
Occupied Japan, 147
plastic, 146
Sears, 147
Sigma, 147
Thailand, 147
Frosty the Snowman, Hong Kong, 73
fruit stacks
Coventry, 174
Enesco, 175
plastic, 174
unknown, 174
frying pan and egg, unknown, 221
frying pans, Brillium, 232
garbage can and
bear, Japan, 83
raccoon, Japan, 119
garbage disposal, plastic, 203
garden lizards, Japan, 152
gas flame, unknown, 202
gas pump and station attendant, Vandor, 3
gas pumps, plastic, 19, 192-193
geese, Japan, 135
ghost and pumpkin, Srp, 76
ghosts,
Japan, 76

JSNY, 76
Srp, 76
Taiwan, 76
Gibson , raccoons, 120
Gibson California Arts, seals, 121
Gifco
mark, 183
onions in bag, 183
gift and
birthday cake, unknown, 223
card, unknown, 216
Giftcorp, cats, 90
Giftcraft
carrots, 185
mark, 185
gifts
Inarco, 216
Ron Gordon Designs, 216
unknown, 216
gingham dog and calico cat
Brayton, 97
Ceramic Arts Studio, 93
giraffe and zebra, Japan, 107
giraffes
Ceramic Arts Studio, 107
Chase Import, 107
Five and Dime, 107
Japan, 107, 124
one-piece, 107
SDD, 107
girl and boy in basket, Japan, 56
girl head
Holt Howard, 64
Lefton, 58
girl on chair, Ceramic Arts Studio, 55
Gladding McBean & Company
Fransician apples, 171
glass
bottles, 13, 210-212
Challinor's forget-me-not, 253
chef heads, Rubicon, 29
coffee pots on metal burner,
unknown, 209
Consolidated Lamp & GlassCo.
bulging 3-petal, 254
cotton bale, 253
cut, unknown, 256
Depression, pink, unknown, 25
Dithridge, sunset, 253
Eagle Glass forget-me-not, 253
Farber Brothers, 255
Japan, 255
Krome Kraft, 255
light bulbs,
Hong Kong, 13
unknown, 13, 226
milk glass, 254, 255
Mt. Washington,
egg, 252
figs, 252
tomato, 252
mugs,
Artmark, 215
Taiwan, 215
non-figural, 249-251
one-piece, 254, 255
painted, 255
Pilgrim Glass Company, 256
radio tubes, unknown,
ST2steins, unknown, 213
U.S. Glass
Iowa, 253
globe (in hand), Aura, 23
GNCO, dogs, 102
gnomes
in shoe,
unknown, 79
Peerless Beer, unknown, 80
sitter, Japan, 144
goats
Brinn's, 108
Rosemeade, 108
Goebel
cacti, 164
Cardinal Tuck, 32
cats,
chickens, 128
chimney sweeps, 34
corn, 180
Dutch couple, 44
dogs, 100
ducks, 132
English toy soldiers, 32
fish, 156
fish condiment, 159
Flower (skunk), 119
flower girls, 57
Friar Tuck, 32
kissing kids, 55
marks, 116, 122, 123, 156, 241
mallards, 132
monks with instruments, 32
Munich condiment, 58
non-figural condiment, 241
nut and squirrel, 122
owls, 141
peppers, 185
pigs, 116
Pinnochio and Cleo, 77
poodles, 11
rabbits (Thumper), 119
seahorses, 162
skunks, 119, 121
squirrel and nut, 122
squirrels, 122
Thumper eyes closed, 119
Thumper eyes open, 119
Goldammer, farmer and wife, 34
Golden Aspen
jugs, 215
mark, 215
Goldilocks, Regal, 68
Golden Girls, Enesco, 58
golf ball on grass and golfer, Enesco, 37
golf balls
metal, 217
wood, 218
golfer and golf ball on grass, Enesco, 37
gondola, Occupied Japan, 190
Good Company
bag of groceries and Daisy Duck, 135
Mickey and Minnie Mouse, 114
Mickey Mouse in car, 114
Pluto, 98
Daisy Duck and bag of groceries, 135
Donald Duck, 135
goose and fox, Ceramic Arts Studio, 107
gophers, Rosemeade, 108
gorillas
Ceramic Arts Studio, 116
Japan, 81, 82
King Kong, 81
SDD, 81
gourds, unknown, 176
graduates
Japan, 60
Lego, 60
Grantcrest
corn as chef, 176
pineapple as chef, 176
grandfather clock and mouse,
unknown, 114
granny
Artmark, 52
in chair, Norcrest, 52
pregnant, 52
with gramps, Japan, 52
with granddaughter, Japan, 52
with yarn, unknown, 52
grapes
hanging, 174
Hong Kong, 174
Japan, 174
unknown, 176
grapes and bananas on tray with cup,
unknown, 175
grasshopper on watermelon, Japan, 152
gravestones
unknown, 232
wood, 232
Greyhound buses
Japan, 188
metal, 188
grill, unknown, 202
grinder and spinning wheel, Enesco,
Grindley
dogs, 97
fish, 155
mark, 155
pheasants, 138
ground hog, Punxsutawney Phil, Mar Mar, 125
guitar and violin as people, unknown, 227
guitars
Fern, 226
Japan, 226
plastic, 226
Relco, 226
unknown, 227
gun and
bullet, unknown, 234
Montana, Parkcraft, 237
hamburger and Big Boy, K. Wolfe, 71
hamburgers
Our Own Import, 222
unknown, 222
ham and Virginia, Parkcraft, 237
hammer and anvil, unknown, 239
hand and globe, Aura, 23
hands, S & P Club, 66
Handy Flame, unknown, 202
Hanna Barbera,Jetsons, 66
Happy and doghouse, unknown, 99
Happy Homer, OMC, 71
Hardy, Oliver and Stan Laurel, 70
harem girl and sultan, Ceramic Arts Studio, 43
harmonicas Japan, 229
Harper Pottery
mark, 120
seals, 120
harps, Japan, 227
Harrington, Betty, A Nod To Abe, 25
Harvestore silos, unknown, 18
hat and
briefcase, unknown, 220
pipe, unknown, 220

purse, unknown, 220
hatchet and stump, unknown, 239
hatching chicks, Japan, 132
hats
on hall tree, 220
MacCarthur's hat and pipe,
unknown, 220
plastic, 220
unknown, 220
Hazel Atlas, ink bottles, 212
head and chef's hat, unknown, 28
hearts and arrow on tray, unknown, 208
heart with arrow, unknown, 234
hedgehogs
Japan, 109
unknown, 109
helmet and football, unknown, 217, 21
hens, (see: chickens)
Hershey's chocolate kisses, unknown, 224
Hiawatha and Minehaha, Ceramic Arts Studio, 38
hickory dickory dock, unknown, 114
hillbillies
Fairway, 61
Holt Howard, 61
Hong Kong, 61
Japan, 61
Norcrest, 61
plastic, 61
unknown, 60, 61
hillbilly and
jug,
Japan, 61
unknown, 60
wife,
Norcrest, 61
unknown, 61
Hillbottom Pottery
ball, 23
Hi Nosey, unknown, 63
hippos
Enesco, 108
Japan, 108
UCGC, 109
unknown, 10
H. Kato
non-figural, 249
Hollywood sign, Five and Dime, 200
Holt Howard
bananas with faces, 167
beets, 178
cats, 90
chickens, 130
chicks in egg cups, 132
Christmas girls, 64
girl's head, 64
hillbilly, 61
marks, 56, 132
partridges, 139
pixies, 77
shakers on springs, 56
strawberries, 173
trees, 163
Hop and Lo, Vernon Kilns, 182
horse (rear end), Japan, 36
horse and
Indian, unknown, 38
jockey, Japan, 231
covered wagon, Japan, 191
horsehead and Tennessee, Parkcraft, 238
horsehead salt and peppers with toothpick, Japan, 110
horseheads
Ceramic Arts Studio, 109
Japan, 110
Juanitaware, 109
Relco, 110
Rosemeade, 109
horses
Howard Johnson's, 110
Japan, 110
Rosemeade, 109
Srp, 109
horses and wagon, wood, 189
hotdog and bun, Knobler, 222
hotdogs, Bernard Studios, 222
Hot N Kold chefs, unknown, 27
hot plate and pot, Japan, 201
huggers
boy and dog, Regal/Van Tellingen, 59
dog and boy, Regal/Van Tellingen, 59
ducks,
Occupied Japan, 135
Regal/Van Tellingen, 135
love bugs, Regal/Van Telllingen, 17
penguins, Japan, 145
pigs, Regal/Van Tellingen, 115
rabbits, Regal/Van Tellingen 118, 119
hula girl and Hawaii, Parkcraft, 236
hula girls, Victoria, 49
Hull, Little Red Riding Hood, 68
Hummel-like children, CNC, 56
hunter and bear, Enesco, 37

IAAC
eskimos, 49
mark, 49
iceboxes
Boldman-Morgan, 203
Lego, 203
plastic, 203
ice cream cones, unknown, 223
ice cream soda and straws, Napco,224
ice cream sodas, plastic, 224
ice cream sundae, unknown, 223
igloo and Alaska, Parkcraft, 23
Imperial Porcelain
Ma and Doc, 34
pigs, 117
Inarco
babies, 53
gifts, 216
squirrels, 123
Indian and
buffalo, Srp, 38
drums, Japan, 37
horse, unknown, 38
Oklahoma, Parkcraft, 237
teepee, Japan, 38
totem pole, Japan, 41
Indians
Ceramic Arts Studio, 38
chalkware, 40
in canoe, Japan, 37
in moccasins, A Quality Product, 39
Japan, 37, 38, 39, 40, 41
Mark L Exclusive, 40
metal, 4].
Napco, 39
nodders, 40
plastic, 41
unknown, 37, 39, 40
Victoria, 38
ink and bottle, unknown, 8
ink bottle and typewriter, El Ranchito, 190
insulators, unknown, 232
Iowa, U.S. glass, 253
irons
Japan, 204
metal, 204
plastic, 204
I walked my feet of in...., unknown, 65
JASCO, Christmas mice, 11
Jetsons in flying saucer, Vandor, 66
Jiggs and Maggie, unknown, 71
jockey and horse, Japan, 231
Jonah and whale, unknown, 69
JSNY
dogs, 100, 102
Eveready batteries, 234
ghosts, 76
witches, 76
JTC, bears, 84
Juanitaware, horseheads, 109
Juan Products
Lenny Lennox, 33
mark, 33
jug .and Kentucky, Parkcraft, 238
jugs
Golden Aspen, 215
Kelvin's, 215
Purinton, 216
Shawnee, 215
Thrifco, 215
unknown, 210
U.S.A., 210
jukebox
Five and Dime, 229
Vandor, 229
Kanga and Roo, unknown, 111
kangaroo
Brinn's, 110
Canada, 111
Ceramic Arts Studio, 16
Dee Bee Company, 112
Disney, 111
Fitz and Floyd, 112
Japan, 110, 111
Palomar, 111
Sutton Ware, 111
Taiwan, 110, 111
unknown, 111, 112
Karol Western, milk cans, 209
Kass Pottery, deer, 95
kayak with eskimos, unknown, 49
Keebler elf, Taiwan, 80
Kelvin
mark, 72
Santa and Mrs. Claus, 72
Kelvin's
jugs, 215
mark, 215
whales, 159
Kenmar
seashells, 161
Kennedy in rocker, 69
Kentucky, Klay Kraft/Milford, 238
Kermit and Miss Piggy, Sigma, 147
kettle and
man, Japan, 65
unknown, 65
witch, Fitz and Floyd, 76
kettles, wood, 214
king and queen of diamonds, unknown, 219
King Kong and Empire State Building, Five and Dime, 81
Klay Kraft
mark, 238
states, 238
Knobler
bun and hotdog, 222
hotdog and bun, 222
locomotives on tray, 188
mark, 188
musical notes, 226
owls, 141
toaster, 203
Kreiss, devil and angel, 75
Krome Kraft, glass shakers with metal tray, 255
kumquats, unknown, 170
K. Wolfe Studios
Big Boy and hamburger, 71
dog and eskimo, 48
hamburger and Big Boy, 71
Mackinac Bridge, 198
Nixon and Agnew, 69
Ladybugs,
Duncan Ceramics, 151
Vallona Starr, 151
V.G., 150
lantern and candle, Japan, 225
lantern candy container, Victory Glass Company, 226
Lantz, Walter
Wally and Windy, Napco, 67
Woody and Winnie, 145
Woodpecker, Napco, 145
Woody Woodpecker, unknown, 145
Laurel and Hardy, Beswick, 70
lawnmower, plastic, 19
Leeds China Company, Donald Duck, 13
Lefton
cattle, 93
children's heads, 55
girl's head, 58 mark, 73
Santa and Mrs. Claus, 72, 73
snow people, 235
Lego
babies, 53
cats, 9
cattle, 93
children praying, 5
dogs, 102
graduates, 60
iceboxes, 203 mark, 201
pheasants, 137
stove, 201
legs, Elbee Art, 66
lemons
chalkware, 170
Empress, 170
on tree, Japan, 169
Lenny Lennox, Juan Products, 33
leopards,
Ceramic Arts Studio, 112
Vandor, 112
lettuce as people, Japan, 176
Liberty Bell, unknown, 230
Liberty Bell and Pennsylvania, Parkcraft, 237
light bulbs
Hong Kong, 13
Rite-Lite, 13
Taiwan, 13
unknown, 226
lighthouse and Delaware, Parkcraft, 238
Lincoln and Illinois, Parkcraft, 237
lions
Ceramic Arts Studio, 121
Japan, 112
Lipper & Mann
mark, 73
Santa and Mrs. Claus, 73
lipstick and perfume, unknown, 205
Little Jack Horner, Japan, 69
Little Red Riding Hood
Hull, 68
Japan, 68
lizards, Japan, 152
lobster condiment, Japan, 161
lobsters
Japan, 160, 161
unknown, 153
with instruments, Occupied Japan, 160
locomotive and caboose, unknown, 188
locomotive and tender
Charles Products, 187
Japan, 187
metal, 188
unknown, 187
locomotives
Japan, 187
Knobler, 187
metal, 187
log and birds, A Quality Product, 142
lovebugs
Japan, 150
Regal/Van Tellingen, 17, 150
LP gas tanks, plastic, 194-195
Lucite, non-figural, 247
Ludwig Von Drake and Donald Duck, Inarco, 134
Lugenes, stove, 201
lunch box, plastic, 216
Lu Ray, 244
Luzianne Mammy,
F & F, 31
F & F impostor, 31
mark, 31
Ma and Doc, Imperial Porcelain, 34
MacArthur's hat and pipe, unknown, 220
Mackinac Bridge
Japan, 198
K. Wolfe, 198
Made for each other, unknown, 51
Maggie and Jiggs, unknown, 71
Magic Chef
milk glass, 27
plastic, 27
magician, the, unknown, 69
magnetic
cats, 89
skunks, 21
pigs, 21
maid, Enesco, 35
maid and bellhop, Germany, 35
mail box, plastic, 216
Maine, Klay Kraft/Milford, 238
mallards
Goebel, 132
Rosemeade, 140
mandolins, unknown, 227
Mapco
mark, 12
Oyster and pearls, plastic, 12
maple syrup pail and Vermont, Parkcraft, 238
Marilyn Monroe,
half body, Clay Art, 69
in cake, Clay Art, 69
Mark L Exclusive, Indians, 40
Marks
Applause, 233
Armita, 76
Artmark, 94
AVSCO, 195
Bernard Studios, 240
Beswick, 14
Ceramarte, 8
Ceramic Arts Studio, 45, 46, 47, 98, 104
Charls Products, 188
Chase, 139
Chikaramachi, 242
Clay Art, 233
CNC, 56
Commodore, 73
Decoware, 212
Delfts Blue, 45
Disney, 207
Elbee Art, 49
El Ranchito, 190
Empress, 170
Enesco, 20, 168, 202
Exclusive, 136
F & F, 31
Fitz and Floyd, 88, 153
Five and Dime, 82, 200
Francie, 226
Giftcraft, 185
Goebel 116, 122, 123, 156, 241
Golden Aspen, 215
Harker Pottery, 120
Holt Howard, 56, 132
IAAC, 49
Juan Products, 33
Kelvin, 73
Kelvin's, 215
Klay Kraft, 238, 239
Knobler, 188
Lefton, 73
Lego, 201
Lipper & Mann, 73
Luzianne Mammy, 31
Mapco, 12
Mar Mar, 125
McCoy, 178
Milford Pottery, 238, 239
Miyad, 46
Muth & Sons, 13
Noritake, 241, 244
Our Own Import, 222
Palomar, 103
Parkcraft, 238
Parksmith, 189
rooster, 186
Rosemeade, 83, 88
Royal, 40
Royal Winton, 244
Sarsaparilla, 82, 200
Seymour Mann, 124
Shafford, 27
Spin Original, 214
Thrift Ceramics, 122
Vallona Starr, 151
Vandor, 66
Van Tellingen, 53
Vernon Kilns, 182
Wade, 167
Wales, 89
Wedgwood, 71
Wihoa's, 129

Wisecarver, 52, 129
Maserati, Parksmith, 189
Mason jar, unknown, 210
McCoy
cabbage, 178
mark, 178
milk cans, 209
non-figural, 244
salt, pepper, creamer, 215
megaphones, Japan, 218
megaphone and football, Bernard Studios, 218
Meridan
elephants, 105
snails, 153
mermaid and sailor, Regal/Van Tellingen, 33
mermaids
Japan, 76
unknown, 76
metal
Amish buggy and horse, 192
Amish couple, 4S
Amish man with wheelbarrow, 49
Amish person in chair, 49
Amish woman and cow, 34
anvils, 240
banjos, 228
batteries, 234
bicycle with glass shakers, 190
boots, 219
buses, 188
cacti, 161
Campbell's soup cans, 212
cans, 13
carpet sweeper, 204
cars, 188, 189
coffee grinders, 202
Corn Palace, 196
cradles, 240
fish, 158
Ford Rotunda, 13
golf balls, 217
Greyhound buses, 188
Indians, 41
iron, 204
locomotive and tender, 188
locomotives, 188
log cabin, 13
Morman Temple and Tabernacle, 196
Mt. Rushmore, 199
non-figural, 248
oil lamps, 225
pistols, 234
radiators, 189
sewing machine, 204
stoves, 202
streetcars, 188
teapots, 219
tender and locomotive, 188
U.S. Army condiment, 242
mice
Applause, 114
Brinn's, 113
Christmas, 114
cowboy and Indian, 113
Dee Bee Company, 114
Enesco, 113
Good Company, 114
in corn, 113
Jasco, 114
Mickey, 114
Mickey and Minnie, 114
plastic, 114
Rosemeade, 113
Milford Pottery
mark, 238
states, 238
milk can and cow, unknown, 99
milk cans
Karol Western, 209
McCoy, 209
Shawnee, 209, 240
milk glass
steins, 213
Magic Chef, 27
non-figural, 254, 255
Minehaha and Hiawatha, Ceramic Arts Studio, 28
miner's donkey, Japan, 104
miniatures
alligators, 148
angels, 75
apple and books, 171
bears, 85
birthday cake and gift, 223
birthday card and gift, 216
books and apple, 171
buffalo and Indian, 38
cake and slice, 223
cat and dog, 91
coffee cup and donut, 221
cow and pig, 94
dog and cat, 91
donut and coffee cup, 221
ducks, 133
eskimo and igloo, 10
ghost and pumpkin, 76
gift and birthday cake, 223
gift and card, 216
horses, 109
Indian and buffalo, 38
monks, 32
Old woman and shoe, 68
pie and rolling pin, 221
pig and cow, 94
pigs, 117
pumpkin and ghost, 76
rolling pin and pie, 221
shoe and old woman, 6
snow babies, 54
snow people, 235
squirrels, 123
Miss Cutie Pie, Napco, 58
Miss Piggy and Kermit, Sigma, 147
mitt and baseball, unknown, 217
mixer, plastic, 203
Miyad
mark, 46, 99
posh poodles, 99
Scottish children, 46
mocassins with Indians, A Quality Product, 39
modern lambs, Ceramic Arts Studio, 121
monkey and
bananas, 82
drum, 19
stump, 15
monkeys
Goebel, 81
hangers, 82
nodders, 82
sitter, 82
Victoria, 82
monks
Cardinal Tuck, Goebel, 32
Friar Tuck, Goebel, 32
plastic, 32
with instruments, Goebel, 32
Twin Winton, 31
Monroe, C.F., billow pattern, glass, 7
moon with cow, Artmark, 94
moose, unknown, 115
mortar board and
Connecticut, Parkcraft, 238
scroll, unknown, 233
Mother's Little Helper, Enesco 58
Mouse, Mickey and Minnie, 114
mouse and
cheese, unknown, 113, 114
clock, unknown, 114
piano, PY, 114
mountain lions, Rosemeade, 118
Mr. Micawber and Sairey Gamp, Beswick, 70
Mr. Peanut
ceramic, unknown, 184
plastic, unknown, 183-184
Mt. Rushmore
Adrian Forrette, 199
metal, 199
one-piece, 199
plastic, 199
unknown, 199
Mt. St. Helens, 234
Mt. Washington Glass Company
egg, 252
fig, 252
tomato, 252
mugs
Burroughs Mfg. Corp., 214
Burrite, 214
glass, 215
Japan, 214
Spin Original, 214
Takahshi, 214
wood, 13, 214
mule and,
Colorado, Parkcraft, 236
Missouri, Parkcraft, 236
musical notes
Francie, 226
Knobler, 226
Sarsaparilla, 228
unknown, 226
music book and cello, unknown, 228
mushroom and frog
Ceramic Arts Studio, 146
Japan, 146
unknown, 146
mushrooms
Hop and Lo, Vernon Kilns, 175
unknown, 182
with faces, Napco, 182
Muth & Sons
bottles, 13, 211, 212 mark, 13
Napco
bananas, 168
birds, wooden, 142
chicks, 131
cowboy and cowgirl, 39
dogs with hats, 101
Dutch children heads, 55
elves, 75
fish, 154, 155
flower children, 56
fruits and vegetables with skirts, 176
ice cream soda and straws, 224
Indian boy and girl, 39
International kissing couples, 4, 39, 47
Miss Cutie Pie, 58
New Years baby and Father Time, 71
Santa and Mrs. Claus, 73
straws and ice cream soda, 224
wooden birds, 142
NEC, piano and bench, 228
nectarines, unknown, 170
nest and chicken, unknown, 103, 131
New Years baby and Father Time, Napco, 71
New....land Ceramics Corporation, seals, 121
Niko China, before and after couple, 51
Nipper
plastic, 101
unknown, 18, 101
Nippon
non-figural condiment, 240
non-figural shakers, 244
Nissen, Dave, foxes, 106
Nixon and Agnew, K. Wolfe, 69
Noah's Ark, Enesco, 190
nodders-figural
Abe Lincoln, Regal, 25
baby, Sarsaparilla, 53
birds, plastic, 144
boy holding something, Japan, 59
camel, Japan, 59
cats, Japan, 88
clown shaker, 17
dog shaker, 17
ducks, Japan, 133
elephant Japan, 17
Indians, Royal, 40
unknown, 40
pigs, walking, Japan, 116
Raisin Nodder, Aura, 24
snake charmer and cobra, unknown, 43
nodders-non-figural
bears, 82, 83
bird condiment, 144
birds, 144
bullfighter and bull condiment, 36
chickens, 128
ducks, 133
fish, 160
four-eyed people, 51
horse and rider condiment, 144
Indians, 40
monkeys, 82
parakeets, 144
skulls, 17
Norcrest
eskimos, 49
hillbillies, 61
mark, 46
mermaids, 76
pheasants, 2
Scottish people, 46
seals, 121
Smokey the Bear, 85
woman in chair, 52
Noritake
azalea, 244
azalea condiment, 241
mark, 244
non-figural condiment, 15
nut and squirrel
Goebel, 122
Japan, 123
unmarked, 123
nuts and bolts, unknown, 240
nuts on tray, wood, 175
Occupied Japan
apples in basket, 171
boy and girl, 56
bride and groom, 50
cucumbers, l81
ducks, 133, 135
Dutch people, 45
fish, 157
frogs, 147
girl and boy, 56
gondola, 190
Hindus, 44
non-figural, 240
pears, 173
penguins, 145
sprinkling cans, 240
strawberries in basket, 171
teapots, 207
windmill, 200
Octopi, Japan, 162
Oh! Ahhh!, unknown, 62
Ohio (Underplate), Aura, 25
Oil lamps,
Artmark, 225
metal, 225
plastic, 225
Oil well and North Dakota, Parkcraft, 237
Oil wells, glass, 195
Old King Cole (company), Magic Chef, 27
Old King Cole, Japan, 69
Old Salty and Cap'n Pepper
Our Own Import, 32
plastic, 33
wood, 33

Old woman and shoe
Japan, 69
Srp, 68
Olive Oyl and Popeye
Japan, 67
Vandor, 68
OMC, Happy Homer, 71
0nce more for old times sake, Japan,52
one-piece
alligator, unknown, 148
bears, Japan, 85
cabbage, Japan, 177
cactus, unknown, 164
carrots, Japan, 177
cats (black), Enesco, 20
chef and cook, unknown, 20
chicken, Japan, 128
china painting, unknown, 243
corn, Japan, 177
cow, Empress, 94
cow and calf, Japan, 94
cucumbers, Japan, 177
ducks, unknown, 20
dutch couple, unknown
fish, unknown, 160
fruit, unknown. 20
garbage disposal, plastic, 203
giraffe, Five and Dime, 107
hot peppers, Japan, 177
non-figural,
glass, 255
plastic, 246, 248
onions, unknown, 177
oriental man and urn, unknown, 20
peas, unknown, 177, 186
rhinoceros, Craftsman, 120
Styron, plastic, 248
Tappan stove, plastic, 201
zebra, Craftsman, 126
onion, Otagiri, 183
onions, one-piece, 177
onions and artichoke, Japan, 177
Onions in bag, Gifco, 183
0range and California, Parkcraft, 236
oranges
Artmark, 170
chalkware, 169, 170 ~
empress, 170
Honk Kong, 169
in crate, 169
Japan, 173
napkin set, 169
On metal tree, 169
plastic, 169
Taiwan, 169
with faces, 169
0riental lamps, Japan, 225
0riental shoes, unknown, 220
Otagiri
apple, 183
apples, 170
bears, 84
Onion, 183
strawberries, 173
Otters, unknown, 115
0ur Own Import
hamburgers, 222
mark, 222
01d Salty and Cap'n Pepper, 32
sacks, 186
Owl and pussycat, Fitz and Floyd, 89
owls
Agiftcorp, 141
Brinn's, 141
Goebel, 141
Japan, 141
Knobler, 141
metal 141
Shawnee, 141
Taiwan, 141
unknown, 141
Woodsy, 141
ox (Babe)
Japan, 67
Rosemeade, 67
unknown, 67
ox and covered wagon
Ceramic Arts Studio, 191
unknown, 191
oyster and Maryland, Parkcraft, 238
PAC
bears, 86
bullfighter and bull, 36
pack mule and Colorado, Parkcraft,236
pail and Vermont, Parkcraft, 238
Palomar
donkey, 103
elephants, 105
kangaroo, 111
mark, 103
palm tree and coconut or date, unknown, 163
palm trees, unknown, 84
pan with turkey, unknown, 140
parakeets
Ceramic Arts Studio, 142
nodder, 144
plastic, 144
Victoria, 144
Parkcraft
marks, 238
states, 19, 236-238
U.S. flag, 249
Parksmith
car, 189
Maserati, 189
parrots, unknown, 142
partridges, Holt Howard, 139
peas
Japan, 176
one-piece, 177
unknown, 177
peaches with faces, Japan, 171
peanuts
Mr. Peanut, 183-134
plastic, 184
with faces,
Japan, 183
pearls, Mapco, 12
pear and worm, unknown, 172
pears
chalkware, 173
Enesco, 172
Japan, 172 2
plastic, 172, 175
unknown, 172
with faces, Japan, 172
peas
as people, 185
Japan, 185
one-piece, 186
unknown, 185, 186
peek-a-boo babies, Regal/Van Tellingen, 53-54
pelicans
Exclusive, 136
Japan, 136, 137
Rosemeade, 136
penguins
Ceramic Arts Studio, 145
England, 145
Japan, 145
huggers, 145
Occupied Japan, 145
plastic, 145
unknown, 145
pepper mill and salt shaker, alabaster, 216
peppers
Artmark, 180
cherry, Japan, 176
Goebel, 185
hot, 175, 177
Japan, 175, 178, 181
one-piece, 175, 177
plastic, 185
Rosemeade, 181
sweet, 175, 178, 180, 181, 183, 185
unknown, 185
with faces, 183
perfume and lipstick, unknown, 205
pheasant and South Dakota, Parkcraft, 237
pheasants
Grindley, 138
Hors d'oeuvres, 138
Japan, 137
Lego, 137
nodders, 138
Norcrest, 2
Rosemeade, 138
Philgas, unknown, 194, 195
phonograph and record, Vandor, 205
photographer and camera, unknown, 36
piano and
bench,
Five and Dime, 228
NEC, 228
mouse, PY, 114
stool, Japan, 228
pianos, Davis Products, 205, 228
pickles, Arcadia, 182 unknown, 185
picnic table, Japan, 231
pie and rolling pin, unknown, 221
pig and
Arkansas, Parkcraft, 237
cow, Srp, 94
platter, unknown, 117
pig driving car, unmarked, 189
pigs
Ceramic Arts Studio, 116
chef,
Japan, 116
Enesco, 117
farmer, Shawnee, 116
Goebel, 116
Imperial Porcelain, 117
Japan, 116, 117
magnetic, 21
Miss Piggy and Kermit, Sigma,147
nodder, 116
on corn, 117
plastic, 116
pushbutton, 116
Regal, 115, 116
Shawnee, 115, 116
Smiley, Shawnee, 115
Srp, 117
Taiwan, 116
unknown, 116, 117
Van Tellingen, 115
Winnie, Shawnee, 115
with instruments, unknown, 152
Pilgrim Glass Company, 256
pilgrims
Elbee Art, 49
Japan, 49
pilgrim squirrels, Srp, 123
pillow for dog, cat, 98
Pillsbury Doughboy, unknown, 27
pimentos, Japan, 185
pineapple as chef, Grantcrest, 176
pineapples
Enesco, 168
Japan, 168
unknown, 168, 176
with faces, 168, 174
pinecones
Japan, 163
unmarked, 163
pine tree and Maine, Parkcraft, 238
Pinocchio
Enesco, 77
Goebel, 77
Japan, 77
Pinocchio and Cleo, Goebel, 77
pipe and
hat, unknown, 220
holder, Trevewood, 220
pipes, smoking
Fern, 240
Japan, 78, 232
pistol and Montana, Parkcraft, 237
pistols, metal, 234
pixie and
mushroom,
Ceramic Arts Studio, 79
Japan, 78
pixies
An Enterprise Exclusive, 78
as pipes, Japan, 78
as teapots, unknown, 78
at home, Japan, 80
chefs, unknown, 78
condiment, Japan, 14, 78
Enesco, 77, 79
Holt Howard, 77
Japan, 77, 78, 80
On rockets, Relco, 78
playing baseball, Japan, 77
Relco, 78 unknown, 77-80
with dice, Enesco, 77
with moveable eyes, unknown, 77
plane and square, unknown, 239
plastic
apples, 171
Bakelite (non-figural), 247
barbeque with glass chefs, 29
beans, 178
blacks, 21
blanket controls, 205
blenders, 203
bluebirds, 143
camel, 87
Campbell Kids, 27
car, 189
cash register, 206
cats, 92
chefs, 27, 28
children praying, 55
clothes pin with faces, 235 .~
coffee pot, 207
coffee pots, 209
croquet set, 219
dogs, 100
drums, 248
Dutch couple, 44
Fifi and Fido, 91
film containers, 216
flower pot and sprinkling can, 240
frog, 146
garbage disposal, 203
gas pumps, 18, 192-193
guitars, 226
hanging teapot, 208
hillbillies, 61
iceboxes,
ice cream soda, 224
Indians, 41
irons, 203
Kenmore appliances, 246
lawnmower, 19
lunch box, 216
Magic Chef, 27
mail box, 216
mammy and chef,
mixer, 203
monks, 32
Mr. Peanut, 183, 184
Mt. Rushmore, 199
non-figural, 248
oil lamps, 225
one-piece, 246, 24
onions, 183
orange napkin set, 169
oranges, 169
oyster and pearls, 12
parakeets, 144
pearls and oysters, 12
pears, 172, 175
penguins, 145

peppers, 185
pianos, 205, 228
pigs, 116
pushbutton, 21, 116
Red Lobster, 246
Seagram's, 234
sewing machines, 203
sheet metal rolls, 235
ship's wheel, 191
snow people, 235
soup cups, 65
sprinkling can and flower pot, 240
steins, 213
St. Labre Indian School, 248
strawberries, 172
toaster, 203
Tappan stove, 201
telephone, 205
television, 19, 205
vacuum sweepers, 203
Venus de Milo, 71
washing machine, 203
water lilies, 166
windmills, 200
witches, 76
with real seashells, 162
wooden shoes, 219
Pluto
American Bisque, 9
Good Company, 92
Japan, 92
Pluto and doghouse, Good Company, 92
Plymouth Rock, unknown, 234
Poinsettia Studios
black children, 29
fish, 157
pomeranians
Ceramic Arts Studio, 98
Goebel, 99
poodles, Ceramic Arts Studio, 98
posh poodles, Miyad, 99
pot and hot plate, Japan, 201
potatoes
baked, 186
Enesco, 186
Japan, 186
sweet, 186
unknown, 185, 186
praying children
Hong Kong, 55
Japan, 54-55
Lego, 55
Presley, Elvis, unknown, 70
prostitute and cop, metal, 32
pueblo and New Mexico, Parkcraft, 236
pumas, Rosemeade, 118
pump and tire, unknown, 193
pumpkin and ghost, Srp, 76
pumpkins, Japan, 186
Punxsutawney Phil, Mar Mar, 125
Purinton
jugs, 216
non-figural, 244
PY
flowers with faces, 166
mice, 113
seashells, 161
Pyrofax tanks, plastic, 194
quail
Red Wing, 138
Rosemeade, 138
queen and king of diamonds
unknown, 219
rabbit and carrot, Japan, 119, 179
Rabbit and Winnie-the-Pooh, unknown, 84
rabbit driving car, Japan, 189
rabbits
Ceramic Arts Studio, 118
Disney, 119
Enesco, 119
Goebel, 119
Japan, 119
luster, 119
Rosemeade, 118
Regal/Van Tellingen, 118, 119
Thumper, 119
race car and Indiana, Parkcraft, 237
raccoons
Gibson, 120
Japan, 119, 120
Rosemeade, 120
Victoria, 120
raccoon with garbage can, Japan, 119
radiators, metal, 189
radios, Vandor, 205
radio tubes, glass, 205
raft (base), Wisecarver, 24
Raggedy Ann and Andy, unknown, 68
Raisin Nodder Aura, 25
Raisins, Aura, 25
Razorback and Arkansas, Parkcraft, 237
Reach out to Other Collectors, Trish Ceramics, 24
rebel cap and Georgia, Parkcraft, 237
record and phonograph, Vandor, 205
Red Brand fence unknown, 18
Red Lobster, plastic, 246
redware
bartender, drinker and horse, 36
bears, 82
black cats, 11, 20, 89
chicken condiment, 129
dogs, 19, 102
gorillas, 82
Indians, 37
poodle spice set, 19
mammy and cook heads, 29
palm trees, 163
stacked condiment, 129
stove, 201
teapots, 209
whales, 124
Red Wing, Bob White pattern, 138
Reebs, black couple, 51
refrigerator and
chef, unknown, 27
woman, Clay Art, 65
refrigerators, Aluminum Housewares, 202
Regal China
A Nod to Abe, 25
bears, 85
boy and dog, 59
boy and girl, 34
dog and boy, 59
ducks, 135
girl and boy, 34
Goldilocks, 68
Lincoln, Abraham, 25
love bugs, 17, 150
mermaid and sailor, 33
Old MacDonald boy and girl, 34
peek-a-boo babies, 53-54
pigs, 115, 116
rabbits, 118, 119
sailor and mermaid, 33
Tweedle Dum and Tweedle Dee, 230
Regaline Dutch napkin set, 44
Reindeer,
Rudolph the Red-Nosed, Japan, 74
unknown, 74
Relco
donkey, 103
flowers in basket, 165
guitars, 226
horseheads, 110
pixies on rockets, 78
squirrels, 122
zebras, 126
revolver and Montana, Parkcraft, 237
rhinoceros, Craftsman, 120
Rite Lite Corporation, light bulbs, 13
riverboats, Japan, 191
roaster and turkey, Japan, 223
Robinson Crusoe, Japan, 69
robin with babies, China, 143
rock bass, unknown, 155
rocket and boy, Enesco, 59
rocking chairs, Japan, 231
rocking horses, unknown, 110
rocking zebras, unknown, 126
rolling pin and pie, unknown, 221
rolling pins, wood, 214
Ron Gordon Designs gifts, 216
roosters, *(see: chickens)*
Rosemeade
Babe and Paul
bears, 83
begging dogs, 96
bison, 86
bisque color, 88, 136
buffalo, 86
Bunyan, Paul and Babe, 67
cabbage, 181
cats, 88
cattle, 92
chickens, 130
color of bisque, 88, 136
corn, 181
cucumbers, 181
deer, 95
dog heads, 96
dogs, 96
ducks, 96
elephants, 104
fish, 3, 155-156
flamingoes, 136
flickertails, 108
flowers, 167
goats, 108
goldfish, 3
horses, 109
mallards, 140
mark, 88
mice, 113
mountain goats, 108
mountain lions, 118
non-figural, 244
Paul Bunyan and Babe, 67
pelicans, 136
peppers, 181
pheasant hors d'oeuvres, 138
pheasants, 138
pumas, 118
quail, 138
rabbits, 118
raccoons, 120
running rabbits, 118
sailboats, 190
skunks, 121
songbirds, 143
swans, 135
swordfish, 3, 159
tulips, 167
turkeys, 140
wolves, 125
zebras, 126
roses, 165
Royal
ducks, 133
Indian figural nodders, 40
mark, 40
Royal Canadian Mounted Police, unknown, 32
Royal Winton
mark, 244
non-figural, 244
rub a dub dub condiment, unknown, 68
Rubicon barbecue grill and chefs, 29
Rudolph and Santa
Japan, 74
unknown, 74
running rabbits
Ceramic Arts Studio, 118
Rosemeade, 118
sacks, Our Own Import, 186
Saf-T-Cup, unknown, 223
sailboats, Rosemeade, 190
sailor and
Bo Peep, Shawnee, 67
girlfriend, Japan, 33
mermaid, Regal/Van Tellingen, 33
sailors, 33
St. Labre Indian School plastic, 248
Sairey Gamp and Mr. Micawber, Beswick, 70
salt and pepper sacks, Our Own Import, 186
S & P Club, hands, 66
Sandeman Brandy, Wedgwood, 71
saucer and cup, Japan, 16
Sarsaparilla *(see also: Five and Dime, and SDD)*
baby and nodder baby, 53
Empire State Building and King Kong, 81
Hollywood sign, 200
King Kong and Empire State Building, 81
mark, 82, 200
musical notes, 228
nodder baby and non-nodder baby, 53
zebras, 126
scarecrows, unknown, 233
school books and apple, Srp, 171
scroll and mortar board, unknown, 233
SDD *(see also; Five and Dime, and Sarsaparilla.)*
giraffes, 107
Spuds McKenzie, 101
sea captain
plastic, 33
wood, 33
Seagram's, unknown, 234
seahorses
Goebel, 162
Japan, 162
seahorse and coral, cat, 162
seals
Elbee Art, 120
Gibson California Arts, 121
Harper Pottery, 120
Japan, 120, 121
New....land Ceramics Corpora
Norcrest, 121
sea otters, unknown, 115
Sears
frogs, 147
seashells
Japan, 161
Kenmar, 161
PY, 161
unknown, 162
sewing machine
metal, 204
plastic, 204
Seymour Mann
mark, 124
walruses, 124
Shafford
black cats, 89
boy and girl, 57
cats, 89
chefs condiment, 27
mark, 27
Shakespeare's house, Beswick condiment, 14
Shawnee
Bo Peep and sailor, 67
cats, 88
chanteclear roosters, 127
chickens, 127
dogs, 97
ducks, 141
Dutch people, 44
farmer pig, 116

fill hole, 127
flowers in pots, 166
jugs, 215
milk cans, 209, 240
Mugsey, 97
non-figural, 244
owls, 141
pigs, 115, 116
Puss 'n Boots, 8
pots with flowers, 166
sailor and Bo Peep, 67
size of fill hole, 127
Smiley pig, 115
sprinkling cans, 240
sunflowers, 10, 167
Swiss children, 47
wheelbarrows, 240
Winnie Pig, 115
sheep, cas, 121
sheet metal rolls, plastic, 235
shells
glued to glass, unknown, 249
Japan, 161
Kenmar, 161
Mapco, 12
oyster and pearls, 12
PY, 161
unknown, 162
ship, unknown, 189
ship's wheel plastic, 191
shoe and
gnome, unknown, 79
old woman, Srp, 68
shoes
Japan, 219
oriental, unknown, 220
with cats, Lego, 90
wooden,
plastic, 219
wood, 220
shopping couple, Japan, 51
shotgun shells, unknown, 234
shutterbug, Enesco, 37
Sigma, Kermit and Miss Piggy, 147
silver, 251
skillet and fish, unknown, 223
skillets, unknown, 222
skull and cactus,Vandor, 125
skull nodders, A Quality Product, 17
skunks
Ceramic Arts Studio, 122
Flower, 119
Goebel, 119, 121
magnetic, 21
on stumps, 122
Rosemeade, 121
Thrift Ceramics, 122
unknown, 122
slice of cake and cake, unknown, 223
Smiley pig, Shawnee, 115
Smokey the Bear,
Empress, 85
Japan, 85
Norcrest, 85
Twin Winton, 85
Smoking pipes, Fern, 240
Smoos, Japan, 230
Snails
Enesco, 153
Fitz and Floyd, 153
Japan, 153
Meridan, 153
snail shells, unknown, 161
Snap and Pop (Kellogg's), Japan, 71
snow babies, Srp, 54
snowman, Taiwan, 236
snowman and New Hampshire, Parkcraft, 238
snow people
Japan, 235
Lefton, 235
miniatures, 235
plastic, 235
Srp, 235
Snow White and Dopey, unknown, 6
snuff bottles, plastic, 213
socket couple, Japan, 62
soldier (Civil War and cannon),Japan, 32
soldiers
English toy, Goebel, 32
Japan, 32
soup bowl and chicken, Vandor, 130
soup cups, F & F, 65
Southern Potteries
chickens, 127, 128
flowers, 167
speckled trout, unknown, 155
spice set (poodle), Japan, 19
spinning wheel and grinder, Enesco, 206
Spin Original
mark, 214
mugs, 214
spoons and forks as people,
Japan, 221
unknown, 221
sprinkling can and flower pot, plastic, 240
sprinkling cans
Japan, 240
Occupied Japan, 240
Shawnee, 240
Spuds McKenzie, Five and Dime, 101
square and plane, unknown, 239
squash, unknown, 182
squash with faces, Japan, 186
squirrel and nut
Goebel, 122
Japan, 123
unmarked, 123
squirrel and stump, Japan, 123
squirrels
Elbee Art, 122
Goebel, 122
Inarco, 123
Japan, 122, 123
pilgrim, Srp, 123
Relco, 122 Srp, 123
Thrifco, 122
unknown, 123
Srp, Sandy
alligators, 148
angels, 75
bears, 85
black babies, 54
buffalo and Indian, 38
cactus and donkey, 103
cat and dog, 91
cow and pig, 94
dog and cat, 91
donkey and cactus, 103
ducks, 133
eskimo and igloo, 10, 48
ghost and pumpkin, 76
horses, 109
igloo and eskimo, 10, 48
Indian and buffalo, 38
old woman and shoe, 68
pig and cow, 94
pigs, 117
pumpkin and ghost, 76
Santa and Mrs.. Claus, 73
shoe and old woman, 68
snow babies, 54
snow people, 235
squirrels, 123
stage coaches, Japan, 192
Stanfordware, corn, 180
Starke Design
fruit bowl, plastic, 174
stove, plastic, 201
states
Klay Kraft, 238
Milford, 238
Parkcraft, 238
Statue of Liberty and New York State, Parkcraft, 237
steamboat and Mississippi,
Parkcraft, 237
steins, Japan, 213
Sterling
banjos, 228
non-figural, 251
Stoney Mouse, Enesco, 113
stool and piano, Japan, 228
stove
Clover, 201
Enesco, 202
Japan, 201
Lego, 201
Lugenes, 201
metal, 202
plastic, 201
Starke Design, 201
Tappan, 201
unknown, 202
wood, 202
wood burning, 202
stove and
coffee pot, unknown, 202
woodpile, unknown, 202
Strasburg Railroad conductor, 33
strawberries
Arcadia, 171
chalkware, 160
Holt Howard, 171
in basket, Occupied Japan, 171
in bowl with creamer, Japan, 173
on metal tree, Japan, 169
Otagiri, 171
unknown, 171
with faces, 171
straws and ice cream soda, Napco, 224
Streetcar Named Desire, A, unknown, 188
streetcars, metal, 188
strutting pheasant, Rosemeade, 138
street lights, wood, 225
stump and
hatchet, 239
monkey, 15
squirrel, 123
stumps with skunks, unknown, 122
Styron, non-figural, 248
suit,case and alligator, Vandor, 149
suitcases, Japan, 65, 232
sultan and harem girl, Ceramic Arts Studio, 43
sundae, unknown, 223
sunbathers
Japan, 63
three-piece sets, 62-63
unknown, 62-63
with sombrero, 63
Sunday Dinner, Wisecarver, 129
sunfish, Japan, 155
Sunset, Dithridge, 253
Sunshine Biscuit chefs, unknown, 26
Superlon
non-figural, 248
teapot, 208
surfer and wave, unknown, 36
Sutton Ware, kangaroos, 111
Suzette on pillow, Ceramic Arts Studio, 98
swans
Japan, 135
Rosemeade, 135
unknown, 136
with flowers, 135
sweet potatoes, Japan, 186
Takahshi, mugs, 214
Tambo and Tembino, Ceramic Arts Studio, 104
teakettles, unknown, 209
teapot, plastic, hanging, 208
teapots
as people, 209
metal, 219
Occupied Japan, 207
on bull, 94
with faces, 209
teardrop, Tiffin Glass, 256
Tee and Eff, unknown, 70
teepee with Indian boy, Japan, 38
telephone, Japan, 230
telephone and book
Japan, 230
unknown, 16, 205
telephones, plastic, 205
television
plastic, 19, 205
with snow, 205
Tembino and Tambo, Ceramic Arts Studio, 104
tender and locomotive
Charls Products, 187
Japan, 187
metal, 188
unknown, 187
Thai and Thai-Thai, Ceramic Arts Studio, 88
Thames, donkey, 103
then and now refrigerators, Aluminum Housewares, 202
Thrifco Ceramics
jugs, 215
mark, 122, 215
skunks, 122
squirrels, 122
Thumper eyes closed, Goebel, 119
Thumper eyes open, Goebel, 119
thundermug and boy, unknown, 61
thundermugs
for toilet condiment, 231
Japan, 231
unknown, 231
Tiffin Glass Company, teardrop, 256
tiger
one-piece, 124
pushbutton, 124
Whirley Industries, 124
tiger and Sambo, Ceramic Arts Studio, 6
tigers in drums, Japan, 124
tire and Ohio, Parkcraft, 236
tire and pump, unknown, 193
toaster
Knobler, 203
metal and plastic, 203
plastic, 203
Tucker Products, 203
unknown, 203
U.S.A., 203
toaster and waffle iron, unknown, 223
toilet and black boy, Japan, 53
toilet condiment, Japan, 231
tomatoes
as people, Japan, 175
condiment, 186
Japan, 176
unknown, 182, 186
toothbrush and toothpaste, unknown, 231
top hat, unknown, 220
totem pole and Indian, Japan, 41
trailer and car
Aura, 24
unknown, 187
Treasure Craft
fish, 160
seahorses, 162
tree and
hula girl, Japan, 49
Paul Bunyan, Ceramic Arts Studio, 67
trees, Christmas, unknown, 7
Trevewood, pipe and holder, 220

Trico China, butler, 16
Trice Brothers, English cottage condiment, 197
tricycle with bears, unknown, 84
Trish Ceramics, hand and globe, 23
trucks, unknown, 188
trumpets,Japan, 228
tugboat and captain, unknown, 33
tulips
Enesco, 165
Japan, 165
Rosemeade, 167
turkey in pan, unknown, 140
turkey and roaster Japan, 223
turkeys
Chase, 139, 140
Enesco, 140
Japan, 139, 140
Rosemeade, 140
unknown, 139, 140
turnip and cabbage, unknown, 185
turtles, Japan, 148
Tweedle Dum and Tweedle Dee, Regal, 231
Twin Winton, 2 monks, 31
typewriter and ink bottle, El Ranchito, 190
UCAG, corn, 178
umbrella and alligator, Vandor, 149
umbrellas on stand, plastic, 220
U.S. Army condiment, unknown, 242
U.S. flags, Parkcraft, 249
U.S. Navy non-figural, unknown, 249
vacuum sweeper
E.W. Lowe, 203
plastic, 203, 204
Vallona Starr
1ady bugs, 151
mark. 151
Vandor
alligator and suitcase, 149
alligator and umbrella, 149
buzzard and cactus, 142
cactus and,
buzzard, 142
coyote, 125
skull, 125
cow in chair, 94
chicken and soup bowl, 130
coyote and cactus, 125
gas station attendant and pump, 34
Jetsons, 66
jukebox, 229
leopards, 112 mark, 66
Popeye and Olive Oyl, 66
radios, 205
record player and record, 205
skull and cactus, 125
snake nodder, 17
suitcase and alligator, 149
umbrella and alligator, 149
Van Tellingen
bears, 85
boy and dog, 59
dog and boy, 59
ducks, 135
love bugs, 17, 150
mermaid and sailor, 33
peek-a-boo babies, 53-54
pigs, 115
rabbits, 118, 119
sailor and mermaid, 33
Venus de Milo, plastic, 71
Vernon Kilns
Hop and Lo., 182
mark, 182
mushrooms, 182
V.G.
canning jars, 210
ladybugs, 150
Victor Goldman
beans, 178
Victoria
beavers, 87
carrots with faces, 179
fish, 154
Indian drums, 38
Indian nodders, 40
monkeys, 82
nodder ducks, 133
racoons, 120
victrola and clock, unknown, 101
violin and guitar as people, unknown, 227
violins with hands, Japan, 227
vulture and cactus, Vandor, 142
Wade
bramble condiment, 167
mark, 167
waffle iron and toaster, unknown, 223
wagon with Dutch couple, A Quality Product, 45
wagon wheels, Frankoma, 192
wagons and horses, wood, 189
Wales
cats, long black, 89
mark, 89
Wally and Windy, Japan, 67
walnuts, black, 224
walruses
Japan, 124
Seymour Mann, 124
Taiwan, 124
unknown, 124
washing machine, plastic, 203
Washington Nonument, Bakelite, 199
Washington State and apple, Parkcraft, 19
watches, Brillium Metals Corporation, 231
watermelon and
Alabama, Parkcraft, 237
black children, unknown, 60
black head,
chalkware, 59
Japan, 59
grasshoppers, Japan, 152
whole and slice, Japan, 176
watermelons as people, Japan, 174
water wheel and windmill, Japan, 200
wave and surfer, unknown, 36
Wavecrest, billow pattern, glass, 7
Weary and doghouse, unknown, 99
Webb, Paul, Ma and Doc,Imperial Porcelain, 34
Wedgwood
blue jasper, 244
mark, 71
Sandeman Brandy, 71
Wendy and Woody Woodpecker, Napco, 145
whale and Jonah, unknown, 69
whales
Effco, 124
Japan, 124, 125
Kelvins, 159
unknown, 124, 125
wheat and Kansas, Parkcraft, 236
wheelbarrows, Shawnee, 240
Whirley Industries, tiger pushbutton, 124
whiskey jug and Kentucky, Parkcraft, 238
Willing and Valentine Boy Lover, Ceramic Arts Studio, 55
windmill and water wheel, Japan, 200
windmills
Japan, 200
Occupied Japan, 200
plastic, 200
unknown, 200
Windy and Wally, Japan, 67
wine corks, France, 210
Winnie pig, Shawnee, 115
Wisecarver
Afloat on the Mississippi, 24
black busts, 52
chickens, 129
mark, 52
raft with boys, 24
Sunday Dinner, 129
witch and kettle, Fitz and Floyd, 76
witches,
Fitz and Floyd, 76
JSNY, 76 plastic, 76
wolves, Rosemeade, 125
woman and
refrigerator, Clay Art, 65
chair,
Norcrest, 52
unknown, 52
granddaughter, Japan, 52
woman with man grabbing breast, Japan, 61
wood
apples, 170
birds, 142
bottles, 213
bowling pin and ball, 21
bullets, 232
cats, 89, 90
chefs, 29
chopping blocks, 231
coffee pots, 13, 214
cups, 214
decks of cards, 219
golf balls, 218
gravestones, 232
horses and wagon, 189
kettles, 214
mugs, 13, 214
musical shakers, 229
non-figural, 13, 250
nuts in tray, 175
pots, 213
rolling pins, 214
stove, 202
street lights, 225
teapots, 207
wagon and horses, 189
wooden shoes, 219
woodchuck
Mar Mar, 125
Punxutawney Phil, 125
wooden shoes
plastic, 219
wood, 219
wood pile and stove, unknown, 202
Woodsy Owl, Japan, 141
Woody Woodpecker, unknown, 145
Woody and Wendy Woodpecker, Napco, 145
worm and
apple, unknown, 119
pear, unknown, 172
wrench and bolt, unknown, 218, 239
wringer washer, plastic, 203
yarn ball and cat
Japan, 90
unknown, 90
yarn balls with granny base, unknown, 52
zebras,
Ceramic Arts Studio, 125
Craftsman, 126
Japan, 110
Relco, 126
rocking with boy and girl, unknown, 126
Rosemeade, 126
Sarsaparilla, 126
unknown, 126
with musical instruments, Japan, 126

The Complete Salt and Pepper Shaker Book
Price Guide

Determining realistic prices for salt and peppers is an inexact science at best. Age, condition, rarity, desirability, size, decoration, presence or lack of gold trim, manufacturer, importer, country of origin, crossover competition, the state of the economy, and geographical area all come into play. Additionally, values can change very rapidly. Prices of Rosemeade shakers, for example, have increased significantly over the past twelve months, while prices of Occupied Japan shakers appear to have fallen some during the same period.

Values were arrived at by consulting both hobby-specific and general price guides and ads in trade papers; from prices noted at antique malls, shows, and flea markets; and through conversations with collectors and dealers. Auction prices, which are often the result of emotion instead of logic, were not considered.

Results from observations made at shows, malls and flea markets were allowed the greatest influence because these seem to represent the real world more accurately than printed listings including, incidentally, the printed listing below. But I wish to stress that actual tag prices were seldom used as more often than not they are inaccurate. First, there is the "dickering" factor most dealers build in, setting their prices artificially high in order to have some room to come down. Second, many dealers lack the knowledge necessary to realistically assess shaker values.

Also taken into account was how quickly shakers sold, or if they sold at all. For example, if a pair was purchased as soon as a dealer put them out, a note was made and additional research conducted on the assumption the pair was probably priced well below its true market value. Most often this proved to be the case. Pairs that sat there week after week and month after month turned out to be a bit different. In the cases of known manufacturers such as Shawnee and Ceramic Arts Studio, it was generally found that the prices were indeed above the market. With more generic shakers, such as those of most importers and those made in Japan, however, just as often as not the problem wasn't overpricing but simply that the shakers were undesirable. Shakers such as these—plain, simple, poorly painted or generally just not well executed—have been assigned very low values in the price guide.

In several places you will notice the words, *price not determined.* This should not be taken as an indication of rarity or high value. It simply means that at the time I compiled the price guide I lacked sufficient information to determine a realistic value.

Where you see *expensive,* however, this does indicate rarity along with a very high value. In these instances reliable comparative pricing information was too scant to quote a viable price. Since my job is to reflect prices, not set them, no attempt was made to give a dollar-and-cents figure.

Finally, as its name implies, this price guide is exactly that, a guide. Because the market is constantly changing, the best way to use the guide would probably be to compare the value of one set of shakers relative to the value of another, instead of relying on exact dollar amounts. I would also implore you to compare the prices below to those quoted by Ms. Davern and Ms. Guarnaccia in their books, and make your own determination, as the three of us differ greatly on some shakers.

Page	Item	Price
Title page		
	w/gold	$110
	w/o gold	$75
2	wings up	$10
	tails down	$6
	tails curled	$12
3	swordfish	$100
	other	$60
4		$12 each pair
6		$8
7		$200
8	Hamm's	*price not determined*
	ink	$14
10	large sunflower	$55
	small sunflower	$25
	eskimo	$35
11	Adam & Eve	$6
	cats sitting	$20
	cats in shoes	$12
	poodles	$30
12	dogs	$8 each pair
	oyster	$20
13	cabin	$0
	cans	*price not determined*
	Ford rotunda	$110
	wood souvenirs	
	coffee pots	$2
	mugs	$2
	oak & walnut	$8
	beer bottles	
	left	$6
	right	$12
	light bulbs	
	large brown	$12
	small brown	$8
	clear	$10
14	advertising shakers	
	Morton	$0.50
	Northwest	$12
	Shakespeare's house	$38
	Pixies	$45
15	condiment	$55
	monkey	$12
	crocodile	$28
16	CAS animals	
	bears	$32
	kangeroos	$55
	gorillas	$45
	butler	$550
	hangers	$12
	teacup	$14
	phone	$24
17	love bugs	$60
	skulls	$38
	elephant	$225
	snake	$20
18	Esso	$26
	Nipper	$48
	Harvestore	$50
	Red Brand	$45
	kissers	
	eskimo	$14
	cowboy	$14
19	mower	$24
	TV	$14
	Washington	$18
	drummers	$22 each set
	poodle spice set	$18
20	black cats	
	long one-piece	$28 each
	short	$20
	one-piece	
	fruit	$20
	cooks	$22
	Dutch	$22
	ducks	$16
	china man	$22
21	pushbutton	$32
	magnetic	$14 each set
	frogs	$12
23	single	$15
	hand & globe	$18
24	Raisin Nodder	$23
	Ohio	$28
	boys on raft	$33
25	Lincoln	$40
26	chefs	$14
27	Tappan chefs	$8
	condiment chefs	$38
	Magic Chefs (figural)	$14
	Magic Chefs (milk glass)	$12
	Pillsbury doughboys	$16
	Campbell Kids	$65
	Hot N Kold	$35
	refrigerator & chef	$30
28	top left	
	left	$10
	center	$16
	right	$10
	center left	
	left	$14
	center	$12
	right	$65
	bottom left	$18
	bottom center	$12
	top right	$10
	center right	$10
	bottom right	$10
29	top left	$10
	center left	
	left	$5
	right	$4
	bottom left	
	inside	$60
	outside	$55
	top center $45	
	top right $100	
	center right	
	w/stove	$120
	w/o stove	$90
	bottom right	$75
30	top left	$60
	center left	$18
	bottom left	$70
	top center	
	rear left	$95
	rear right	$95
	front	$65
	center	
	inside	$95
	outside	$80
	top right	$28
	center right	$130
	bottom right	
	left	$85
	right	$110
31	top left	
	inside	$80
	outside	$100
	center left	$18
	bottom left	$150
	center	
	stove set	$45
	table set	$26
	top right	
	green (w/stripe)	$260
	red	$85
	bottom right	
	shakers	$24
	cookie jar	$45
32	miniature monks	$10
	Goebel monks	
	w/instruments	$60
	w/books	$65
	Friar Tuck	$35
	Cardinal Tuck	$95
	mounties	$22
	bobbies	$20
	cop & girl	$12
	North-South soldiers	$16
	rebel & cannon	$14
	toy soldiers	
	Goebel	$50
	Japan	$10
	Old Salty & Cap'n	$12
33	Old Salty & Cap'n(full figure)	$8
	Old Salty & Cap'n (busts)	$6
	sailor figures & busts	
	figures	$14
	busts	$28
	Capt. Bob-Lo	$45
	sailor & mermaid	$125
	conductor singles	$6 each
	true pair	$18
	capt. & tug	$18
	Lenny Lennox	$75
34	porters	
	left	$50
	right	$125
	gas pump	$18
	bottom left	$12
	top center	$75
	Regal Old MacDonald	$85
	Amish woman	$10
	Ma & Doc	$110
	chimney sweeps	$45
35	cowboy	$18
	maid & bellhop	$30
	Deary & clowns	
	Deary	$14
	clowns	$12
	German clowns	$30
	top right	
	inside	$10
	outside	$10
	center right	
	left	$18
	center	$16
	right	$10
	CAS clown & dog	$75
36	top left	
	photog & camera	$16
	single camera	$6
	top bullfighter	$26
	bottom bullfighter	$22
	condiment bullfighters	
	left	$85
	right	$75

Page	Item	Price
	horse's patoot	$12
	boxers	$10
	surfer	$22
37	top left	$28
	center left	$12
	bottom left	
	Indian drummer	$26
	Indians in canoe	$28
	top right	
	fisherman	$14
	hunter	$14
	golfer	$16
	center right	
	single	$6
	true pair	$14
	bottom right	
	Indians in canoe	$24
	redware Indians	$12
38	top left	
	Hiawatha & Minehaha	$170
	Wee Indians	$32
	center left	$12
	teepee	$28
	brave & horse	$28
	Indian & buffalo	$35
39	top left	
	left	$14
	center	$12
	right	$14
	Indians in moccasins	$16
	Indians or Eskimos	$22
	bottom left	$16
	top right	$14
	right center (3 Indians)	
	left	$14
	center	$14
	right	$8
	right center (2 Indians)	
	left	$18
	right	$18
	bottom right	
	Indians	$18
	cowboys	$18
40	top left	
	chalkware	$12
	ceramic	$20
	nodders—	
	rectangular bases	$45 each set
	round bases	$65 each set
	top right	
	left	$16
	center	$16
	right	$12
	center right	
	left	$14
	center	$16
	right	$16
	bottom right	
	left	$16
	right	$24
41	top left	
	left	$14
	right	$16
	center left	
	inside	$14
	outside	$16
	bottom left	$24
	top right	$32
	center right	$3
	bottom right	$5
42	top left	
	rear left	$10
	center	$6
	right	$8
	front left	$8
	right	$8
	bottom left	$28 as is, $48 w/lid
	top right	
	wee Chinese	$18
	large Chinese	$28
	bottom right	*price not determined*
43	top left	$12
	bobbing heads	$30
	orientals w/underplate	$14
	bottom left	
	outside	$10
	inside	$10
	top center	$12
	center	
	inside	$10
	outside	$24
	CAS Hindu boys	$100
	top right	$85
	snake charmer	$110
44	top left	$14
	Shawnee Dutch (4-3/4 inches)	
	w/gold	$60
	w/o gold	$40
	Shawnee Dutch (5-1/4 inches)	
	w/gold	$45
	w/o gold	$95
	bottom left	$32
	center	$22
	top right	$14
	center right	$8
	bottom right	
	left	$22
	right	$14
45	top left	$45
	center left	
	CAS wee Dutch	$22
	CAS imposters	$10
	top center	$35
	top right	$12
	three Goebel sets	
	inside	$35
	outside	$30
	front	$30
	Dutch w/sugar	$32
	bottom	
	left	$10
	right	$10
46	top left	
	left	$14
	middle	$12
	right	$18
	center left	
	left	$14
	right	$14
	top right	
	CAS wee Scotch	$25
	CAS wee Swedish	$30
	CAS wee French	$30
47	Shawnee Swiss	
	w/gold	$60
	w/o gold	$35
	bottom left	$16
	top right	$22
	Carribean Islander	$14
	right center	
	Spanish couple	$14
	Scottish couple	$14
	unmarked Spanishcouple	$22
48	top left	$8
	three pair of Mexicans	
	left	$6
	center	$10
	right	$6
	two pair of Mexicans	
	left	$6
	right	$6
	bottom right	$18
	top center	$12
	two eskimo sets	
	left	$10
	right	$14
	top right	$30
	Srp Eskimos	$35
	Srp Eskimo & igloo	$30
	bottom right	$38
49	top left	
	left	$10
	center	$12
	right	$10
	Eskimos in vessel	$22
	Amish folks	
	left	$6
	center	$6 each pair
	right	$6
	pilgrims	
	left	$10
	right	$10
	hula girls	$16
	hula girl & palm	$14
50	top left	$14
	bottom left	$16
	top right	$12
	bottom right	$12
51	top left	
	left	$12
	center	$14
	right	$12
	Niko China couple	$18
	what's his is hers	$18
	four-eyed nodders	$45 each set
	top right	$22
	made for each other	$14
	Reebs black couple	$150
52	top left	$26
	bench sets	
	left	$16
	right	$16
	bottom right	$16
	granny & yarn	$22
	four busts	
	left	$10
	right	$10
	five seated ladies	$16 each pair
	top right	$8
	granny & granddaughter	$14
53	top left	$18
	Peek-A-Boo Babies	
	small	$175
	large	$375
	top right	
	rear inside	$16
	rear outside	$16
	front	$8
54	top left	$95
	Srp snow babies	$35
	girl & boy	$14
	tyke in cradle	$18
	top center	$16
	Clay Art babies	$15
	Enesco heads	$10 per pair
	candy cane babies	$18
	Srp babies	$40
	praying kids	$14 each pair
55	top left	$14 each pair
	center left	
	left	$14
	center	$8
	right	$14
	bottom left	$12
	bottom center	$12
	Goebel kissers	$35
	CAS willing & boy	$80
	CAS bashful & boy	$80
	CAS kids in chairs	$45 each set
56	flower children	$12
	Hummel-like	$12
	basket set	$14
	spring set	$12
	top right	$16
	OJ boy & girl	$14
	Enesco boy & girl	$12
57	Amish kids	$14
	farmer boy & girl	$8
	New York World's Fair	$16
	bottom left	$10
	reversible boy & girl	$16
	Goebel flower girls	$55 each set
	top left	$14
	Shafford singles	$7 each
58	golden girls	
	w/tray	$18
	w/o tray	$12
	mother's little helper	$12
	bottom left	$20
	black boy on toilet	$35
	Miss Cutie Pie	$20
	Goebel	
	condiment set	$175
	shakers	$75
59	top left	$65
	black boys	$10
	Van Tellingen huggers	
	left	$50
	right	$65
	center	$16
	boy-wagon-dogs	$35
	CAS blacks	
	gator & black boy	$140
	elephant & black boy	$150
	future astronaut	$14
	watermelon sets	
	left	$50
	right	$65
60	watermelon kids	$100
	bums w/neckties	
	inside	$18
	outside	$12
	graduates	$12 each pair
	bum children	$10
	hillbilly & jug	$14
61	top left	$10
	center left	$12
	bottom left	$14
	center	$10
	top right	$10
	center right	$6
	bottom right	
	rear left	$12
	rear center	$8
	rear right	$8
	front	$8
62	Oh!-Ahhh	$22
	socket couple	$18
	sunbather	$12
	boy & thunder mug	$14
	OUCH!	$32
63	top left	$10
	center left	
	rear left	$14
	rear right	$14
	front left	$10
	front right	$12
	bottom left	
	left	$10
	right	$10
	top right	$10
	center right	$14
64	Adam & Eve	$8
	bottom left	$10
	top right	$16
	bottom right	$10
65	woman & refrigerator	$16
	seated cannibal & dinner	$18
	standing cannibal & dinner	$14
	top right	$12
	soup cups	$45
	luggage and feet	
	luggage	$12
	feet	$2
	bottom right	
	left	$4
	center	$6
	right	$4
66	top left	
	left	$2
	center	$4
	right	$2
	legs	$6
	Jetsons	*expensive*
	Popeye & Olive Oyl	$40
67	top left	$80
	Wally & Windy	$40
	Rosemeade Paul Bunyan	$75
	bottom left	$30
	top right	
	rear left	$35
	rear right	$30

Page	Item	Price
	front	$30
	CAS Paul & tree	$65
	Shawnee Bo Peep & sailor	
	w/gold	$50
	w/o gold	$20
68	Hull Little Red Riding Hood	
	3-1/4 inches	$50
	4-1/2 inches (not shown)	$400
	5-1/2 inches (not shown)	$110
	Japan LRRH	$30
	Regal Goldilocks	
	left	$135
	right	$135
	Snow White & Dopey	$60
	Srp old woman & shoe	$30
	bottom	$25
	Black Sambo	$350
	Mowagli	$80
	bottom right	$75
69	top left	
	old woman	$22
	Jack Horner	$22
	center left	
	Robinson Crusoe	$22
	Bobby Shaftoe	$22
	bottom left	$22
	top center	$24
	Jonah & whale	$85
	Nixon & Agnew	$15
	Marilyn Monroe	$22
	JFK	$45
	Marilyn Monroe cake	$22
70	top left	$35
	Laurel & Hardy	$110
	Gamp & Micawber	$40
	bottom left	$20
	Dempsey & Tunney	$24
	Blue Bonnet Sue	$22
	Tee & Eff	$28
71	top left	
	shakers	$45
	decanter	$50
	Big Boy & burger	$25
	Japan Big Boys	$175
	Happy Homer	$35
	Snap & Pop	$28
	Maggie & Jigs	*expensive*
	Venus	$10
	Father Time	$18
	Father Time & baby	$16
72	top left	
	left	$12
	center	$10
	right	$10
	right center	
	inside	$10
	outside	$10
	bottom right	$20
73	top left	
	left	$10
	center	$12
	right	$10
	center left	
	in chairs	$14
	right	$12
	bottom left (Noel)	$14
	Srp Santa & Mrs. Claus	$35
	Claus's on skis	$16
	top right	
	left	$12
	right	$14
	napkin set	$16
74	top left	$10
	center left (3 pair)	
	unmarked heads	$10
	Santa & bag	$12
	rocking horse	$10
	center left (1 pair)	$4
	bottom left	
	chalkware Santas	$4
	Japan Santas	$10
	top center	$16
	top right	$45
	center right	
	inside	$12
	outside	$10
	bottom right	$18
75	angel & bow	$30
	center left	
	elves	$10
	angels	$16
	bottom left	$8
	top right	$10
	center right	$12
	bottom right	$10
76	ghosts	
	left	$10
	middle	$10
	right	$8
	ghost & pumpkin	$25
	witch & pot	$16
	three mermaids	
	left	$14
	middle	$16
	right	$14
	four mermaids	
	left rear	$16
	center left	$16
	center right	$14
	right	$16
77	Pinocchio	
	outside	$40
	inside	$40
	front	$60
	center left	$12
	bottom left	*price not determined*
	top right	
	left	$8
	right	$8
	moveable eyes	$12
	pixies & dice	$10
	Holt Howard pixies	$8
	pixies & vegetables	$12
78	play ball!	$16
	pixies cooking	$16
	pixie & mushroom	$14
	pipe pixies	$8
	teapot pixies	$14
	rocket pixies	$20
	cue ball pixies	$8
	kissing pixies	*pr. not determined*
	Each shaker is 3-1/8	$8
	pixie condiment	$32
79	top left	$8 each pair
	center left	
	left	$8
	right	$10
	oak sprites	
	left	$28
	right	$28
	pixie & mushroom	$38
	Gnome in shoe	$16
	bottom right	$8
80	Keebler elf	$14
	brewery gnomes	$18
	pixies at home	$12
	pixies w/feathers	$8
	elf & eggs	$20
	bottom right	$8
81	Goebel monkeys	$35
	top right	$10
	King Kong	$18
82	macho ape	$10
	center left	
	monkey sitter	$18
	monkey hanger	$12
	bottom left	
	bears	$32
	monkeys	$32
	monkey & bananas	$10
	top right	$6
	armadillos	$14
	bottom right	$12
83	all Rosemeade	
	bears shown	$38 per pair
	bottom left	$8
	three nodders	$35 each pair
	one nodder	$40
	bear & fish	$12
	unmarked sitter	$14
	center right	
	as is	$10
	w/bike	$16
	bear & garbage can	$10
84	top left	$12
	unmarked 2-3/4 inch bears	$8
	bears with bibs	$16
	bottom left	$12
	top right	$6
	Christmas bears	$12
	Winnie-the-Pooh	$28
	bears on tricycle	$16
85	Smokey shovel up	$18
	Smokey shovel down	$18
	Van Tellingen bears	$25
	bottom	
	chefs	$8
	Smokey	$12
	sitters	$10
	others	$8
	three Smokeys	
	left	$20
	middle	$18
	right	$20
	one-piece rocker	$22
86	top left	
	left	$10
	middle	$10
	right	$10
	nesting bears	$18
	beavers	
	inside	$12
	outside	$18
	two camels	
	left	$12
	right	$12
	Rosemeade bison	$75
	unmarked bison	$14
	top right	$275
	CAS camels	$95
	intertwined camels	$25
87	one plastic camel	$8
	two plastic camels	
	left	$10
	right	$8
	Cat-a-Lac	$100
	top right	
	rear left	$10
	rear right	$12
	front left	$10
	front right	$8
	2-3/4 inch cats	$8
	drinking cats	$14
	bottom right	
	rear left	$8
	rear middle	$12
	rear right	$10
	front	$8
88	Puss 'N Boots	
	as shown	$28
	w/gold	$40
	w/gold & fancy paint	$85
	CAS gray cats	$35
	Thai & Thai-Thai	$50
	Rosemeade cats	
	rear row	$45 each pair
	middle row	$35 each pair
	front	$35
	owl & pussycat	$30
	nodder cat	$85
	cats in baskets	$14
	bottom right	
	left	$8
	right	$8
89	top left	
	rear left	$40
	rear right	$40
	front	$35
	center left	
	left	$12
	middle	$10
	right	$18
	bottom left	
	s & p	$18
	teapot	$48
	cream & sugar	$38
	wallpocket	$42
	top right	
	left	$5
	right	$5
	long black cats	$22
90	top left	
	black cats	$10
	cats on shoes	$8
	white cats	$6
	wood cat chefs	
	small	$2
	medium	$4
	large	$5
	bottom left	$10
	top right	$6
	animals in dresses	$24
	three cats & toys	
	left	$10
	middle	$10
	right	$10
	bottom right	
	inside	$8
	outside	$8
91	top left	$6 each pair
	CAS cats	
	outside	$65
	inside	$55
	bottom left	$25
	Fifi & Fido	$20
	bottom right	$10
92	CAS cat & dog	$30
	cats & birds	
	cats	$10
	birds	$8
	kitty salt & pepper	$14
	Rosemead cattle	
	heads	$40
	full figure	$110
	Japan cattle	$10
93	top left	$8 each pair
	2-5/8 inch cattle	$10
	bottom left	$10
	top right	$10
	Elsie & Elmer	$95
	Elsie sitters	
	left	$55
	right	$15
94	top left	$18
	bull w/teapots	$10
	one-piece cows	$16
	lounging cow	$15
	cow & moon	$10
	Regal feedsacks	$135
	top right	$8
	Srp cow & pig	$35
	long one-piece	$22
95	top left	
	cows	$16
	bears	$16
	plastic	$6
	CAS deer	$75
	Rosemeade	
	resting fawns	$55
	top right	$8
	Rosemeade jumping fawns	$60
	Kass-like fawns	$14
	bottom right	
	rear left	$10
	rear right	$12
	front	$8
96	top left	$12
	Rosemeade dog heads	
	chihuahuas (not shown)	$60
	all others	$35
	top right	$8
	Rosemeade	
	begging dogs	$50 each pair
97	top left	$14
	center left	
	w/o gold	$40

w/gold $85
range size Mugsey
inside $140 per pair
outside $60
top right $40
center right $48
bottom right
Billy & Butch $45
spaniels $35
98 top left
inside $45
other $40
bottom left
scotties $40
pomeranians $45
two Plutos
left $28
right $28
center right $40
bottom right $16
99 center left $14 each pair
posh poodles $22
top right
left $10
right $14
center right
dog $12
cow $12
dog & hydrant
condiment $30 each set
100 top left $16
dog & house $12
Goebel dogs
left $40
right $50
dalmatians $10
top right $8
dogs in basket $12
one-piece dog $14
4-1/8 inch dogs $10
Christmas pooches $10
101 top left
left $50
right $35
Nipper & clock
Nipper $50
clock $14
Spuds $18
bottom left
scotties $14
fish $12
top right $10
center right $12
bottom right $10
102 top left $8
center left
left $12
middle $10
right $12
bottom left
left $6
middle $10
right $8
top right
rear left $8
rear middle $8
rear right $10
front $6
center right
outside $10
inside $10
103 top left $12
two dogs
inside $8
outside $10
two sitters
dog in chair $12
hen on nest $10
bottom left $12
top center $55
top right $22
center right $20
bottom right $24
104 top left $16

donkey on farm $16
chalkware donkey &
elephant $12
Dem & Rep $50
Tambo & Tembino $95
two elephants w/trunks up
CAS $30
imposters $8
top right $10
Rosemeade elephnts $60 ea. pr.
105 top left $12
Meridan elephants $12
redware Dumbos $14
American Bisque Dumbos
large $22
small $14
top right
front $12
rear $12
snazzy foxes $14
bottom right $16 each pair
106 top left $8 per pair
Blossom Valley foxes $35
top right $12
bottom
inside $14
outside $6
107 top left $55
center left
left $24
middle $10
right $10
bottom left $120
top right
left $18
center $28
right $20
center right $8
bottom right $14
108 Brinn's goats $18
Rosemeade goats $110
Rosemeade gophers
rear left $60
rear right $50
front $45
top right $8
dressed hedgehogs $12
three hippos
rear inside $12
rear outside $10
front $10
bottom right
left $12
right $12
109 top left $10
hippos & frogs
hippos $12
frogs $14
Rosemeade ponies $60
Rosemeade horse heads $45
Srp rearing horses $35
Juanitaware horse heads $14
CAS horse heads $38 each pair
110 top left
left $10
middle $14
right $8
center left $12
bottom left $16
top right $12
zebras $8
saddle toothpick
& heads $18
kangeroos
far left $14
left $10
right $18
far right $14
111 top left $55
dapper kangaroo standing $95
dapper kangaroo lying $95
Kanga & Roo $60
center right
rear left $8

rear middle $14
rear right $10
front $10
top right $14
bottom right $10
112 top right
left $18
middle $14
right $18
Fitz & Floyd
salt & pepper $20
cream & sugar $30
cookie jar $75
Japan lions $12
Vandor leopards $15
CAS leopards $150
CAS lions $100
113 top left $18
Stoney Mouse $8
Rosemeade mice $75
top right $12
lady & gentleman mouse $18
cowboy & Indian mouse $16
114 top left
left $12
right $12
Applause Mickey & Minnie $14
bottom left $22
hickory dickory dock $12
mouse & piano $18
Good Company
Mickey in car $18
Mickey & Minnie $14
plastic Christmas mice $10
bottom right $100
115 top left $10
sea otters $14
Van Tellingen pigs
3-5/8 inches (not shown) $55
4-1/2 inches $130
6-1/2 inches $250
table size Shawnee pigs
Smileys
as shown $38
w/gold $70
Smiley & Winnie
as shown $48
with gold $95
range size Shawnee pigs
Smileys
as shown $75
w/gold $140
Smiley & Winnie
as shown $100
w/gold $160
116 top left
as shown $24
w/gold $50
CAS pigs $30
Regal one-piece $100
bottom left $14
Goebel pigs $40 each pair
figural nodder pigs $200
Taiwan pigs $8
one-piece plastic chef pigs $10
ceramic chef pig heads $12
117 top left $8
pigs on corn $10
Imperial Porcelain pigs $100
Srp pigs $25
top right
inside $8
outside $8
Enesco farmer pigs $12
gentlemen pigs $16
2-1/2 inch unmarked pigs $12
roasted pig $10
bottom right
rear inside $8
rear outside $8
front $6
118 top left $95
center left $75
bottom left $40/ true pair

top right $35
center right $65
Huggie rabbits $18
119 top left $45
center left
Goebel Thumper
eyes closed $130
eyes open $140
Goebel Flower $85
Enesco Bambi &Thumper $70
bottom left $10
top right
rabbit & carrot $10
apple & worm $10
raccoon & garbage can $12
bottom right
left $10
right $12
120 top left $85
center left
left rear $12
right rear $12
front $8
top right $30
center right $16 each pair
bottom right
rear $18
front $22
121 center left
left rear $12
right rear $12
left front $10
right front $10
bottom left $10
top right
inside $12
outside $10
CAS lambs $32
Goebel skunks $75
Rosemeade Skunks
inside $60
outside $75
122 top left $8
skunks on stumps $6
CAS mother & baby $45
top center $8
center
squirrels $8
beavers $10
squirrel hangers $18
top right
left $45
right $55
123 squirrels & nuts
squirrels $8
nuts $6
squirrels or skunks $6
bottom left $10
top center $35
top right $25
squirrel & stump $10
three squirrels
left $12
middle $16
right $10
bottom right
left $8
right *price not determined*
124 top left $15
center left $12
bottom left
giraffes $10
tigers $10
top right
left $8
right $12
three walruses
left $8
center $16
right $10
bottom right
rear left $8
rear right $8
front left $10

Page	Item	Price
	front right	$10
125	top left	
	left	$8
	right	$10
	Rosemeade wolves	$110
	cactus & skull	$16
	cactus & coyote	$20
	Punxsutawney Phil	$8
	CAS zebras	$110
126	top left	$45
	center left	$10
	bottom left	
	rear	$14
	front inside	$8
	front outside	$16
	long one-piece zebra	$40
	center right	
	rear	*price not determined*
	front	$8
	bottom right	$14
127	top left	$45
	center left	$45
	Blue Ridge chickens	$75
	Shawnee chickens	
	as shown	$40
	w/gold & fancy paint	$130
128	top left	$75
	center left	
	left	$40
	center	$35
	right	$40
	bottom left	$8
	top right	$40
	chicks & eggs	$12
	stacked condiment	$18
	bottom right	$22
129	top right	
	chickens	$15
	cookie jar	$85
	white chickens	$8
	bottom right	
	front	$12
	rear	$8
130	top left	
	rear left	$8
	rear center	$10
	rear right	$10
	front left	$12
	front right	$25
	two chickens	
	large	$10
	small	$8
	Rosemeade chickens	
	left	$60
	center	$50
	right	$85
	bottom	
	rear left	$8
	rear center	$10
	rear right	$8
	front	$6
	top right	$16
	center right	
	Japan	$14
	Holt Howard	$10
	Vandor chicken	$14
131	top left	
	left	$8
	center	$8
	right	$12
	chickens on nests	
	left	$12
	right	$14
	soft sculpture chickens	$6
	bottom left	
	left	$12
	center	$10
	right	$8
	top right	$10
	center right	
	left	$8
	right	$8
	bottom right	$12
132	top left	$14
	center left	$12
	Rosemeade ducks	
	left	$40
	right	$50
	top right	$12
	Holt Howard chicks	$22
	Goebel mallards	$35
133	top left	
	left	$10
	right	$10
	Srp ducks	$20
	OJ ducks	$12
	top right	$10
	long-billed ducks	$24
	two nodders	
	left	$50
	right	$35
	Peppy & Salty	$14
134	top left	
	outside	$40
	inside	$35
	center left	$135
	bottom left	$10
	top right	
	rear outside	$75
	rear inside	$20
	front	$25
	right center	
	left	$14
	right	$10
	bottom right	
	left	$14
	right	$8
135	top left	
	ducks	$12
	geese	$8
	center left	$40 each pair
	bottom left	$14
	top right	$8
	Japan swans w/flowers	$16
	plain Japan swans	$8
	bottom right	
	left	$50
	right	$50
136	top left	
	big	$12
	little	$8
	center left	$14
	bottom left	
	left	$12
	right	$16
	top right	
	pelicans	$60
	flamingoes	$50
	center right	$50
137	top left	$16
	center left	$12
	bottom left	
	inside	$28
	outside	$16
	top right	$25
	center right	
	front	$20
	rear	$24
	bottom right	
	rear	$14
	center	$18
	front	$12
138	top left	$16
	center left	
	rear	$120
	front left	$35
	front right	$60
	bottom left	
	shakers	$75
	hors d'oeuvres	$125
	top right	
	left	$38
	right	38
	Red Wing bob white	$20
	bottom right	
	inside	$70
	outside	$35
139	top	$6
	center left	$22
	bottom left	
	rear left	$28
	rear right	$18
	front	$18
	four turkeys	
	rear left	$16
	rear right	$16
	front left	$14
	front right	$12
	bottom right	
	rear left	*price not determined*
	rear right	$18
	front	*price not determined*
140	top left	
	ducks	$40
	turkeys	$60
	three turkeys	
	left	$14
	center	$14
	right	$12
	3-1/2 inch turkeys	$18
	bottom left	
	left	$12
	right	$12
	top right	$16
	center right	$22
	roast turkey	$8
141	top left	
	rear left	$12
	rear center	$12
	rear right	$14
	front	$12
	owls & ducks	
	left	$95
	center	$32
	right	16
	2-1/2 inch owls	$8
	bottom left	$45
	plated metal owls	$4
	bottom center	$8
	top right	$8
	Agiftcorp owls	$10
	dressed owls	$10
	Woodsy Owl	$50
142	top left	$22
	CAS parakeets	$50
	birds on log	$18
	bottom left	$12
	top center	$6
	just below top center	
	left	$12
	right	$8
	wooden birds	$6
	Rosemeade	$45
	3-inch parrots	$12
	bottom right	$18
143	Rosemeade songbirds	
	bluebirds	$35
	robins	$35
	goldfinches	$35
	chickadees	$35
	unmarked ceramic bluebirds	$10
	plastic bluebirds	$4
	bottom left	
	left	$2
	right	$2
	top right	$12
	center right	*price not determined*
	bottom right	$16
144	plastic parakeets	$8
	two birds on bases	
	left	$14
	right	$6
	bottom left	
	left	$75
	right	$55
	top right	$45
	center right	$12
	bottom right	
	left	$16
	right	$14
145	top left	$38
	two Woodys	
	outside	$35
	inside	$25
	Woody & Winnie	$100
	bottom left	
	left	$18
	center	$8
	right	$10
	top center	$40
	metal penguins	$8
	top right	$6
	center right	
	left	$10
	center	$10
	right	$16
	bottom right	$14
146	top left	$8
	center left	
	plastic	$12
	ceramic	$14
	CAS frog	$40
	top right	$10
	bone china frogs	$12
	two frogs & mushrooms	
	left	$10
	right	$12
	bottom right	
	left	$14
	right	$10
147	top left	
	left	$16
	right	$8
	Japan tray frogs	$16
	Thailand frogs	$8
	bottom left	$10
	top right	
	inside	$20
	outside	18
	center right	
	hobos	$20
	musicians	$18
	bottom right	$14
148	top left	
	outside	$6
	inside	$6
	just below top left	
	left	$14
	right	$6
	Oravitz turtles	$16
	bottom left	$16
	top right	
	left	$12
	right	$12
	long one-piece	$25
	Srp alligators	$30
	bottom right	$12 each pair.
149	top left	
	front	$12
	rear	$12
	center left	$14
	bottom	
	w/suitcase	$18
	w/umbrella	$18
	top left	$16
	center left	$12
150	top left	$16
	center left	
	bees	$10
	ladybugs	$12
	bottom right	$12
	Van Tellingen lovebugs	
	3-5/8 inches	$60
	4-3/8 inches	$150
	ladybugs	$10 each pair
151	top left	
	left	$14
	center	$8
	right	$10
	3-1/4 inch bugs	$10
	bottom left	
	left	$16
	right	$10
	bug band	$18 each shaker
	center right	$12 each pair

Page	Item	Price
	bottom right	
	inside	$16
	outside	$12
152	top left	$18 each pair
	center left	$14
	garden lizards	$22
	top right	
	bugs	$14
	pigs	$12
	grasshoppers & watermelon	$30
153	top left	
	rear	$10
	front left	$10
	front right	$12
	bottom	
	snails	$10
	lobsters	$10
	top right	
	rear left	$10
	rear right	$10
	front	$10
154	top left	$10
	center left	
	rear left	$12
	rear right	$12
	front	$10
	bottom left	$10
	top right	$10 each pair
	just below top right	
	rear	$16
	center	$14
	front left	$10
	front right	$10
	Oravitz fish	$8
	bottom right	
	rear	$10
	front left	$14
	front center	$10
	front right	$16
155	top left	
	left	$14
	right	$14
	Grindley fish	$10
	bottom left	
	northerns	$100
	muskies	$100
	top right	$25
	center right	$25 each pair
	bottom right	
	walleye	$100
	trout	$110
	perch or smallmouth	$100
156	top left	
	left	$85
	right	$110
	CAS fish	
	inside	$40
	outside	$45
	four fish	
	left	$14
	center left	$12
	center right	$10
	right	$14
	bottom left	$18
	top right	$8
	two Goebel fish	
	front	$35
	rear	$35
	bottom right	$45
157	top left	
	rear left	$12
	rear right	$8
	front	$10
	center left	
	inside	$12
	outside	$8
	bottom left	
	inside	$14
	outside	$8
	top right	
	left	$8
	center	$24
	right	$6
	center right	
	left	$12
	right	$12
	bottom right	
	rear left	$8
	rear right	$8
	front	$8
158	angel fish	$14 each pair
	chalkware fish	$2
	three fish (standing & swimming)	
	rear left	$10
	rear right	$12
	front	$12
	bottom left	
	left	$10
	center	$8
	right	$12
	metal fish	$4
	bottom right	
	front	$6
	rear	$6
159	top left	$100
	center left	
	rear left	$14
	rear right	$10
	front	$16
	bottom left	
	outside	$10
	inside	$8
	top right	$14
	center right	
	left	$16
	right	14
	fish w/base	$16
	Goebel condiment	$110
160	Treasure Craft fish	$6
	two nodders	
	left	$40
	right	$30
	fish & creel	$14
	bottom left	$22
	top right	*price not determined*
	center right	$10
	bottom right	$10
161	lobsters	
	rear	$12
	front left	$12
	front right	$12
	center left	$65
	bottom left	
	rear left	$10
	rear right	$12
	front	$14
	center right	
	left	$12
	center	$8
	right	$14
	bottom right	$12
162	top left	$6
	center	
	inside	$14
	outside	$10
	far outside	$12
	Treasure Craft	
	squirrels	$6
	seahorses	$8
	top right	$65
	bottom right	$45 each pair
163	palm & coconut	$10
	bottom left	
	inside	$18
	outside	$10
	top right	
	inside	$12
	outside	$8
	bottom right	$8
164	top left	$8
	unmarked cactus	$6
	Goebel cactus	$40
	bottom left	
	left	$12
	center	$14
	right	$12
	top right	
	rear left	$12
	rear center	$12
	rear right	$12
	front	$10
	center right	$18
	bottom right	$8
165	top left	$18
	boxed tulips	$6
	bottom left	
	left	$16
	right	$24
	top right	$35
	tulips w/bees	$10
	bottom right	$18
166	plastic flowers	$12
	tray w/ceramic flowers	$22
	Elbee Art roses	$14
	bottom left	
	left	$14
	right	$12
	top right	$14
	center right	$18
	bottom right	
	w/gold	$35
	w/o gold	$15
167	Shawnee sunflowers	
	stove set	$55
	table set (not shown)	$25
	Rosemeade flowers	
	rear left	$35
	rear right	$35
	front	$45
	bottom right	$45
	top right	$55
	bananas	$16
168	top left	
	strawberries	$6
	bananas	$6
	fruit w/dresses	
	w/dresses	$16 each shaker
	bananas	$14
	two pineapples	
	left	$12
	right	$16
	bottom left	
	inside	$22
	outside	$22
	top right	
	left	$8
	right	$22
	bottom right	$16
169	top left	$8
	plastic orange tree	$12
	top right	$16
	plastic & chalkware oranges	
	plastic	$6
	chalkware	$6
	oranges w/faces	$18
	S & P tree	$24
170	top left	
	left	$10
	center	$6
	right	$10
	chalkware citrus fruit	$6 per pair
	bottom left	$8
	top right	$4
	center right	$10
	bottom right	
	nectarines	$10
	apples	$10
171	top left	$22
	Arcadia tray set	$22
	3-3/4 inch apples	$8
	bottom left	$16 each set
	top right	$12
	center right	$30
	bottom right	
	left	$22
	right	$22
172	top left	
	rear	$18
	center	$16
	front	$18
	center left	
	left	$8
	center	$8
	right	$10
	bottom left	$14
	top right	$10
	center right	
	left	$16
	center	$12
	right	$16
	bottom right	$8
173	top left	$12
	Otagiri strawberries	$10
	strawberries w/faces	$18
	bottom left	
	left	$10
	center	$12
	right	$10
	top right	
	rear	$14
	center left	$10
	center right	$10
	front left	$8
	front right	$8
	center right	
	left	$6
	center	$10
	rear	$14
	bottom right	
	bananas	$10
	strawberries	$22
174	top left	
	left	$10
	center	$10
	right	$8
	watermelon halves	$10
	watermelons people	$20
	hanging grapes	$8
	top right	$14
	center right	
	rear	$10
	front left	$8
	front right	$8
	bottom right	
	left	$10
	center	$12
	right	$16
175	top left	$35
	center left	$8
	bottom left	$20
	top right	$6
	center right	
	cucumbers	$12
	sweet peppers	$10
	hot peppers	$12
	bananas	$12
	mushrooms & pears	
	mushrooms	$8
	pears	$10
	bottom right	
	left	$8
	center	$16
	right	$22
176	top left	
	pineapples	$6
	gourds	$12
	cherry peppers	$8
	grapes	$12
	center left	
	tomatoes	$8
	watermelon	$16
	bottom left	$20
	top right	$16 each shaker
	six w/faces	
	rear left	$24
	rear center	$24
	rear right	$28
	front left	$22
	front center	$18
	front right	$20
	three being held up	
	left	$20
	center	$20
	right	$20
	bottom right	
	left	$16

center $22
right $20
177 top left $20
center left
rear left $18
rear center $20
rear right $20
front left $20
front right $16
bottom left
rear left $18
rear right $18
center left $20
center $20
center right $18
front left $20
front right $18
top right $80
bottom right *price not determined*
178 asparagus $12
beans $6
McCoy cabbage $25
three veggies
beets $10
cabbage $12
peppers $10
two veggies
cabbage $12
celery $12
corn $14
179 top left
left $8
center $12
right $8
Carson carrots w/faces $22
Oravitz carrots w/faces $22
bottom left $14
hobby cer. veggies $14 each pair
bottom center $16
top right $120
center right $35
bottom right $14
180 top left $18
Goebel corn $45
bottom left $10
corn w/faces $28
three veggies $6 each pair
181 Rosemeade vegetables
peppers $15
cabbage $15
cucumbers $15
corn $25
corn & peppers
corn $10
peppers $10
corn & celery
corn $10
celery $12
bottom left
corn $10
carrots $16
three veggies
cucumbers w/faces $20
peppers $12
plain cucumbers $12
OJ cucumbers $12
bottom right $24
182 top left $12
bottom left $22
four veggies
beets $12
squash $14
tomatoes $10
eggplant $14
Hop & Lo *expensive*
two mushrooms
w/faces $14
plain $10
183 top left $6
Otagiri
apple $10
onion $10
top right
pepper $12
peanut $14
center right $12
bottom right $16
184 three Mr. Peanuts $16 each set
large plastic Mr. Peanut $22
peas $10
top right
large $28
small $18
center right $35
bottom right $8
185 top left $20
three veggies
carrots $14
cauliflower & beets $6 each
front $12
bottom left
rear left $12
rear center $10
rear right $10
front $8
top right
inside $20
outside $6
Goebel $35
pepper $45
potatoes $6
186 top left
rear left $12
rear right $10
front $14
potatoes & sacks
potatoes $8
sacks $10
tray sets
beets $35
pumpkins $16
bottom left $12
top right $45
center right
rear left $32
rear right $32
front $16
bottom right $20
187 top left $14
center left
left $20
right $12
bottom left
left $12
right $14
top right
left $14
right $12
bottom right
left $14
right $12
188 metal trains & trolley
rear left $6
rear right $6
bottom left $14
top right $45
metal Greyhounds
left $35
right $35
ceramic automobiles $12
bottom right $4
189 top left
radiators $4
cars $6
bottom left $18
top right $22
car & rabbit $16
bottom right
car & pig $16
horses & wagon $8
ship $24
190 top left $6
Noah's ark $55
gondola $18
top right $110
center right
aircraft carrier $18
typewriter & ink $16
191 top left $16
clipper ships $14
ship's wheel $8
bottom $65
top right $8
Connestoga condiment $28
bottom right
left $22
right $14
192 top left $12
wagons & stagecoaches
wagons $6
stagecoaches $18
horse & wagon $24
bottom left $45
top right $10
center right $10
bottom right $30
193 top left $28
center left
left $28
right $28
bottom left
Gulf $28
Sohio $28
Phillips 66 $28
top center $32
top right $30
two pumps
Conoco $28
Texaco $28
three pumps
Richfield $36
Atlantic $36
Shell $28
bottom right
inside $20 as is
outside $16
194 top left $12
center left $20
bottom $20 each pair
top right $16
Pyrofax $26
195 top left $18
top right $18
196 top left $16
Empire State Building
inside $8
outside $8
top right
ceramic $18
metal $10
center right $8
bottom $16
197 Hearst Castle $18
cottage condiment $55
bottom left $12
top right
left $14
right $12
gingerbread houses $18
barns & silos
left $14
right $16
Harvestore silos
left $50
right $50
lighthouses $16 each pair
198 top left *price not determined*
lighthouse & cabin $24
Japan outhouses
inside $8
outside $12
bottom left
outhouses $8
toilets $8
cedar outhouses $2
two Mackinac bridges
rear $18
front $22
top right $10
center right
no. 1 $75
any other $65
199 top left $28
center left $48
one-piece Mt. Rushmore $16
center right
rear left $10
rear center $12
rear right $10
front $6
bottom right
rear left $20
rear center $20
rear right $18
front $6
200 top left $14
bottom left
rear left $18
rear center $16
rear right $10
front $12
top right $18
bottom right $20
201 top left $24
bottom left $22
top right
left $10
center $14
right $32
bottom right
rear left $10
rear right $16
front $12
202 top left $12
bottom left *price not determined*
top center $14
two woodburners
left $10
right $14
stoves & coffee grinders
left $6
right $6
top right $12
handy flames $14
bottom $20
203 top left $12
plastic washer $22
Filter Queen vacuums $20
bottom left $14
top center $12
Naylor toaster $14
Lilly toasters
left $16
right $16
top right $24
center right $16
bottom right $18
204 top left $8
center left $8
plastic irons
outside $12
inside $16
top right $16
three metal sets
sewing machines $4
sweepers $4
irons $4
bottom right $24
205 top left *price not determined*
teenagers' icons
telephone $12
lipstick & perfume $16
telephone & book $14
bottom left
left $14
center $12
right $12
top right $14
pop-up TV $14
piano & TV
piano $10
TV $10
bottom $22
top right
rear $15

Page	Item	Price
	front	$15
	radio tubes	$35
206	left	$12
	top right	
	spring works	$45
	doesn't work	$28
	bottom right	$28
207	top left	$18
	wood coffee pots	$2
	Disneyland	
	inside	$10
	outside	$8
	plastic coffee pot	$14
208	top left	$10
	center left	
	rear left	$16
	rear right	$20
	front	$12
	bottom left	
	inside	$8
	outside	$12
	top right	$12
	center right	
	left	$14
	right	$12
	bottom	
	left	$8
	center	$12
	right	$8
209	top left	
	left	$4
	center	$6
	right	$10
	center left	$22
	bottom left	
	left	$14
	center	$24
	right	$18
	top right	
	inside	$16
	outside	$8
	center right	$8
	bottom right	$8 each pair
210	top left	
	rear left	$6
	rear center	$8
	rear right	$8
	front	$6
	center left	$14
	bottom left	
	inside	$8
	outside	$4
	ceramic Mason jars	$14
	3 bottles, 1 jar	
	rear left	$6
	rear center	$12
	rear right	$8
	front	6
	bottom left	$4
	top right	
	outside	$24
	inside	$16
	Coke bottles	22
	bottom right	*price not determined*
211	top left	
	7-Up	$8
	Sealtest	$18
	Coors Golden Export	$18
	beer in two sizes	
	left	$16
	right	$16
	bottom left	
	left	$8
	right	$14
	top right	
	left	$6
	center	$14
	right	$6
	center right	
	left	$15
	center	$12
	right	$14
	bottom right	
	left	$8
	center	$16
	right	$10
212	top left	$12
	center left	$0.50
	bottom left	$4
	top center	$26
	Campbell cans	$8
	top right	$4
	Higgins ink	$16
213	top left	$12
	wood bottles& pots	$2 each pair
	metal minis	*price not determined*
	bottom left	$4
	milk glass shakers	$6
	three sets	
	left	$2
	center	$4
	right	$2
	top right	$14
	Japan steins	$10
	bottom right	
	left	$8
	right	$4
214	top left	$2 each pair
	center left	$6
	bottom left	
	inside	$6
	outside	$6
	top right	
	inside	$0.50
	outside	$1
	center right	$3 each pair
	bottom right	
	inside	$3
	outside	$4
215	top left	$1
	Oklahoma mugs	$1
	Vegas/New Mexico	
	left	$4
	right	$6
	center right	
	salt & pepper	$5
	creamer	$4
	bottom right	$45
216	top left	$16
	pepper mill & shaker	$10
	plastic lunch-box, as is	$14
	if perfect	$22
	top right	
	left	$14
	center	$6
	right	$8
	Thornburg gifts	$16
	bottom center	$36
	film canisters	$1
	plastic mailbox	$22
217	top left	$10
	bottom left	
	left	$12
	middle	$10
	right	$14
	top right	$25
	center right	
	left	$10
	right	$14
	bottom right	$22
218	top left	
	football	$10
	golf	$16
	wood bowling ball & pin	$4
	top right	
	rear	$10
	front	$14
	megaphones	$8
	four go-togethers	
	megaphone & football	$12
	cannon & cannon balls	$14
	bolt & wrench	$12
	bowling ball & pin	$14
	plastic bowling condiment	$24
219	top left	$20
	center left	$12
	bottom right	$4
	top right	$14 each pair
	shoes as shown	$12
	w/unshown holder	$16
	metal	
	teapots	$3
	boots	$2
	bottom right	$10
220	top left	$6
	two hats	
	inside	$10
	outside	$10
	four accessories	
	left rear	$10
	rear right	$14
	front left	$16
	front right	$12
	bottom left	
	rear left	$12
	rear right	$14
	front	$12
	top center	$8
	top right	$12
	center right	$20
	bottom right	$12
221	top left	$14
	center left	
	left	$16
	center	$14
	right	$14
	bottom left	$14
	top right	$12
	center right	$14 each set
	bottom right	
	left	$22
	right	$20
222	top left	$16
	bread	$20
	hotdog go-together	$16
	bottom left	
	inside	$8
	outside	$8
	top right	
	pans	$8
	hamburgers	$8
	cheese	$10
	bottom right	$10
223	top left	
	left	$35
	center	$18
	right	$12
	five food-related	
	left rear	$20
	center rear	$24
	right rear	$26
	left front	$10
	right front	$20
	two cooked items	
	left	$12
	right	$8
	bottom left	$6
	top right	$6
	center right	$14
	bottom right	$40
224	top left	$12
	center left	
	left	$14
	right	$16
	bottom right	
	shakers	$8
	bank	$14
	top right	$1
	center right	$12
	bottom right	$16
225	top left	$12
	colonial oil lamps	
	w/o box	$10
	w/box	$12
	two oil lamps	
	inside	$10
	outside	$8
	bottom left	
	left	$8
	center	$8
	right	$8
	top right	$10
	center right	$6
	bottom right	$8
226	top left	$16
	center left	$14
	bottom left	
	left	$12
	center	$12
	right	$14
	top right	
	left	$16
	center	$14
	right	$8
	bottom right	$12 each
227	top left	$6
	bottom left	$10
	top right	$8 each set
	center right	$12
	bottom right	
	inside	$12
	outside	$12
228	top left	$12
	center left	$14 each
	bottom left	
	left	$14
	center	$16
	right	$16
	top right	
	inside	$8
	outide	$8
	center right	
	rear outside	$10
	rear inside	$10
	front	$35
	accordians	$8
229	top left	$10
	center left	$6
	bottom left	
	left	$18
	right	$20
	center right	$6
	bottom right	$25
230	top left	$4
229	top left	$10
	center left	$6
	bottom left	
	left	$18
	right	$20
	center right	$6
	bottom right	$25
230	top left	$4
	telephone & binoculars	
	telephone	$12
	binoculars	$12
	books	$8
	bottom left	$4
	top right	*expensive*
	center right	$38
	bottom right	$10
231	top left	$22
	chairs by fire	
	chairs	$10
	fireplaces	$10
	bottom left	$8
	top left	$6
	center right	
	sofa & chair	$18
	jockey & horse	$18
	toothpaste & brush	$12
	bottom right	
	toilet condiment	$26
	thundermugs	$8
232	top left	$20
	two wood sets	
	tombstones	$2
	bullets	$4
	Pepper Tate & Salty O'Day	$10
	bottom left	$14
	top right	
	comb & brush	$14
	jar & atomizer	$18
	suitcases	$14
	Indian pieces	
	left	$16
	center	$10

right $8
bottom right $12
233 top left
scarecrows $12
gram & gramps $10
cap & scroll $22
Autumn Serenade $4
top right $18
bottom right *price not determined*
234 top left $35
top center $8
gun & bullet $14
shotgun shells $12
batteries $12
rebel cannons $10
top right $6
broken heart $6
plastic shakers $10
225 top left $18
center left $12
bottom left $10
top right
left $6
right $6
center right $35
bottom right
left $10
center $10
right $10
226 top left $8
Parkcraft States $16 per set
238 Milford Pottery states $10
239 hatchet & stump $14
clever & cutting block $14
planes & squares $12 per pair
wrench & bolt $14
bellows $6
bottom right
bowling ball & pin $14
hammer & anvil $12
bread & butter $14
240 top left
pipes $8
bolts $8
hammer & anvil $14
center left $4 per set
bottom left $8
top right $10
plastic wateringcan & flower pot $4
metal cleaning accessories
left $4
right $4
three Shawnee sets
wheelbarrows $14
milk cans $16
sprinkling cans $12
241 Azalea condiment $75
Goebel condiments
left $120
right $140
242 top left $65
center left $40
top right $55
center right $15
bottom right $35
243 top left $18
center left $15
bottom left
left $12
right $20
top right
as is $0
good condition $22
center right (per pair)
chartruese $22
yellow $12
turquoise $12
ivory $12
cobalt $16
rose $22
red $18
green $12
gray $22
dark green $22
medium green $38
bottom right $30
244 top left $10
Purinton $20
Shawnee $30
bottom left $30
top center $10
McCoy $8
Wedgwood $60
top right $40
Noritake Azalea
inside $35
outside $35
245 top left $20
center left
left $15
center $30
right $15
bottom left
rear left $30
rear center w/sugar $45
rear right $15
front inside $15
front outside $25
top right $35
center right $10
bottom right $10
246 top left
inside $6
outside $8
plastic kitchen shakers
left $2
center $2
right $2
pocket one-piece $3
bottom left
inside $2
outside $4
top right $2
center right
rear left $8
rear right $8
front left $2
front right $8
bottom right
left $4
right $4
247 top left $2
center left
left $2
right $2
bottom left $4
top right $6
sailboats $6
stacked plastic $6
bottom right $28
248 top left $8
Superlon $8
St. Labre $0.25
bottom left $6
top center $12
three different media
left $6
center $4
right $6
two plastic $8 each pair
top right $6
center right $6
bottom right $2 per pair
249 top left $22
U.S. Navy $10
chrome tops w/plastic bases $8
bottom left $6
OJ on tray $15
top right $4
S90 $14
shells & handptd
left $2
right $8
bottom right $2 each pair
250 top left $2
center left
rear left $2
rear center $2
rear right $2
front left $4
front right $2
bottom left $3
top center $4
top right *price not determined*
boxed glass $6
Greece $12
bottom right $1 each pair
251 top left $12
bottom left
left $4
right $6
top right $18
center right $6
bottom right $16
252 center left $120 each
bottom left $80 each
top right $80 each
center right $100 each
bottom right $100 each.
253 top left
left $35
right $25
center left $90 pair
bottom left $80
top right $50
center right $40
bottom right
left $80 pair
right $75 pair
254 top left $130
center left $300 pair
bottom left *price not determined*
top right $60
center right $20
bottom right
inside $15
outside $18
255 top left $45
center left *price not determined*
bottom left $22
top right $20
milk glass w/holder $20
painted clear glass $6
bottom right $6
256 top left $8
4-7/8 inch cutglass shakers $22
three empty pair
left $14
center $25
right $14
three full pair
left $3
center $3
right $3
top right 5
center right *price not determined*
bottom right $12